Industrial Engineering and Production Management

Industrial Engineering and Production Management

Edited by Courtney Hoover

CLANRYE
INTERNATIONAL
www.clanryeinternational.com

Clanrye International,
750 Third Avenue, 9th Floor,
New York, NY 10017, USA

ISBN: 978-1-63240-584-5

Cataloging-in-publication Data

Industrial engineering and production management / edited by Courtney Hoover.
    p. cm.
Includes bibliographical references and index.
ISBN 978-1-63240-584-5
1. Industrial engineering. 2. Production management. 3. Production engineering. 4. Production control. I. Hoover, Courtney.
T56.24 .I53 2017
658.5--dc23

For information on all Clanrye International publications
visit our website at www.clanryeinternational.com

Printed in the United States of America.

Contents

Preface

Industrial Engineering is a vast field of study. It involves the optimization of various complex process associated with industrial output. Production management is a sub-set of Industrial Engineering and is primarily concerned with the production of goods. This elaborate book traces the progress and conjunction of this field and highlights some of the key concepts and applications. It presents researches and studies performed by experts across the globe. Those with an interest in industrial engineering and production management would this book helpful. It will serve as a reference for graduate and post graduate students.

Various studies have approached the subject by analyzing it with a single perspective, but the present book provides diverse methodologies and techniques to address this field. This book contains theories and applications needed for understanding the subject from different perspectives. The aim is to keep the readers informed about the progress in the field; therefore, the contributions were carefully examined to compile novel researches by specialists from across the globe.

Indeed, the job of the editor is the most crucial and challenging in compiling all chapters into a single book. In the end, I would extend my sincere thanks to the chapter authors for their profound work. I am also thankful for the support provided by my family and colleagues during the compilation of this book.

Editor

Challenges and future perspectives for the life cycle of manufacturing networks in the mass customisation era

D. Mourtzis[1]

Abstract Manufacturers and service providers are called to design, plan, and operate globalised manufacturing networks, addressing to challenges of increasing complexity in all aspects of product and production life cycle. These factors, caused primarily by the increasing demand for product variety and shortened life cycles, generate a number of issues related to the life cycle of manufacturing systems and networks. Focusing on the aspects that affect manufacturing network performance, this work reviews the exiting literature around the design, planning, and control of manufacturing networks in the era of mass customisation and personalisation. The considered life cycle aspects include supplier selection, initial manufacturing network design, supply chain coordination, complexity, logistics management, inventory and capacity planning and management, lot sizing, enterprise resource planning, customer relationship management, and supply chain control. Based on this review and in correlation with the view of the manufacturing networks and facilities of the future, directions for the development of methods and tools to satisfy product–service customisation and personalisation are promoted.

Keywords Manufacturing systems and networks · Design · Planning · Mass customisation

This article is part of a focus collection on "Robust Manufacturing Control: Robustness and Resilience in Global Manufacturing Networks".

✉ D. Mourtzis
mourtzis@lms.mech.upatras.gr

[1] Laboratory for Manufacturing Systems and Automation, University of Patras, 26500 Patras, Greece

1 Introduction

Mass production (MP) has been the established manufacturing paradigm for nearly a century. MP initially answered to the need of the continuously increasing population around the globe, with a gradual improvement in its living standards, especially in the developed world, for goods and commodities. However, since the 1980s and with the beginning of the new millennium, a saturation of the market towards mass produced products is observed. In 2006, Chryssolouris states that: "It is increasingly evident that the era of MP is being replaced by the era of market niches. The key to creating products that can meet the demands of a diversified customer base, is a short development cycle yielding low cost and high quality goods in sufficient quantity to meet demand" [1]. Currently, the need for increased product variety is intensifying, and customers in many market segments request truly unique products, tailored to their individual taste. Companies are striving to offer product variety while trying to produce more with less [2] (i.e. maximise their output while minimising the use of materials and environmental footprint), while the landscape that they must operate in, inflicted by the economic recession, has become more complex and dynamic than ever.

In the mass customisation (MC) paradigm, the establishment of which is still an ongoing process, instead of treating customers merely as product buyers, a producer must consider them as integrated entities in the product design and development cycle. In this customer-driven environment that is shifting towards online purchases and market globalisation, the underlying manufacturing systems and chains are heavily affected. Owing to its multidisciplinary nature, the manufacturing domain in general lacks of unified solution approaches [3]. The management of the co-evolution of product, process and production on a strategic and operational level is a huge challenge. Market globalisation broadens the target audience of a product, while at the same time it constitutes supply strategies and logistics' more difficult to manage. Adding to that, the Internet, one of the primary enablers of globalisation, allowed online customisation and purchasing, leading to new disruptive purchasing models. In their turn, these models affected long-established businesses that could not form an online presence fast and succumbed to the competition. Moreover, the economic recession highlighted the need for quick adaptation to demand; companies that could not adapt to the new requirements suffered economic losses and their viability was challenged. Simultaneously, the decreasing product costs and the increase in purchasing power in developing countries generated new markets and destabilised demand. Finally, the emergence of new materials, new forms of production, and key enabling technologies constitute new diversified product features and processes feasible, as well as they allow the interconnection between ICT systems, humans, and engineering/manufacturing phases.

It becomes apparent that manufacturers and service providers are presented with numerous external and internal drivers and challenges [4] that have a visible impact on the smooth operation of the entire value-adding network down to each individual manufacturing facility [5]. A root cause for these problems is that while the MC paradigm proposes a set of practices and solutions for tackling these issues, its practical implementation is still considered as work in progress in terms of effectiveness of coordination and collaboration between stakeholders, design and planning of networks and facilities, and execution and control efficiency [6]. An enabling solution for realising a cost-effective implementation of MC is to properly configure easily adaptable manufacturing networks, which are capable to handle the complexity and disturbances that modern production requirements inflict [7]. Support systems for the design, planning, and control with inherent robustness are necessary in order for companies to withstand the antagonism through sustainable practices. Technology-based business approaches comprise a major enabler for the realisation of robust manufacturing systems and networks that offer high value-added, user-oriented products and services. These qualities are critical for companies in order to master variety and maintain their viability [8]. Significant work has been conducted on this field, yet a focused review of the literature regarding the influence of MC practices on different aspects of the manufacturing network life cycle is missing. In particular, the lack of dedicated reviews on the challenging issues of design, planning, and operation of manufacturing networks in the framework of MC forms the motivation for conducting this work [9].

Towards bridging this gap in academic approaches, this work reviews the existing literature related to the basic aspects of a manufacturing network from its design,

planning, and control life cycle perspectives within the general MC landscape, targeting to the understanding of the current situation and identification of future developments. For the scope of the paper, areas of supplier selection, initial manufacturing network design, supply chain coordination, complexity, logistics management, inventory and capacity planning and management, lot sizing, enterprise resource planning (ERP), customer relationship management (CRM), and supply chain control are reviewed. The purpose is to establish an overview of the current status of academic research and pinpoint the challenges that have yet to be addressed by academic work. Departing from that, major drivers and enabling technologies are identified, as well as concepts that can lead to a more sustainable implementation of MC are proposed.

The review is based on structured search in academic journals and books, which were retrieved primarily from Scopus and Google Scholar databases, using as keywords the main fields of interest of the study, namely: evolution of manufacturing paradigms, issues in MC and personalisation environments, the role of simulation for manufacturing, methods and technologies related to product and production complexity, and inventory management and capacity planning, among others. Academic peer-reviewed publications related to the above fields were selected, ranging over a period of 30 years, from 1984 to 2015, with only a few notable exceptions. Sciences that were considered in the search were: engineering, management, business, and mathematics. The review was carried out in three stages: (1) search in scientific databases with relevant keywords, (2) identification of the relevant papers after reading their abstract, and (3) full-text reading and grouping into research topics. Indicatively, the frequency of results from a search with the keywords "mass customisation" or "product personalisation" in the abstract, title, and keywords of the article as obtained by the Scopus database is depicted in Fig. 1.

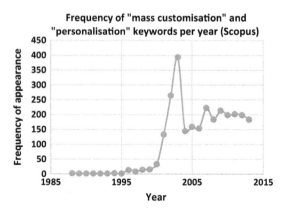

Fig. 1 Frequency of appearance of the keywords "mass customisation" and "personalisation" in the abstract, title, and keywords of the article

The above figure also visualises the increase of interest on these topics by the scientific community, and the establishment of MC as a distinct field of research. The trend resembles a typical hype cycle. In the beginning, the abstract concept of MC is born from the realisation that product variety is increasing. Then, key enabling technologies, such as the rise of the Internet, web-based collaboration means, and flexible manufacturing systems act as a trigger in the spread of MC, quickly reaching a peak during late 1990s and early 2000s. Until then, most studies are concerned with management and strategic issues of MC, failing to address critical MC implementation issues. Afterwards, researchers realised that a series of sub-problems ought to be tackled first, leading to research indirectly associated with MC (e.g. investigation of product family modelling techniques). Nevertheless, MC is here to stay, therefore, research interest on complete MC solutions starts appearing after 2005 and continues up to the current date.

The rest of the paper is structured as follows. Section 2 presents the evolution of manufacturing paradigms and discusses the recent shift towards customer-centred manufacturing. Section 3 performs a literature review on major topics related to the life cycle of manufacturing networks, together with the latest advances in ICT for supporting the design, planning, and control of manufacturing networks. Section 4 summarises the challenges that need to be addressed, aided by a generic view of the manufacturing landscape of the near future. Finally, Sect. 5 concludes the paper.

2 Evolution of manufacturing and current challenges

2.1 Evolution of manufacturing paradigms

Over time, manufacturing paradigms, driven by the pressure of the environment in which they operate, change in character and evolve in patterns (Fig. 2). The various patterns witnessed up to now can be roughly correlated to movements between three stages: (1) craft shops that employ skilled artisans, (2) long-linked industrial systems using rigid automation, and (3) post-industrial enterprises characterised by flexible resources and information intensive intellectual work [10]. Prevailing manufacturing paradigms are, in chronological order of appearance, the following: craft production, American production, mass production, lean production, mass customisation, and global manufacturing. Apart from American production, all other paradigms are still "operational" today in different industrial sectors [11].

By studying these notable transitions, which are attributed to the pressure applied by social needs, political factors, and advances in technology, it is noticeable that factory systems and technologies have been evolving in two directions. Firstly, they increased the versatility of the allowable products' variety that they produced. This resulted in numerous production innovations, design technology advances, and evolution in management techniques. Secondly, companies have extended factories like tools and techniques. Factories emerged from firms that introduced a series of product and process innovations that made possible the efficient replication of a limited number of designs in massive quantities. This tactic is widely known as economies of scale [12]. Factory systems replaced craft modes of production as firms learned how to rationalise and product designs as well as standardise production itself [13]. Although factory organisations provided higher worker and capital productivity, their structure made it difficult to introduce new products or processes quickly and economically, or to meet the demands of customers with distinctive tastes; factory-oriented design and production systems have never completely replaced craftsmanship or job shops even if the new technologies continue to appear. The result, in economic, manufacturing, and design concepts, has been a shift from simple economies of scale, as in the conventional MP of a limited number of products, to economies of scope and customer integration [14]. It is clear that MP factories or their analogues are not appropriate for all types of products or competitive strategies. Moreover, they have traditionally worked best for limited numbers of variants suited to mass replication and mass consumption. The craft approach offers a less efficient process, at least for commodity products, but remains necessary for technologies that are still new or emerging and continues to serve specific market niches, such as for tailoring products for individual needs and luxury or traditional items. A categorisation of the different production concepts based on the indicators system reconfigurability, demand volatility, and product complexity is depicted in Fig. 3.

Today, issues introduced by the shift of business models towards online purchasing and customisation [15] must be tackled in cost-efficient and sustainable ways in order for companies to maintain their competitiveness and create value [16]. To respond to consumer demand for higher product variety, manufacturers started to offer increased numbers of product "options" or variants of their standard product [17]. Therefore, practice nowadays focuses on strategies and methods for managing product, process, and production systems development that are capable of supporting product variety, adaptability, and leanness, built upon the paradigms of MC and product personalisation. The currently widespread MC is defined as a paradigm for "developing, producing, marketing and delivering affordable goods, and services with enough variety and

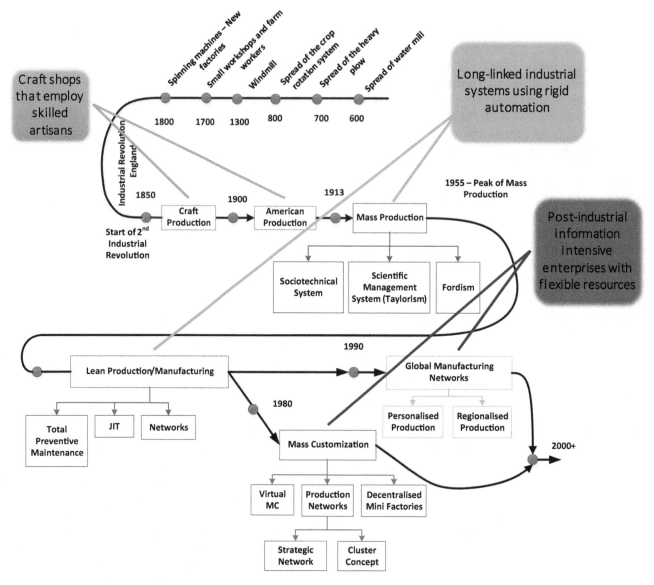

Fig. 2 Evolution of manufacturing paradigms (adapted from [11])

customisation that nearly everyone finds exactly what they want" [17]. This is achieved mostly through modularised product/service design, flexible processes, and integration between supply chain members [18, 19]. MC targets economies of scope through market segmentation, by designing variants according to a product family architecture and allowing customers to choose between design combinations [20]. At the same time, however, MC must achieve economies of scale, in a degree compared to that of MP, due to the fact that it addresses a mass market. Another significant objective for companies operating in an MC landscape is the achievement of economies of customer integration in order to produce designs that the customers really want [14]. On the other hand, personalised production aims to please individual customer needs through the direct integration of the customer in the design

of products. The major differences between the prominent paradigms of MP, MC, and personalisation in terms of goals, customer involvement, production system, and product structure are depicted in Fig. 4.

A research conducted in the UK related to automotive products revealed that 61 % of the customers wanted their vehicle to be delivered within 14 days [21], whereas consumers from North America responded that they could wait no longer than 3 weeks for their car, even if it is custom built [22]. Such studies point out the importance of responsiveness and pro-activeness for manufacturers in product and production design.

During the last 15 years, the number of online purchases has increased and recent surveys show that 89 % of the buyers prefer shopping online to in-store shopping [23]. Web-based and e-commerce systems have been

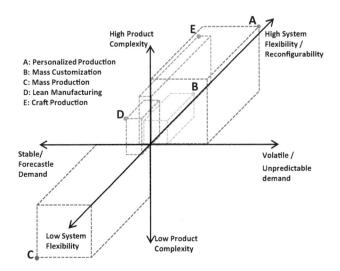

Fig. 3 Characterisation of production paradigms based on demand structure, product complexity, and product flexibility

	Mass production	Mass customization	Personalization
Goal	Economy of Scale	Economy of scope	Value differentiation
Customer involvement	Buy	Choose	Design
Production System	Dedicated Manufacturing System (DMS)	Reconfigurable Manufacturing System (RMS)	On Demand Manufacturing System
Product Structure	Common parts	Common parts / Custom parts	Common parts / Custom parts / Personalized Parts

Fig. 4 Differences between production paradigms (adapted from [20])

implemented and have proved to be very effective in capturing the pulse of the market [24]. These web-based toolkits aim at providing a set of user-friendly design tools that allow trial-and-error experimentation processes and deliver immediate simulated feedback on the outcome of design ideas. Once a satisfactory design is found, the product specifications can be transferred into the firm's production system and the custom product is subsequently produced and delivered to the customer [25]. Still online 2D and 3D configurators do not solve practical issues such as the assembly process of these unique variants. Although proposed approaches include e-assembly systems for collaborative assembly representation [26] and web-based collaboration systems [27], the research in this area needs to be expanded in order to provide tools for assembly representation and product variant customisation. An additional constraint is that globalised design and manufacturing often

require the variants for local markets to be generated by regional design teams, which use different assembly software and source parts from different supply bases [20]. The incorporation of the customers' unique tastes in the product design phase is a fairly new approach to the established ways of achieving product variety and entails significant reorganisation, reconfiguration, and adaptation efforts for the company's production system. Variety is normally realised at different stages of a product life cycle. It can be realised during design, assembly, at the stage of sales and distribution, and through adjustments at the usage phase. Moreover, variety can be realised during the fabrication process, e.g. through rapid prototyping [28].

It should finally be noted that naturally, even if the trends dictate a shift towards personalised product requirements, it should always be considered that forms of production such as MP cannot be abandoned for commodities and general-purpose products, raw materials, and equipment. After all, paradigms are shaped to serve specific market and economical situations.

2.2 Globalisation

Globalisation in manufacturing activities, apart from its apparent advantages, introduces a set of challenges. On the one hand, a globalised market offers opportunities for expanding the sphere of influence of a company, by widening its customer base and production capacity. Information and communication technologies (ICT) and the Internet have played a significant role to that [29]. On the other hand, regional particularities greatly complicate the transportation logistics and the identification of optimum product volume procurement, among other. Indicatively, the difficulty in forecasting product demand was highlighted as early as in 1986 by the following observation from Intel laboratories: when investigating the match between actual call off and the actual forecast, they estimated that supply and demand were in equilibrium for only 35 min in the period between 1976 and 1986 [30, 31]. Enterprises started locating their main production facilities in countries with favourable legislation and low cost of human labour [32]; thus, the management of the supply chain became extremely complex, owing primarily to the fact that a great number of business partners have to mutually cooperate in order to carry out a project, while being driven by opportunistic behaviours. Thus, manufacturing networks need to properly coordinate, collaborate, and communicate in order to survive [33].

On a manufacturing facility level, the impact of supply chain uncertainties and market fluctuations is also considerable. The design and engineering analysis of a complex manufacturing system is a devious task, and the operation of the systems becomes even harder when flexibility and reconfigurability parameters must be incorporated [34].

The process is iterative and can be separated into smaller tasks of manageable complexity. Resource requirements, resource layout, material flow, and capacity planning are some of these tasks [1], which even after decomposition and relaxation remain challenging [35]. In particular, in the context of production for MC businesses, issues such as task-sequence-dependent inter-task times between product families are usually ignored, leading to inexact, and in many cases non-feasible, planning and scheduling. Even rebalancing strategies for serial lines with no other inter-dependencies is challenging, leaving ample room for improvement in order for the inconsistencies between process planning and line balancing to be minimised [20].

From a technological perspective, the increased penetration of ICT in all aspects of product and production life cycles enables a ubiquitous environment for the acquisition, processing, and distribution of information, which is especially beneficial for a globalised paradigm. With the introduction of concepts like cyber physical systems (CPS) and Internet of things (IoT) in manufacturing [36], new horizons are presented for improving awareness, diagnosis, prognosis, and control. Also, the relatively new paradigm of agent-based computation provides great potential for realising desirable characteristics in production, such as autonomy, responsiveness, distributiveness, and openness [37].

3 Manufacturing networks life cycle and mass customisation

In this section, the recent advances and the challenges presented during the life cycle of a manufacturing network are discussed. A typical modern manufacturing network is composed of cooperating original equipment manufacturer (OEM) plants, suppliers, distribution centres, and dealers that produce and deliver final products to the market [38]. The topics discussed include supplier selection, supply chain coordination, initial network configuration, manufacturing network complexity, inventory management, capacity planning, warehousing, lot sizing, ICT support tools, and dynamic process planning, monitoring, and control. These topics are in line with the life cycle phases of a manufacturing network as reported in [39] (Fig. 5).

3.1 Supplier selection

The building blocks of any manufacturing network are the cooperating companies. The significance of the selection of these stakeholders (supplier, vendors) has been indicated as early as in 1966 as stressed in [40] and is known as the supplier selection problem. This decision-making problem is highly challenging since it goes beyond simple comparison of component prices from different suppliers. It is often decomposed into sub-problems of manageable complexity, such as formulation of criteria for the selection, qualification of partners, final selection, and feedback verification. A comprehensive literature review on the issue of supplier selection in agile manufacturing chains is included in [41]. In Fig. 6, the decomposition of the supplier selection problem into small more manageable problems is presented, together with indicative methods for solving these sub-problems.

The supplier selection problem becomes even more complicated in the era of MC since a certain level of adaptability and robustness is necessary when operating within a volatile and rapidly changing environment. The

Fig. 5 Manufacturing network life cycle

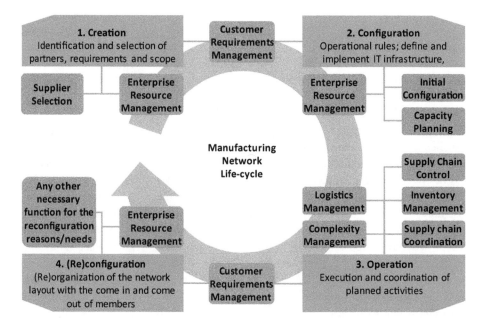

Fig. 6 Supplier selection problem, its decomposition into small more manageable problems, and indicative methods for solving them (adapted from [51])

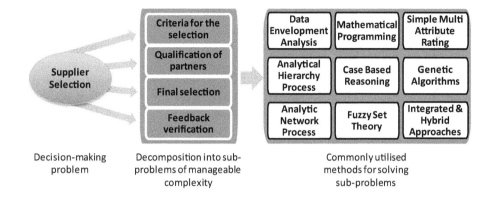

Decision-making problem   Decomposition into sub-problems of manageable complexity   Commonly utilised methods for solving sub-problems

most commonly used criteria in supplier selection studies include quality and performance [42]. However, when having to deal with unpredictability and fluctuating demand, which are common in MC, additional factors need to be considered such as management compatibility, transparency of operations, strategic direction, reliability, and agility [43]. While trying to adhere to eco-friendliness directives, frameworks like the one proposed in [44] incorporate environmental footprint criteria to green supply chain design. Moreover, several other criteria may be relevant according to the design and planning objectives of a niche supply chain, which could be identified using data mining methods [45].

The Internet and web-based platforms are used in recent years in order to counterbalance uncertainty, monitor altering parameters (e.g. weather in supply routes), and proactively adapt to changes [46]. Moreover, several proposed supplier selection models incorporate the relative importance of the supplier selection factors depending on the types of targeted MC implementation, e.g. for the component-sharing modularity type of MC, the requirements for selecting suppliers would not be the same as the component-sweeping modularity implementation type [47]. Like in the case of a stable low variety production, the analytic hierarchy process (AHP) is commonly used as a means to solve the multi-criteria decision-making problem of supplier selection. Incorporating uncertain information about the real world, essentially extending the Dempster–Shafer theory, the authors in [48] propose the D-AHP method for solving the supplier selection problem. The suggested D numbers preference relation encapsulates the advantages of fuzziness and handles possible incomplete and imprecise information, which is common in human-driven systems such as supply chains. Similarly, a combined analytic hierarchy process—quality function deployment (AHP–QFD) framework is described by [49] that handles uncertain information, selects suppliers, and allocates orders to them. A multi-criteria decision-making method to support the identification of business-to-business

(B2B) collaboration schemes, especially for supplier selection is proposed in [50].

3.2 Supply chain coordination

The literature on organisational knowledge creation points out that "coordination" plays an important role in combining knowledge from stakeholders [52], while it also mediates the relationship between product modularity and MC [53]. A report on coordination mechanisms for supply chains was compiled in [54].

Concerning coordination in supply chains, in general, two topologies are studied, namely the centralised and the decentralised one [11] (Fig. 7). In the first, the coordination decisions are taken by a central body, often the leading supply chain OEM, whereas in the second, each member independently makes its own operational decisions. The decentralised topology has been proven to improve the performance in the context of MC [38, 55]. A supply chain that is commissioned to provide a variety of customised products requires a total systems approach to managing the

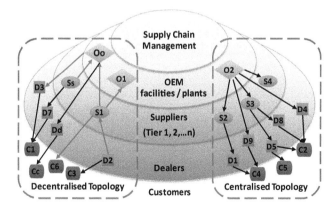

Fig. 7 Centralised and decentralised supply chain topologies. In a centralised topology, material and information move only downstream. In the decentralised one, material/information can be transferred both upstream and downstream to better serve customisation, personalisation, and/or regionalisation [6]

entire flow of information, materials, and services in fulfilling customer demand [56]. Further incentives have to be provided to the members, so as to entice their cooperation through the distribution of the benefits of the coordination for instance.

The need for adaptation to the new MC requirements has led to the definition of a novel framework for autonomous logistics processes. The concept of autonomous control "describes processes of decentralised decision-making in heterarchical structures, and it presumes interacting elements in non-deterministic systems, and possess the capability and possibility to render decisions independently [57]". However, regardless the topology, the alignment of the objectives of the different collaborating organisations in order to successfully carry out projects, optimise system performance, and achieve mutual profits is indispensable [58]. While an action plan suffices for the coordination of a centralised supply chain, it is inadequate with a decentralised one [59] since entities tend to exhibit opportunistic behaviour. Nevertheless, in terms of overall network performance, decentralised topologies have shown great benefits for serving the mass customisation paradigm [6, 7].

3.3 Initial manufacturing network configuration

The initial manufacturing network configuration must consider the long-term needs of cooperation and often determines its success. In a constantly changing environment, the configuration of the manufacturing network must be, therefore, flexible and adaptable to external forces. The problem has been extensively addressed in the literature using approaches classified in two main categories, namely approximation (artificial intelligence, evolutionary computation, genetic algorithms, tabu search, ant colony optimisation, simulated annealing, heuristics, etc.) and optimisation techniques (enumerative methods, Lagrangian relaxation, linear/nonlinear integer programming, decomposition methods, etc.) and their hybrids [60, 61] (Fig. 8).

Focusing on agile supply chains, a hybrid analytic network process mixed-integer programming model is proposed in [62] with uttermost aim the fast reaction to customer demands. Fuzzy mathematical programming techniques have been employed to address the planning problems for multi-period, multi-product supply chains [63]. A coloured Petri Nets approach for providing modelling support to the supply chain configuration issue is included in [64]. A dynamic optimisation mathematical model for multi-objective decision-making for manufacturing networks that operated in a MC environment is suggested in [65].

Still, the accuracy of planning ahead in longer horizons is restricted. The incorporation of unpredictable parameters in the configuration through a projection of the possible setting of the network in the future may lead to unsafe results.

3.4 Inventory management/capacity planning/lot sizing

Inventories are used by most companies as a buffer between supply chain stages to handle uncertainty and volatile demand. Prior to the 1990s, where the main supply chain phases, namely procurement, production, and distribution, were regarded in isolation, companies maintained buffers of large inventories due to the lack of regulatory mechanisms and feedback [66]. The basis for manufacturing and inventory planning was relatively safe forecasts. However, in the era of customisation the basis is actual orders and the pursuit is minimisation of inventories. These requirements constitute inventory management and capacity planning functions very important for a profitable MC implementation.

In complex distributed systems such as modern manufacturing companies with a global presence, the question of optimal dimensioning and positioning of inventory emerges as a challenging research question. Various strategies for inventory planning have been reported based on how

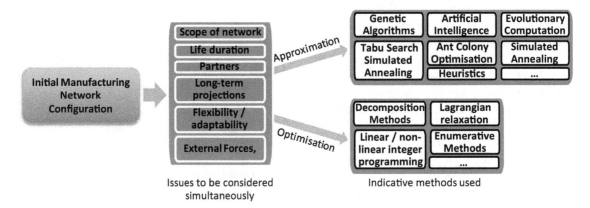

Fig. 8 Issues to be considered during the initial manufacturing network configuration and indicative methods used

Fig. 9 Methods (indicative) used for solving the capacitated lot sizing problem (adapted from [70])

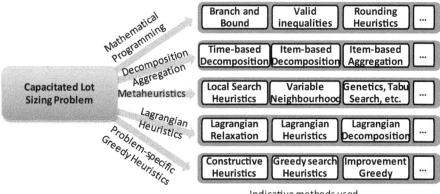

Indicative methods used

the underlying demand and return processes are modelled over time, thus making a distinction between constant, continuous time-varying, and discrete time-varying demand and return models [67]. Integrated capacity planning methods encompassing stochastic dynamic optimisation models over volatile planning horizons exhibit high performance in the context of MC and personalisation [68]. The DEWIP (decentralised WIP) control mechanism was proposed in [69], focusing on establishing control loops between work centres for adjusting the WIP levels dynamically. Its performance was assessed against other well-accepted systems such as LOOR, Conwip, and Polca. Methods used for solving the capacitated lot sizing problem are indicatively shown in Fig. 9.

In particular, in just-in-time (JIT) environments, MC impacts the amount of inventory that needs to be carried by firms that supply many part variants to a JIT assembly line. In addition, the supply of parts is performed either on constant order cycles or more commonly under non-constant cycles [71]. The goal chasing heuristic, pioneered within the Toyota production system, seeks to minimise the variance between the actual number of units of a part required by the assembly line and the average demand rate on a product-unit-by-product-unit basis, while applying penalties for observed shortages or overages [72]. Of course, information sharing and partner coordination systems are a prerequisite for JIT procurements. For instance, DELL, which achieved a highly coordinated supply chain to respond to MC, communicated its inventory levels and replenishment needs on an hourly basis with its key suppliers and required from the latter to locate their facilities within a 15-min distance from DELL facilities [73]. Another consideration during inventory management is the type of postponement applied in a company. Studies have shown that postponement structures allow firms to meet the increased customisation demands with lower inventory levels in the case of time postponement (make-to-order), or with shorter lead times in the case of form postponement [74]. Also, an assemble-to-order process, a variation of

form postponement, does not hold inventory of the finished product, while in form postponement, finished goods inventory for each distinct product at the product's respective point of customisation is kept [75]. An indicative example is given in the case of Hewlett Packard, where using form postponement, the company achieved the postponement of the final assembly of their DeskJet printers to their local distribution centres [76].

3.5 Logistics management

Logistics can play a crucial role in optimising the position of the customer order decoupling point and balance between demand satisfaction flexibility and productivity [77]. In a customer-centric environment, the supply chain logistics must be organised and operated in a responsive and at the same time cost-effective manner. Customisation of the bundle of product/services is often pushed downstream the supply chain logistics, and postponement strategies are utilised as an enabler for customisation [78]. Maintaining the product in a neutral and non-committed form for as long as possible, however, implicates the logistics process. Traditional logistics management systems and strategies need to be revisited in the context of customisation, since distribution activities play a key role in achieving high product variety, while remaining competitive. Most OEMs form strategic alliances with third-party logistic (TPLs) companies. The introduction of TPLs in the supply chain serves two purposes. First, it acts as a means of reducing the complexity of management for an OEM through shifting the responsibilities of transportation, and in many times customisation, to the TPLs [79]. Second, it extends the customisation capabilities as TPLs can actively implement postponement strategies [80]. Postponement strategies with logistics as an enabler are located at the bottom of Fig. 10 and can serve all types of customisation, from plain shipment to order up to extremes of engineer-to-order or personalisation.

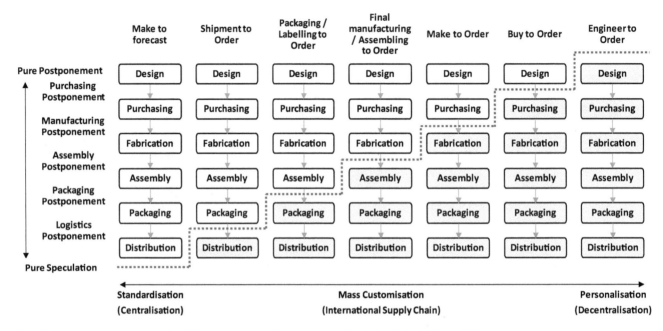

Fig. 10 Postponement strategies for different supply chain structures and logistics (adapted from [81])

Moreover, the management of logistics is a process inherently based on communication and collaboration. Developments of either function-specific or all-in-one ICT solutions targeted on logistics are analysed in [6, 82]. Tools for warehouse and transportation management, ERP, supply chain management (SCM), and information sharing are reported under the umbrella of e-logistics. The concept of virtual logistics is also proposed for separating the physical and digital aspects of logistics operations [83], having Internet as an enabling means to handle ownership and control of resources.

3.6 Supply chain control

The information transferred from one supply chain tier to the next in the form of orders is often distorted, a phenomenon known as the bullwhip effect. In particular, when customer demand is volatile such as the case is in MP, the bullwhip effect misguides upstream members of the supply chain in their inventory and production decisions [84]. Nevertheless, the performance of the supply chain is highly sensitive to the control laws used for its operation. The application of the wrong control policy may have as a result the amplification of variance instead on its minimisation. Dynamic modelling approaches have been proposed to manage supply chains, accounting for the flow of information and material, to capture the system dynamics [85]. Multi-agent approaches for modelling supply chain dynamics are proposed in [86]. Software components known as agents represent supply chain entities (supplier, dealers, etc.), their constituent control elements (e.g. inventory policy), and their interaction protocols (e.g. message types). The agent framework utilises

a library of supply chain modelling components that have derived after analysis of several diversified supply chains. For instance, a novel oscillator analogy in presented in [87] for modelling the manufacturing systems dynamics. The proposed analogy considers a single degree of freedom mass vibrator and a production system, where the oscillation model has as input forces, while the manufacturing system has demand as excitation. The purpose is to use this simple oscillator analogy to predict demand fluctuations and take actions towards alignment.

Another necessity in supply chain control is the traceability of goods. Traceability methods, essential for perishable products and high-value shipments, exploit the radio frequency identification (RFID) technology during the last years [88, 89]. A traceability system that traces lots and activities is proposed by Bechini et al. [90]. The study examines the problem from a communication perspective, stressing the need to use neutral file formats and protocols such as XML (extended markup language) and SOAP (simple object access protocol) in such applications. The emerging technology of IoT can provide ubiquitous traceability solutions. Combining data collection methods based on wireless sensor network (WSN) with the IoT principles, the method proposed in [91] can support the traceability of goods in the food industry. In a similar concept, the role of an IoT infrastructure for order fulfilment in a collaborative warehousing environment is examined in [92]. The IoT infrastructure is based on RFID, ambient intelligence, and multi-agent system, and it integrates a bottom-up approach with decision support mechanisms such as self-organisation and negotiation protocols between agents based on a cooperation concept.

Supply chains formed for servicing customisation are more complex as structures and less predictable in their dynamic behaviour than stable traditional supply chains. Recent complexity studies deal with the emerging aspects of increasing complexity of manufacturing activities and the dynamic nature of supply chains [93]. The importance of managing the complexity in supply chains is evident, as recent studies depict that lower manufacturing network complexity is associated with reduced costs and overall network performance [38, 94]. A complete and comprehensive review of complexity in engineering design and manufacturing is presented in [95–97].

3.7 Simulation and ICT support systems for manufacturing networks life cycle

Robust and flexible ICT mechanisms are rendered necessary for improving performance in each of the previous life cycle aspects of supply chains and for bridging inter- and intra-enterprise collaboration environments. Digital enterprise technologies in general represent an established, new synthesis of technologies and systems for product and process development and life cycle management on a global basis [98] that brings many benefits to companies. For instance, the benefits offered by the adoption of virtual engineering through the life cycle of production are shown in Fig. 11 [99]. To manage the huge portfolio of products and variety, as well as tracking the expanding customer base, ERP and CRM suites are necessary tools. Additionally, cloud technology is already revolutionising core manufacturing aspects and provides ample benefits for supply chain and manufacturing network life cycle. Cloud technology and the IoT are major ICT trends that will reshape the way enterprises function in the years to come [100, 101].

3.8 Simulation for manufacturing network design

Literature on ICT-based systems for improving manufacturing networks is abundant and highlights the need for increased penetration of ICT systems in design, planning,

and operation phases. A simulation-based method to model and optimise supply chain operations by taking into consideration their end-of-life operations is used to evaluate the capability of OEMs to achieve quantitative performance targets defined by environmental impacts and life cycle costs [102]. A discrete event simulation model of a capacitated supply chain is developed and a procedure to dynamically adjust the replenishment parameters based on re-optimisation during different parts of the seasonal demand cycle is explained [103]. A model is implemented in the form of Internet-enabled software framework, offering a set of characteristics, including virtual organisation, scheduling, and monitoring, in order to support cooperation and flexible planning and monitoring across extended manufacturing enterprise [58]. Furthermore, the evaluation of the performance of automotive manufacturing networks under highly diversified product demand is succeeded through discrete event simulation models in [55] with the use of multiple conflicting user-defined criteria such as lead time, final product cost, flexibility, annual production volume, and environmental impact due to product transportation. Finally, the application of the mesoscopic simulation approach to a real-world supply chain example is illustrated utilising the MesoSim simulation software [104, 105].

Existing simulation-based approaches do not tackle the numerous issues of manufacturing network design in a holistic integrated manner. The results of individual modules used for tackling network design sub-problems often contradict each other because they refer to not directly related manufacturing information and context (e.g. long-term strategic scheduling vs. short-term operational scheduling), while harmonising the context among these modules is challenging. This shortcoming hinders the applicability of tools to real manufacturing systems as it reduces the trustworthiness of results to the eyes of the planner among other reasons.

3.9 Enterprise resource planning

An ERP system is a suite of integrated software applications used to manage transactions through company-wide

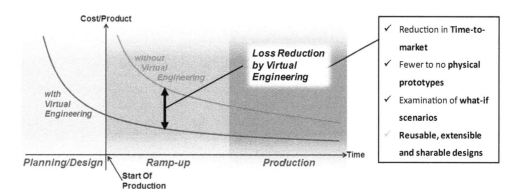

Fig. 11 Increased efficiency through virtual engineering approaches (adapted from [99])

business processes, by using a common database, standard procedures, and data sharing between and within functional areas [106]. Such ICT systems entail major investments and involve extensive efforts and organisational changes in companies that decide to employ them. ERP systems are becoming more and more prevalent throughout the international business world. Nowadays, in most production distribution companies, ERP systems are used to support production and distribution activities and they are designed to integrate and partially automate financial, resource management, commercial, after-sale, manufacturing, and other business functions into one system around a database [107].

A trend, especially in the mid-market, is to provide specific ERP modules as services. Such need generates the challenge for ERP system providers to offer mobile-capable ERP solutions. Another issue is the reporting and data analysis, which grows with the information needs of users. Research in big data analytics and business intelligence (BI) should become more tightly integrated with research and applications of ERP.

3.10 Customer relationship management

In Internet-based retailing, which is the preferred business model followed in MC, customer information management is a necessity. In particular, exploiting consumer data, such as purchase history, purchasing habits, and regional purchasing patterns, are the cornerstone of success for any company active in MC. In business-to-business and business-to-customer, CRM suites are thus indispensable. According to Strauss and Frost [108], CRM involves, as a first step, research to gain insight so as to identify potential and current customers. In a second step, customer information is used to differentiate the customer base according to specific criteria. Finally, the third step involves customised offerings for those customers that are identified as "superior" from the previous phase, enabling thus, the targeted offering of customised products. During the first step of identification of customers, market research and consumer behaviour models are used. In a second phase, for establishing differentiation techniques, data mining and KPIs assessment are used. Finally, for fine-tuning customisation options, information such as price, variants, promotions, and regions are examined [109].

As Internet becomes ubiquitous in business, CRM has been acknowledged as an enabler for better customisation since it offers management of the new market model less disruptively. Internet-enabled CRM tools also bring the customer closer to the enterprise and allow highly responsive customer-centred systems without significant increase in costs [110]. e-CRM implementations have been assessed in the study [111]. Noticeably, most major CRM

suite vendors have started providing cloud-based services, a business model that suits SMEs that cannot afford huge ICT investments. Based on the balanced scorecard method, the study in [112] assessed e-CRM performance using 42 criteria in a number of companies. The results show that a successful CRM implementation is associated with tangible outcomes, such as improvements in financial indicators, customer value, brand image, and innovation. Finally, the latest generation of CRM tools, referred to as social CRM, exploit social networking technology to harness information about customer insights and engagement.

3.11 Cloud computing and manufacturing

A comprehensive definition of cloud computing is provided by the National Institute of Standards and Technology: "a model for enabling ubiquitous, convenient, on-demand network access to a shared pool of configurable computing resources (e.g. networks, servers, storage, applications, and services) that can be rapidly provisioned and released with minimal management effort or service provider interaction" [113]. Several applications have been reported in recent years where a cloud infrastructure is used to host and expose services related to manufacturing, such as: machine availability monitoring [114], collaborative and adaptive process planning [115], online tool-path programming based on real-time machine monitoring [116], manufacturing collaboration and data integration based on the STEP standard [117], and collaborative design [118].

The benefits of cloud for improving manufacturing network performance are numerous (Table 1). Cloud can offer increased mobility and ubiquitous information to an enterprise since the solutions it offers are independent of device and location. Moreover, computational resources are virtualised, scalable, and available at the time of demand. Therefore, the intensive costs for deploying high-performance computing resources are avoided. In addition to that, purchasing the application using the model software as a service is advantageous for SMEs who cannot afford the huge investments that commercial software suites entail [119]. However, there are some considerations also (Table 1). A main challenge for the adoption of cloud in manufacturing is the lack of awareness on security issues. This major issue can be addressed using security concepts and inherently safe architectures, such as privately deployed clouds. The security concept must include availability of ICT systems, network security, software application security, data security, and finally operational security. Considerable funding is spent by the global security software market, in order to alleviate security issues. Recent reports show that the expenditure on cloud security is expected to rise 13-fold by 2018 [120, 121]. Moreover, there is the possibility of backlash from

Table 1 Benefits and drawbacks of cloud technology for manufacturing

Benefits	Drawbacks
Increased mobility that allows decentralised and distributed SCM	Lack of standardisation and protocols create hesitation in adoption of Cloud solutions
Ubiquitous access to information context empowering decision-making	Security and lack of awareness on security issues, especially in SMEs, that are part of supply chains/clusters
Device and location independent offering context-sensitive visualisation of crucial data relevant to the mfg. network	Privacy issues generate legal concerns, identity management, access control, and regulatory compliance
Hidden complexity permits the diffusion of ICT solutions even to traditional, averted by disruptive solutions, sectors	Dependence on the cloud provider (provider stops providing services, absence of contracts/regulation)
Virtualised and scalable on-demand computational resources (problems of varied computational complexity)	Loss of control over data (assuring smaller companies that their data are not visible by anyone in the supply chain, but the owner is challenging)
Low cost for SMEs that cannot afford huge ICT investments and lack the know-how to maintain them	

entrenched ideas, manufacturing processes and models caused by the hesitation for the adoption of innovative technology. Finally, the lack of standardisation and regulation around cloud hinders its acceptance by the industry [122].

4 Challenges for future manufacturing

MC provides a set of enabling concepts and methods for providing the customer with products they desire and for organising production resources and networks to realise these products. However, on a practical strategic, tactical, and operational level, the tools for the realisation of MC are under development and refinement and a number of issues related to the design of manufacturing networks and their management are still not tackled in a holistic integrated manner. Several particular challenges need to be addressed as described below. Possible solutions are also proposed in the context of supporting a more efficient implementation of MC and personalisation.

4.1 Challenges for the manufacturing network life cycle

Regarding supplier selection, existing frameworks that handle both selection of suppliers, order allocation, and capacity planning are rare in the literature. Therefore, inconsistencies between the design phase and the actual implementation of the supply chain are a common issue. The problem most commonly treated jointly with supplier selection is the order allocation problem, as reported in the works of [123–125] among other. Moreover, several studies point out the difficulties of coordination between large networks of stakeholders. Potential solutions in novel approaches to tackling the issues generated in supply chain coordination for the procurement of customised products

are proposed, such as in [126], where organisation flatness is proposed as a mediator for enhancing MC capability. Flatness in cross-plant and cross-functional organisation alleviate the need to decisions to pass through multiple layers of executives, simplifying coordination and information sharing [127]. Among the several challenges for configuring robust manufacturing networks to satisfy MC are the need for frameworks that handle the entire order fulfilment life cycle (from product design to delivery), methods to allow easy modelling and experimentation of what-if scenarios and deeper examination of the impact of product variety on the performance of manufacturing networks. On the field of SCM, identifying the benefits of collaboration is still a big challenge for many. The definition of variables, such as the optimum number of partners, investment in collaboration, and duration of partnership, are some of the barriers of healthy collaborative arrangements that should be surpassed [128]. Available solutions for lot sizing are following traditional approaches and are not able to address the increasing complexity of problems arising in the modern manufacturing network landscape. The economic order quantity (EOQ), established for more than 100 years, still forms the basis of recent lot sizing practices. In setups of complex and changeable products, the problem of lot sizing becomes extremely complex. Nevertheless, the optimality of inventory and capacity planning is often neglected due to increased complexity of the supply chain problems which comes with higher priority. For instance, in multi-agent manufacturing systems, each agent resolves inventory issues in its domain partition level, without clear global optimisation overview [37]. Furthermore, the broader role of logistics capabilities in achieving supply chain agility has not been addressed from a holistic conceptual perspective [129]. Therefore, an open research question is the relationship between logistics capabilities and supply chain agility. Regarding ERP suites, apart from their apparent benefits, the reported successful implementations of ERP systems are limited when considering implementation costs and disruptions caused in production [130]. One reason for the low success rates in ERP implementations is attributed to the organisation changes needed for the industry that disrupt normal flow of business. Another reason is that production planning, a core function handled by currently deployed closed-loop MRPII (manufacturing resource planning) and ERP suites, is performed through the fundamental MRP (material requirements planning) logic [1, 131]. This leads to the generation of low-detail shop-floor schedules, assuming infinite production capacity and constant time components, thus leading to inflated lead times [132]. Challenges on the technological level of ERP systems include delivery of software as a service, mobile technology, tightly integrated business intelligence, and big data analytics [133, 134]. Challenges in the field of product data management (PDM) are related to the efficiency of these systems with regard to studying factors that affect the accessibility of product data, for instance, the nature of data in different timeframes of a development, the relationship between the maturity of the data, and the probability of them being modified [135]. The deployment and tight integration of product life cycle management (PLM) tools must also be considered since they bring an abundance of benefits against current manufacturing challenges. Yet these benefits are still not appreciated by many industrial sectors, mainly due to the following reasons: (1) they are complex as a concept and understanding their practical application is difficult, (2) they lack a holistic approach regarding the product life cycle and its underlying production life cycle and processes, and (3) the gap between research and industrial implementation is discouraging [136]. Concerning CRM, although data rich markets can exploit the feedback of consumers through social networks to identify user polarity towards a product–service, improve its design, and refine a product service system (PSS) offering, only few initiatives have tapped that potential.

Further challenges that are related indirectly to the previous aspects are discussed hereafter. Concerning individual disparate software modules, it is often observed that they contradict each other because they refer to not directly related manufacturing information and context. The harmonisation, both on an input/output level and to the actual contents of information, is often a mistreated issue that hinders the applicability of tools to real-life manufacturing systems. Limitations of current computer-aided design (CAD) tools include: the complexity of menu items or commands, restricted active and interactive assistance during design, and inadequate human–computer interface design (focused on functionality) [137]. To fulfil the needs of modern manufacturing processes, computer-aided

process planning should be responsive and adaptive to the alterations in the production capacity and functionality. Nowadays, conventional computer-aided process planning (CAPP) systems are incapable of adjusting to dynamic operations, and a process plan, created in advance, is found improper or unusable to specific resources [138]. Highlighted challenges for life cycle assessment (LCA) are modularisation and standardisation of environmental profiles for machine tools, as well as modelling of "hidden flows" and their incorporation in value stream mapping tools [139, 140]. Regarding knowledge management and modelling, reusable agent-oriented knowledge management frameworks, including the description of agent roles, interaction forms, and knowledge description, are missing [141]. Moreover, ontologies used for knowledge representation have practical limitations. In case an ontology is abstract, its applicability and problem-solving potential may be diminished. On the other hand, in the case of very specific ontologies, reasoning and knowledge inference capacities are constrained [142]. Furthermore, in the turbulent manufacturing environment, a key issue of modern manufacturing execution systems is that they cannot plan ahead of time. This phenomenon is named decision myopia and causes undoubtedly significant malfunctions in manufacturing [143]. In the field of layout design and material simulation, some commercial software can represent decoupling data from 3D model and export them in XML or HTML format. While this is an export of properties, it cannot fully solve the interoperability and extensibility issues since the interoperability depends on how the different software and users define contents of data models [144]. Concerning material flow simulation, it can be very time-consuming to build and verify large models with standard commercial-off-the-shelf software. Efficient simulation model generation will allow the user to simplify and accelerate the process of producing correct and credible simulation models [145]. Finally, while the steady decline in computational cost renders the use of simulation very cost-efficient in terms of hardware requirements, commercial simulation software has not kept up with hardware improvements.

4.2 Solutions for addressing the challenges in the future manufacturing landscape

A view of the manufacturing system of the near future that incorporates the latest trends in research and ICT developments and can better support MC is shown in Fig. 12. It is envisioned that, fuelled by disruptive technologies such as the IoT and cloud technology, entities within supply chains will exchange information seamlessly, collaborate more efficiently, and share crucial data in real time. Data acquisition, processing, and interpretation will be

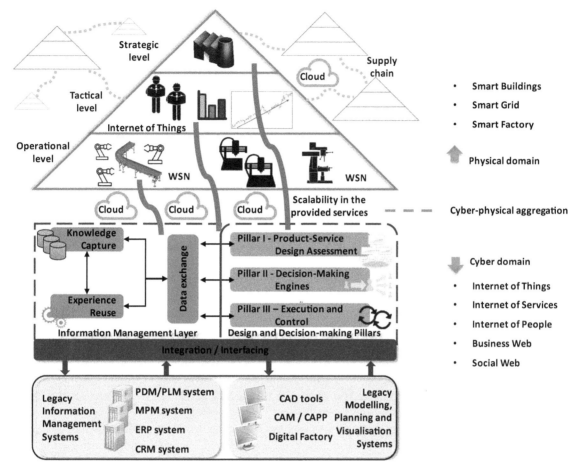

Fig. 12 View of manufacturing in the near future

supported by wireless sensor networks. The information will be available on demand and on different degrees of granularity empowered by big data analytics. Drilling down to specific machine performance and zooming out to supply chain overview will be practically feasible and meaningful. The distinction between the physical and the digital domains will become less clear. Besides, physical resources are already considered as services under the cloud manufacturing paradigm. A tighter coupling and synchronisation between the life cycles of product, production, resources, and supply chains will be necessary, while the distinction between cyber and physical domains will become hazier. A discussion on potential directions for adhering to this view of manufacturing is provided hereafter.

New technologies and emerging needs render traditional SCM and manufacturing network design models obsolete. To support manufacturing network design, planning, and control, a framework that integrates, harmonises, processes, and synchronises the different steps and product-related information is needed. The framework will be capable of supporting the decision-making procedure on all organisation levels in an integrated way, ranging from the overall management of the manufacturing network, down to the shop-floor scheduling fuelled by big data analytics, intuitive visualisation means, smart user interfaces, and IoT. An alignment and coordination between supply chain logistics and master production schedules with low-level shop-floor schedules is necessary for short-term horizons. The framework needs not be restricted on a particular manufacturing domain; since it is conceived by addressing universal industrial needs, its applicability to contemporary systems is domain-independent. The constituents of the framework are described hereafter.

The system will be supported by automated model-based decision-making methods that will identify optimum (or near-optimum) solutions to the sub-problems identified above, such as for the problem of the configuration of manufacturing networks capable of serving personalised product–services. The method must consider the capabilities of the manufacturing network elements (suppliers of different tiers, machining plants, assembly plants, etc.) and will indicate solutions to the warehouse sizing problem, to the manufacturing plant allocation, and to the transportation logistics. The decision support framework requires interfacing with discrete event simulation models of manufacturing networks and assessment of multiple conflicting and user-defined performance indicators.

The joint handling of order allocation, supplier selection, and capacity planning is necessary to alleviate inconsistencies between the supply chain design and implementation phases under a flatness concept. The incorporation of the entire order fulfilment life cycle is additionally envisioned, enhanced with methods that allow easy modelling and experimentation on what-if scenarios. The relationship between logistics capabilities and supply chain agility can also be revealed through this holistic view of the constituents of the supply chain.

Regarding SCM, collaboration concepts based on cloud computing and cloud manufacturing are a game changer. Through the sharing of both ICT as well as manufacturing resources, SMEs can unleash their innovation potential and thus compete more easily in the global market.

Further to that, the measurement and management of the manufacturing network complexity should be considered as a core strategic decision together with classical objectives of cost, time, and quality. Handling a variety of market excitations and demand fluctuations is the standard practice even today in many sectors, while this trend is only bound to intensify. In parallel, a risk assessment engine should correlate complexity results and leverage them into tangible risk indicators. Complexity can then be efficiently channelled through the designed network in the less risky and unpredictable manner.

To address the increasing complexity of problems arising in the modern manufacturing network landscape, the lot sizing and material planning need to be tightly incorporated to the production planning system. The consideration of capacitated production constraints is needed in order to reflect realistic system attributes. A shared and distributed cloud-based inventory record will contain information related to MRP and ERP variables (e.g. projected on-hand quantities, scheduled order releases and receipts, changes due to stock receipts, stock withdrawals, wastes and scrap, corrections imposed by cycle counting, as well as static data that describe each item uniquely). This record should be pervasive and contain dataset groups relevant to intradepartmental variables, as well as datasets visible only to suppliers and relevant stakeholders, in order to increase the transparency of operations.

The mistreated issues of deployment and tight integration of PLM, ERP, and CRM tools must also be tackled through interfacing of legacy software systems and databases for seamless data exchange and collaboration. Software as a service PLM, ERP, and CRM solutions available to be purchased per module will be the ideal ownership model since it allows greater degree of customisation of

solutions, more focused ICT deployment efforts, and reduced acquisition costs. CAD/CAM, PDM, and MPM (manufacturing process management) systems and databases will be interfaced and interact with digital mock-ups of the factory and product–services solutions as well for synchronising the physical with the digital worlds. In addition, the knowledge capturing and exploitation is pivotal in the proposed framework. Product, process, and production information is acquired from production steps and is modelled and formalised in order to be exploited by a knowledge reuse mechanism that utilises semantic reasoning. This mechanism is comprised of an ontological model that is queried by the knowledge inference engine and allows the retrieval of knowledge and its utilisation in design and planning phases. The developments should also mediate the deeper examination of the impact of product variety on the performance of manufacturing networks.

In parallel, there is an urgent need of standardisation and harmonisation of data representation for manufacturing information, for example: the product information (BoM, engineering-BoM and manufacturing-BoM [146]), the manufacturing processes (bill of processes—BoP) including the manufacturing facilities layout, the associated relations (bill of relations—BoR), and related services (Bill of Services—BoS) should be pursued through a shared data model. Moreover, the product complexity needs to be assessed based on functional product specifications using, for instance, design structure matrices (DSM) [147], which incorporate components (BoM), the required manufacturing and assembly processes (BoP) including sequences/plans, relationships (BoR), and the accompanying services (BoS). The complexity of the product in relation to the manufacturing network and service activities (impact on delivery time and cost, and effect on the overall reliability) will be quantified and will be incorporated in the decision-making process.

Last but not least, it should be noted that the components of the proposed framework must be offered following a software as a service delivery method and not as a rigid all-around platform. The framework should act as a cloud-based hub of different solutions, offering web-based accessibility through a central "cockpit" and visualisation of results through common browser technology and hand-held devices (tablets, smartphones, etc.).

5 Conclusions

The ability to customise a product/service is offered to consumers for many years now, while truly unique products will be requested in the near future by users around the globe [148] using the Internet as a means of integration in the design process. In addition, the shortening of life cycles and time to market, increased outsourcing, manufacturing at dispersed sites, and the diverse cooperation in networks increase the complexity of production [149]. Agility, reconfigurability, and synchronisation from process up to supply chain levels are necessary in order for companies to respond effectively to the ever-changing market needs [150]. Driven by the ever-increasing need to reduce cost and delivery times, OEMs are called to efficiently overcome these issues by designing and operating sustainable and efficient manufacturing networks.

This work reviewed the existing literature related to the basic aspects of a manufacturing network life cycle within the MC landscape. The focus was to study existing practices and highlight the gaps in the current approaches related to these aspects of manufacturing network design, planning, and operation. Afterwards, the identification of future directions of academic and industrial research is proposed. Departing from that, major drivers and enabling technologies are identified and concepts that can lead to a more sustainable implementation of MC are proposed.

Summing up, the theoretical foundations of MC have been laid for many years now [150]. Still, there is an apparent gap between the theoretical and the actual application of MC, and bridging this gap is a challenging task that needs to be addressed. A safe conclusion reached is that the complexity generated in manufacturing activities due to the exploding product variety requires a systematic approach to be considered during the design, planning, and operating of the entire manufacturing system [5]. All in all, piecemeal digitalisation of manufacturing network is not a viable option; revisiting of the entire supply and manufacturing network life cycle is essential for sustainability. The pursuit for a smoother, more efficient, more rewarding, and eco-friendly manufacturing is ongoing.

References

1. Chryssolouris G (2006) Manufacturing systems: theory and practice, 2nd edn. Springer, New York
2. Chryssolouris G, Papakostas N, Mavrikios D (2008) A perspective on manufacturing strategy: produce more with less. CIRP J Manuf Sci Technol 1(1):45–52
3. Tolio T, Ceglarek D, ElMaraghy HA, Fischer A, Hu SJ, Laperrière L, Newman ST, Váncza J (2010) SPECIES—co-evolution of products, processes and production systems. CIRP Ann Manuf Tech 59(2):672–693
4. ElMaraghy H, Schuh G, ElMaraghy W, Piller F, Schönsleben P, Tseng M, Bernard A (2013) Product variety management. CIRP Ann Manuf Tech 62(2):629–652
5. Chryssolouris G, Papakostas N (2014) Dynamic Manufacturing Networks in a Globalised Economy. In: 8th International conference on digital enterprise technology—DET 2014. Keynote Presentation, 25–28 Mar 2014, Stuttgart, ISBN: 9783839606971
6. Doukas M, Psarommatis F, Mourtzis D (2014) Planning of manufacturing networks using an intelligent probabilistic approach for mass customised products. Int J Adv Manuf Tech 74(9):1747–1758
7. Mourtzis D, Doukas M, Psarommatis F (2015) A toolbox for the design, planning, and operation of manufacturing networks for mass customization. J Manuf Syst 36:274–286
8. Santoso T, Ahmed S, Goetschalckx Shapiro A (2005) A stochastic programming approach for supply chain network design under uncertainty. Eur J Oper Res 167(1):96–115
9. Da Silveira G, Borenstein D, Fogliatto FS (2001) Mass customisation: literature review and research directions. Int J Prod Econ 72(1):1–13
10. Gunasekaran A, Ngai EWT (2012) The future of operations management: an outlook and analysis. Int J of Prod Econ 135(2):687–701
11. Mourtzis D, Doukas M (2014) The evolution of manufacturing systems: From craftsmanship to the era of customisation. In Modrak V, Semanco P, (eds) Design and management of lean production systems. IGI Global, ISBN13: 9781466650398, http://www.igi-global.com/book/handbook-research-design-management-lean/84172/
12. Chandler AD (1990) Scale and scope. Harvard University Press, Cambridge
13. Hounshell A (1984) From the American system to mass production, 1800–1932: the development of manufacturing technology in the United States. Johns Hopkins University Press, Baltimore
14. Piller FT, Möslein KM (2002) From economies of scale towards economies of customer interaction: value creation in mass customisation based electronic commerce. In: 15th Bled electronic commerce conference, TU München, pp 214–228
15. Thirumalai S, Sinha K (2011) Customisation of the online purchase process in electronic retailing and customer satisfaction: an online field study. J Oper Manag 29:477–487
16. Ueda K, Takenaka T, Váncza J, Monostori L (2009) Value creation and decision-making in sustainable society. CIRP Ann Manuf Tech 58(2):681–700
17. Koren Y (2009) The global manufacturing revolution, 2nd edn. Wiley, Hoboken
18. Fogliatto FS, da Silveira GJC (2008) Mass Customisation: a method for market segmentation and choice menu design. Int J Prod Econ 111(2):606–622
19. Fogliatto FS, Da Silveira G, Borenstein D (2012) The mass customisation decade: an updated review of the literature. Int J Prod Econ 138(1):14–25
20. Hu SJ, Ko J, Weyand L, ElMaraghy HA, Kien TK, Koren Y, Bley H, Chryssolouris G, Nasr N, Shpitalni M (2011) Assembly system design and operations for product variety. CIRP Ann Manuf Tech 60(2):715–733
21. Holweg M, Disney SM, Hines P, Naim MM (2005) Towards responsive vehicle supply: a simulation-based investigation into automotive scheduling systems. J Oper Manag 23(5):507–530
22. Elias S (2002) New car buyer behaviour. 3DayCar Research Report. Cardiff Business School
23. Knight K (2007) Survey: 89 % of consumers prefer online shopping. Bizreport. http://www.bizreport.com/2007/12/survey_89_of_consumers_prefer_online_shopping.html
24. Helms M, Ahmadi M, Jih W, Ettkin L (2008) Technologies in support of mass customisation strategy: exploring the linkages between e-commerce and knowledge management. Comput Ind 59(4):351–363

25. Franke N, Keinz P, Schreier M (2008) Complementing mass customisation toolkits with user communities: how peer input improves customer self-design. J Prod Innov Manag 25(6):546–559

26. Chen L, Song Z, Feng L (2004) Internet-enabled real-time collaborative assembly modelling via an E-assembly system: status and promise. Comput Aided Des 36(9):835–847

27. Feldmann K, Rottbauer H, Roth N (1996) Relevance of assembly in global manufacturing. CIRP Ann Manuf Tech 45(2):545–552

28. Janardanan VK, Adithan M, Radhakrishnan P (2008) Collaborative product structure management for assembly modelling. Comput Ind 59(8):820–832

29. Warren NA (2005) Internet and globalisation. In: Gangopadhyay P, Chatterji M (eds) Economics of globalisation. Ashgate, USA

30. Oliver J, Houlihan JB (1986) Logistics management—the present and the future. In: Proceedings of 1986 BPICS conference, p 91–99

31. Wilding R (1998) The supply chain complexity triangle: uncertainty generation in the supply chain. Int J Phys Distrib Logist Manag 28(8):599–616

32. Gereffi G (1999) International trade and industrial upgrading in the apparel commodity chain. J Int Econ 48(1):37–70

33. Váncza J, Monostori L, Lutters D, Kumara SR, Tseng M, Valckenaers P, Van Brussel H (2011) Cooperative and responsive manufacturing enterprises. CIRP Ann Manuf Tech 60(2):797–820

34. Wiendahl HP, ElMaraghy HA, Nyhuis P, Zäh MF, Wiendahl HH, Duffie N, Brieke M (2007) Changeable manufacturing—classification, design and operation. CIRP Ann Manuf Tech 56(2):783–809

35. Tolio T, Urgo M (2007) A rolling horizon approach to plan outsourcing in manufacturing-to-order environments affected by uncertainty. CIRP Ann Manuf Tech 56(1):487–490

36. Chui M, Loffler M, Roberts R (2010) The internet of things. McKinsey Quarterly, New York

37. Monostori L, Váncza J, Kumara SRT (2006) Agent-based systems for manufacturing. CIRP Ann Manuf Tech 55(2):697–720

38. Mourtzis D, Doukas M, Psarommatis F (2013) Design and operation of manufacturing networks for mass customisation. CIRP Ann Manuf Tech 63(1):467–470

39. Cunha PF, Ferreira PS, Macedo P (2008) Performance evaluation within cooperate networked production enterprises. Int J Comput Int Manuf 21(2):174–179

40. Weber CA, Current JR, Benton WC (1991) Vendor selection criteria and methods. Eur J Oper Res 50(1):2–18

41. Wu C, Barnes D (2011) A literature review of decision-making models and approaches for partner selection in agile supply chains. J Purch Suppply Manag 17(4):256–274

42. Tracey M, Tan C (2001) Empirical analysis of supplier selection and involvement, customer satisfaction, and firm performance. Supply Chain Mana Int J 6(4):174–188

43. Ellram LM (1990) The supplier selection decision in strategic partnerships. J Purch Mater Manag 26(4):8–14

44. Lee AHI et al (2009) A green supplier selection model for high-tech industry. Expert Syst Appl 36(4):7917–7927

45. Ni M, Xu X, Deng S (2007) Extended QFD and data-mining-based methods for supplier selection in mass customisation. Int J Comput Integr Manuf 20(2-3):280–291

46. Hou J, Su D (2007) EJB-MVC oriented supplier selection system for mass customisation. J Manuf Tech Manag 18(1):54–71

47. Hou J, Su D (2006) Integration of web services technology with business models within the total product design process for supplier selection. Comput Ind 57(8–9):797–808

48. Deng X, Hu Y, Deng Y, Mahadevan S (2014) Supplier selection using AHP methodology extended by D numbers. Expert Syst Appl 41(1):156–167

49. Scott J, Ho W, Dey PK, Talluri S (2015) A decision support system for supplier selection and order allocation in stochastic, multi-stakeholder and multi-criteria environments. Int J Prod Econ 166:226–237

50. Tan PS, Lee S, Goh AES (2012) Multi-criteria decision techniques for context-aware B2B collaboration in supply chains. Dec Support Syst 52(4):779–789

51. Ho W, Xu X, Dey PK (2010) Multi-criteria decision making approaches for supplier evaluation and selection: a literature review. Eur J Oper Res 202(1):16–24

52. Nonaka I (1994) A dynamic theory of organizational knowledge creation. Organ Sci 5(1):14–37

53. Ahmad S, Schroeder RG, Mallick DN (2010) The relationship among modularity, functional coordination, and mass customization. Eur J Innov Manag 13(1):46–61

54. Kanda AA, Deshmukh SG (2008) Supply chain coordination: perspectives, empirical studies and research directions. Int J Prod Econ 15(2):316–335

55. Mourtzis D, Doukas M, Psarommatis F (2012) A multi-criteria evaluation of centralised and decentralised production networks in a highly customer-driven environment. CIRP Ann Manuf Tech 61(1):427–430

56. Chase RB (1998) Production and operations management: manufacturing and services. Irwin McGraw Hill, New York

57. Windt K, Böse F, Philipp T (2008) Autonomy in production logistics: identification, characterisation and application. Robot Comput Int Manuf 24(4):572–578

58. Mourtzis D (2011) Internet based collaboration in the manufacturing supply chain. CIRP J Manuf Sci Tech 4(3):296–304

59. Li X, Wang Q (2007) Coordination mechanisms of supply chain systems. Eur J Oper Res 179(1):1–16

60. Jain AS, Meeran S (1999) Deterministic job-shop scheduling: past, present and future. Eur J Oper Res 113(2):390–434

61. Mula J, Peidro D, Mandronero MD, Vicens E (2010) Mathematical programming models for supply chain production and transport planning. Eur J Oper Res 204:377–390

62. Wu C, Barnes D, Rosenberg D, Luo X (2009) An analytic network process-mixed integer multi-objective programming model for partner selection in agile supply chains. Prod Plan Control 20(3):254–275

63. Liang TF, Cheng HW (2009) Application of fuzzy sets to manufacturing/distribution planning decisions with multi-product and multi-time period in supply chains. Expert Syst Appl 36:3367–3377

64. Zhang L, You X, Jiao J, Helo P (2008) Supply chain configuration with co-ordinated product, process and logistics decisions: an approach based on Petri nets. Int J Prod Res 47(23):6681–6706

65. Yao J, Liu L (2009) Optimisation analysis of supply chain scheduling in mass customisation. Int J Prod Econ 117(1):197–211

66. Thomas GJ, Griffin PM (1996) Coordinated supply chain management. Eur J Oper Res 196(94):1–15

67. Akçalı E, Çetinkaya S (2011) Quantitative models for inventory and production planning in closed-loop supply chains. Int J Prod Res 49(8):2373–2407

68. Lanza G, Peters S (2012) Integrated capacity planning over highly volatile horizons. CIRP Ann Manuf Tech 61(1):395–398

69. Lödding H, Yu KW, Wiendahl HP (2003) Decentralised WIP-oriented manufacturing control (DEWIP). Prod Plan Control 14(1):42–54

70. Buschkühl L, Sahling F, Helber S, Tempelmeier H (2008) Dynamic capacitated lot-sizing problems: a classification and review of solution approaches. OR Spectr 32(2):231–261

71. Aigbedo H (2007) An assessment of the effect of mass customisation on suppliers' inventory levels in a JIT supply chain. Eur J Oper Res 181(2):704–715

72. Monden Y (1983) Toyota production system: a practical approach to production management. Industrial Engineering and Management Press, ISBN: 978-0898060348

73. Berman B (2002) Should your firm adopt a mass customisation strategy? Bus Horiz 45(4):51–60

74. Zinn W, Bowersox DJ (1988) Planning physical distribution with the principle of postponement. J Bus Logist 9(2):117–136

75. Su JCP, Chang YL, Ferguson M (2005) Evaluation of postponement structures to accommodate mass customisation. J Oper Manag 23(3–4):305–318

76. Lee HL, Billington C, Carter B (1993) Hewlett Packard gains control of inventory and service through design for localisation. Interfaces 23(4):1–11

77. Rudberg M, Wikner J (2004) Mass customisation in terms of the customer order decoupling point. Prod Plan Control 15(4):445–458

78. Pagh JD, Cooper MC (1998) Supply chain postponements and speculation strategies: how to choose the right strategy. J Bus Logist 19(2):13–33

79. Mikkola JH, Skjott-Larsen T (2004) Supply chain integration: implications for mass customisation, modularisation and postponement strategies. Prod Plan Control 15(4):352–361

80. van Hoek RI (2000) The role of third-party logistics providers in mass customisation. Int J Logist Manag 11(1):37–46

81. Biao Y, Burns ND, Backhouse CJ (2004) Postponement: a review and an integrated framework. Int J Oper Prod Manag 24(5):468–487

82. Helo P, Szekely B (2005) Logistics information systems. Ind Manag Data Syst 105(1):5–18

83. Clarke MP (1998) Virtual logistics: an introduction and overview of the concepts. Int J Phys Distr Logist Manag 28(7):486–507

84. Lee HL, Padmanabhan V, Whang S (1997) Information distortion in a supply chain: the bullwhip effect. Manag Sci 43(4):546–558

85. Perea E, Grossmann I, Ydstie E, Tahmassebi T (2000) Dynamic modeling and classical control theory for supply chain management. Comput Chem Eng 24:1143–1149

86. Swaminathan JM, Smith SF, Sadeh NM (1998) Modelling supply chain dynamics: a multiagent approach. Dec Sci 29(3):607–632

87. Alexopoulos K, Papakostas N, Mourtzis D, Gogos P, Chryssolouris G (2008) Oscillator analogy for modelling the manufacturing systems dynamics. Int J Prod Res 46(10):2547–2563

88. Kelepouris T, Pramatari K, Doukidis G (2007) RFID-enabled traceability in the food supply chain. Ind Manag Data Syst 107(2):183–200

89. Wang LC, Lin YC, Lin PH (2007) Dynamic mobile RFID-based supply chain control and management system in construction. Adv Eng Inf 21(4):377–390

90. Bechini A, Cimino M, Marcelloni F, Tomasi A (2008) Patterns and technologies for enabling supply chain traceability through collaborative e-business. Inf Softw Technol 50(4):342–359

91. Xu DL, He W, Li S (2014) Internet of things in industries: a survey. IEEE Ind Inf 10(4):2233–2243

92. Reaidy PJ, Gunasekaran A, Spalanzani A (2015) Bottom-up approach based on internet of things for order fulfilment in a collaborative warehousing environment. Int J Prod Econ 159:29–40

93. Duffie NA, Roy D, Shi L (2008) Dynamic modelling of production networks of autonomous work systems with local capacity control. CIRP Ann Manuf Tech 57(1):463–466

94. Perona M, Miragliotta G (2004) Complexity management and supply chain performance assessment. A field study and a conceptual framework. Int J Prod Econ 90(1):103–115

95. Efthymiou K, Pagoropoulos A, Papakostas N, Mourtzis D, Chryssolouris G (2012) Manufacturing systems complexity review: challenges and outlook. Proc CIRP 3:644–649

96. Chryssolouris G, Efthymiou K, Papakostas N, Mourtzis D, Pagoropoulos A (2013) Flexibility and complexity: Is it a trade-off? Int J Prod Res 51(23–24):6788–6802

97. ElMaraghy W, ElMaraghy H, Tomiyama T, Monostori L (2012) Complexity in engineering design and manufacturing. CIRP Ann Manuf Tech 61(2):793–814

98. Maropoulos P (1994) Digital enterprise technology–defining perspectives and research priorities. Int J Comput Int Manag 16(7–8):467–478

99. Manufuture (2004) A vision for 2020, report of the high-level group, Nov 2004. http://www.manufuture.org/documents/manufuture_vision_en%5B1%5D.pdf

100. Bughin J, Chui M, Manyika CJ (2010) Big data, and smart assets: ten tech-enabled business trends to watch. McKinsey Quarterly, McKinsey Global Institute, New York

101. Xu X (2012) From cloud computing to cloud manufacturing. Robot Comput Int Manuf 28(1):75–86

102. Komoto H, Tomiyama T, Silvester S, Brezet H (2011) Analyzing supply chain robustness for OEMs from a life cycle perspective using life cycle simulation. Int J Prod Econ 134:447–457

103. Grewal CS, Enns ST, Rogers P (2010) Dynamic adjustment of replenishment parameters using optimum-seeking simulation. In: Simulation conference (WSC), proceedings of the 2010 winter conference, pp 1797–1808

104. Umeda S, Zhang F (2010) A simulation modelling framework for supply chain system analysis. In: Simulation conference (WSC), proceedings of the 2010 winter conference, pp 2011–2022

105. Hennies T, Reggelin T, Tolujew J, Piccut PA (2014) Mesoscopic supply chain simulation. J Comput Sci 5(3):463–470

106. Aloini D, Dulmin R, Mininno V (2012) Modelling and assessing ERP project risks: a petri net approach. Eur J Oper Res 220(2):484–495

107. Mourtzis D, Papakostas N, Mavrikios D, Makris S, Alexopoulos K (2012) The role of simulation in digital manufacturing-applications and outlook. Int J Comput Int Manuf 28(1):3–24

108. Strauss J, Frost R (2001) E-marketing, 2nd edn. Prentice-Hall, Englewood Cliffs

109. Vrechopoulos AP (2004) Mass customisation challenges in Internet retailing through information management. Int J Inf Manag 24(1):59–71

110. McIntosh RI, Matthews J, Mullineux G, Medland AJ (2010) Late customisation: issues of mass customisation in the food industry. Int J Prod 48(6):1557–1574

111. Kimiloglu H, Zarah H (2009) What signifies success in e-CRM? Mark Intell Plan 27(2):246–267

112. Woodcock N, Green A, Starkey M (2011) The customer framework, social CRM as a business strategy. J Database Mark Cust Strategy Manag 18:50–64

113. Mell P, Grance T (2009) Perspectives on cloud computing and standards. Information Technology Laboratory, National Institute of Standards and Technology (NIST), Gaithersburg

114. Mourtzis D, Doukas M, Vlachou K, Xanthopoulos N (2014) Machine availability monitoring for adaptive holistic scheduling: A conceptual framework for mass customisation. In: 8th International conference on digital enterprise technology—DET 2014. 25–28 Mar, Stuttgart, ISBN: 9783839606971

115. CAPP-4-SMEs. Collaborative and adaptive process planning for sustainable manufacturing environments—CAPP4SMEs. EC

Funded Project, 7th Frammework Programme. Grant Agreement No.: 314024

116. Tapoglou N, Mehnen J, Doukas M, Mourtzis D (2014) Optimal tool-path programming based on real-time machine monitoring using IEC 61499 function blocks: a case study for face milling. In: 8th ASME 2014 international manufacturing science and engineering conference, 9–13 June 2014, Detroit, Michigan

117. Valilai OF, Houshmand M (2013) A collaborative and integrated platform to support distributed manufacturing system using a service-oriented approach based on cloud computing paradigm. Robot Comput Int Manuf 29(1):110–127

118. Wu D, Thames JL, Rosen DW, Schaefer D (2012) Towards a cloud-based design and manufacturing paradigm: looking backward, looking forward. In: 32nd Computers and information in engineering conference ASME proceedings, Chicago, Illinois, 12–15, Aug, pp 315–328

119. Lewis G (2010) Basics about cloud computing. Software Engineering Institute, Carnegie Mellon University, Pittsburgh

120. Technavio Insights (2014) Global cloud security software market. http://www.trendmicro.com/cloud-content/us/pdfs/business/reports/rpt_technavio-global-security-software-market.pdf

121. MindCommerce (2014) Cloud security 2014: companies and solutions, July 2014. http://www.mindcommerce.com/cloud_security_2014_companies_and_solutions.php

122. Tao F, Zhang L, Venkatesh VC, Luo Y, Cheng Y (2011) Cloud manufacturing: a computing and service-oriented manufacturing model. Proc Inst Mech Eng Part B J Eng Manuf 225(10):1969–1976

123. Kannan D, Khodaverdi R, Olfat L, Jafarian A, Diabat A (2013) Integrated fuzzy multi criteria decision making method and multi-objective programming approach for supplier selection and order allocation in a green supply chain. J Clean Prod 47:355–367

124. Mafakheri F, Breton M, Ghoniem A (2011) Supplier selection-order allocation: a two-stage multiple criteria dynamic programming approach. Int J Prod Econ 132(1):52–57

125. Amin SH, Razmi J, Zhang G (2011) Supplier selection and order allocation based on fuzzy SWOT analysis and fuzzy linear programming. Expert Syst Appl 38(1):334–342

126. Yinan Q, Tang M, Zhang M (2014) Mass customization in flat organization: the mediating role of supply chain planning and corporation coordination. J Appl Res Tech 12(2):171–181

127. Zhang M, Zhao X, Qi Y (2014) The effects of organizational flatness, coordination, and product modularity on mass customization capability. Int J Prod Econ 158:145–155

128. Ramanathan U (2014) Performance of supply chain collaboration—a simulation study. Expert Syst Appl 41(1):210–220

129. Gligor DM, Holcomb MC (2012) Understanding the role of logistics capabilities in achieving supply chain agility: a systematic literature review". Supply Chain Manag Int J 17(4):438–453

130. Panorama Consulting (2014) ERP Report. Denver, Colorado: A Panorama Consulting Solutions Research Report

131. Olhager J (2013) Evolution of operations planning and control: from production to supply chains. Int J Prod Res 51(23–24):6836–6843

132. Plossl GW (2011) Orlicky's material requirements planning, 3rd edn. McGraw-Hill, New York

133. Su CJ (2009) Effective mobile assets management system using RFID and ERP technology. In: WRI international conference on communications and mobile computing, pp 147–151

134. Schabel S (2009) ERP—mobile computing. Thesis, Wien: Universität, Wien

135. Chan E, Yu KM (2007) A concurrency control model for PDM systems. Comput Ind 58(8–9):823–831

136. Schuh G, Rozenfeld H, Assmus D, Zancul E (2008) Process oriented framework to support PLM implementation. Comput Ind 59(2–3):210–218

137. Nee AYC, Ong SK, Chryssolouris G, Mourtzis D (2012) Augmented reality applications in design and manufacturing. CIRP Ann Manuf Tech 61:657–679

138. Wang L (2013) Machine availability monitoring and machining process planning towards cloud manufacturing. CIRP J Manuf Sci Tech 6(4):263–273

139. Brondi C, Carpanzano E (2011) A modular framework for the LCA-based simulation of production systems. CIRP J Manuf Sci Tech 4(3):305–312

140. Sihn W, Pfeffer M (2013) A method for a comprehensive value stream evaluation. CIRP Ann Manuf Tech 62(1):427–430

141. Dignum V (2006) An Overview of Agents in Knowledge Management. In: Umeda M (ed) Proceedings of INAP-05. Springer, pp 175–189

142. Mourtzis D, Doukas M (2014) Knowledge capturing and reuse to support mass customisation: a case study from the mould making industry. Proc CIRP 2:123–128

143. Valckenaers P, Brussel H, Verstraete P, Hadeli GB (2007) Schedule execution in autonomic manufacturing execution systems. J Manuf Syst 26(2):75–84

144. Shariatzadeh N, Sivard G, Chen D (2012) Software evaluation criteria for rapid factory layout planning, design and simulation. Proc CIRP 3:299–304

145. Lee JY, Kang HS, Kim GY, Noh SD (2012) Concurrent material flow analysis by P3R-driven modelling and simulation in PLM. Comput Ind 63(5):513–527

146. Xu HC, Xu XF, HE T (2007) Research on transformation engineering BOM into manufacturing BOM based on BOP. Appl Mech Mater 10–12:99–103

147. Eppinger SD, Browning TR (2012) Design structure matrix methods and applications. ISBN: 9780262017527

148. Zhang LL, Lee CKM, Xu Q (2010) article is part of a focus collection. Comput Ind 61(3):213–222

149. Simchi-Levi D (2010) Operation rules. MIT Press Ltd, Cambridge

150. Pine BJ (1992) Mass customisation: the new frontier in business competition. Harvard Business School Press, Boston

2

On the use of multi-agent systems for the monitoring of industrial systems

Nafissa Rezki[1] · Okba Kazar[2] · Leila Hayet Mouss[1] · Laid Kahloul[2] ·
Djamil Rezki[1]

Abstract The objective of the current paper is to present an intelligent system for complex process monitoring, based on artificial intelligence technologies. This system aims to realize with success all the complex process monitoring tasks that are: detection, diagnosis, identification and reconfiguration. For this purpose, the development of a multi-agent system that combines multiple intelligences such as: multivariate control charts, neural networks, Bayesian networks and expert systems has became a necessity. The proposed system is evaluated in the monitoring of the complex process Tennessee Eastman process.

Keywords Multivariate process · Hotelling T^2 control chart · Multi-agent system · Bayesian network · Neural network

✉ Nafissa Rezki
nafissa_rezki@yahoo.fr

Okba Kazar
okbakazar@yahoo.fr

Leila Hayet Mouss
hayet_mouss@yahoo.fr

Laid Kahloul
kahloul2006@yahoo.fr

Djamil Rezki
drezki@yahoo.fr

[1] LAP: Laboratory, Industrial Engineering Department, University of Batna, 05000 Batna, Algeria

[2] LINFI Laboratory, Computer Science Department, University of Biskra, 07000 Biskra, Algeria

Introduction

The process monitoring is a critical task in all industrial plant. It can be realized by the use of three principal approaches (Venkatasubramanian et al. 2003): (1) the analytical methods based on mathematics models. These methods compare the real-system outputs to the mathematical model outputs, (2) the methods based on knowledge (Stamatis 2003; Dhillon 2005) that use the human knowledge [risk analysis, failures modes effects and critically analysis (FMECA), decision trees], and (3) the data-based methods that focus on statistic development of the process. The last kind of the method uses, generally, the control charts [(Page 1954), cumulative SUM (CUSUM) (Roberts 1959)] or exponentially weighted moving average (EWMA) (Alt et al. 1985) for the fault detection in the industrial process.

Currently, the manufacturing processes become more and more complex and multivariate. In these systems, the operator recuperates a vast data amount to be analysed. The high volume of data and the big number of process variables make the operator task fastidious. To avoid such problems, the data-based methods are more suitable for the process monitoring. The multivariate control charts [Hotelling T^2 control chart, multivariate CUSUM (MCUSUM), multivariate EWMA (MEWMA)] have been used for the control of multivariate process and have proved their adequacy to reduce the complexity of such process monitoring. Moreover, the monitoring of a multivariate process is a complex task, and it can be devised into four subtasks which are: the detection of abnormal situation, the diagnosis of the faults, the identification of variables that involved in the faults and finally the reconfiguration of the process (Venkatasubramanian et al. 2003).

Many researches have used the control charts for process monitoring (Yu-Chang et al. 2015; Xia 2015; Ehsan and Sadigh 2014; Vijayababu and Rukmini 2014; Assareh et al. 2013). To identify the variables that make an out-of-control in T^2, a decomposition of the statistic T^2 into independent terms has been suggested by Jing et al. (2008). The "MYT approach" has been applied by Mani and Cooper (1999) for the variables identification. The "MYT approach" has a big disadvantage which is the number of T^2 decompositions. For a process with p variables, the number of decompositions is $p!$. To reduce this number and to identify the relationship among the variables, the Bayesian networks have been applied for variables identification by Friedman (2000), Li et al. (2006), Li and Shi (2007), Sylvain (2007).

In this paper, we regroup all the tasks of the multivariate process monitoring in one approach. Our contribution is to determine the best combination of multivariate control charts, neural networks, Bayesian networks, expert systems. The result of this research is a multi-agent system that applied to a multivariate process monitoring. This multi-agent system uses: multivariate control chart for abnormal detection, neural network for faults diagnosis, Bayesian network for variables identification and expert system for reconfiguration task.

The rest of this paper is organized as follows: the process monitoring approach is presented in "The proposed multi-agent system" section with the monitoring algorithm. In "Application of the proposed model on the Tennessee Eastman process" section, a case study of simulated Tennessee Eastman process (TEP) (Downs and Vogel 1993) is employed to illustrate the validity of the proposed approach, including the detection by multivariate control charts executor agent (MCCEA), diagnosis by diagnosis artificial neural network agent (DANNA), identification by Identification Bayesian network agent (IBNA) and the reconfiguration by reconfiguration agent (RA). Finally, conclusions and future works are suggested.

The proposed multi-agent system

The proposed multi-agent system uses a multiple intelligences that are: multivariate control chart, neural network, Bayesian network and expert system in a multi-agent system. The multivariate control charts (T^2 control chart, MEWMA ...) can detect successfully the instability of the process, but it cannot diagnosis the fault that appeared in the process and cannot identify the causes of the instability. In this paper, we use an artificial neural network for the faults diagnosis. The neural networks have demonstrated their ability in the classification of similar faults. The neural networks take time in

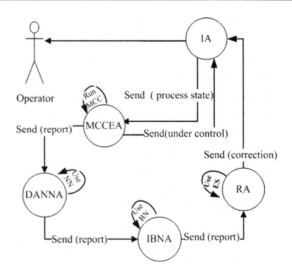

Fig. 1 The agents diagram

the training phase, and then, the classification will be done quickly. After detecting the instability using T^2 control chart, and the diagnosis using neural network, the Bayesian network proposed by Sylvain (2007) is used in the identification task. To realize a complete monitoring system for multivariate process and simplify the reconfiguration task to the operators that are not specializing in the realm, we developed an expert system that assures the process correction. The following paragraphs will describe each of these used agents. The agent diagram of the proposed approach is shown in Fig. 1. In this diagram, the actual agent types are represented by circles. People that must interact with the system are represented by the unified modelling language (UML) actor symbol.

The interface agent

The interface agent (IA) is a reactive agent which represents the interface for the human user access; hence, it receives the request from the users (monitoring the process state). Besides this, the IA transforms the agent's responses to the users. The IA receives a request from the user about the process state, and it sends a message to the MCCEA. If the process is under control, the IA will display to the operator the decision of the MCCEA. In the other case, when the process is out of control, the IA waits the response from the RA and displays it to the user.

The multivariate control chart executor agent

This agent is responsible on the execution of the multivariate control charts [T^2 control chart (Hotelling 1947), multivariate CUSUM (MCUSUM) (Pignatiello and Runger 1990), multivariate EWMA (MEWMA) (Lowry et al. 1992)]. The control charts (T^2) control chart, MEWMA and MCUSUM can successfully detect the process instability, but it cannot give any information about the fault that appeared in the process and the variables that are responsible about the process instability. The use of one chart for process monitoring is not sufficient to detect all out of control situation. So, to monitor successively the process, we suggest to use a software agent that can execute simultaneously a set of multivariate control charts and detect easily the process instability. These different control charts are utilized in the design and implementation of the MCCEA.

The diagnosis artificial neural network agent

We use the neural networks in the diagnosis task because it demonstrated its efficiency in the resolution of classification problem. In addition, the neural networks—after the learning step—has a short response time and a good classification rate. We create a classical multilayer perceptron (MLP), with three layers: (1) the input layer: the number of neurons in this layer is the number of the process parameters, (2) the output layer: in this layer, the number of the neurons represents the number of classes (faults of the process), (3) the hidden layer: it is generally known that the number of neurons in this layer is problematic research. We carried out a set of tests, and we find that the optimal number is equal to: (*number of neurons in the input layer + the number of neurons in the output layer*)/2. This neural network is used in the implementation of the DANNA. So, in our system the DANNA is responsible for the diagnosis task. When the process is out of control, DANNA receives report from the MCCEA. Its principle objective is to find the fault that appeared in the process. After, it sends a report to the IBNA.

The identification Bayesian network agent

The IBNA receives report from DANNA about the fault that appeared in the process. It builds a Bayesian net using the causal decomposition algorithm of T^2 proposed by Sylvain (2007). It finds the variables involved in the fault. This agent simplifies the variable identification in the process. After, it sends report to the RA.

The reconfiguration agent

For the objective, to regroup all the process monitoring tasks (detection, diagnosis, identification and reconfiguration) in one system, we add the RA which helps the operator to reconfigure the process after its failure. It receives report from the IBNA about the variables that involved in the fault. It must propose a reconfiguration plan

to the operator, to maintain the process. Also, it sends its reconfiguration plan to the IA. This agent has been developed using an expert system technology.

The proposed monitoring algorithm

Start
    Get data from data base
    Create the MCCEA
    MCCEA runs the controls charts
    If(MCCEA-decision=stable-process)**Then**
    MCCEA sends report to the IA
    Else
    Create the DANNA
    Create the IBNA
    Create RA
    DANNA creates the ANN using MLP
    IBNA creates the Bayesian net
    For (i=1 to number of observations) Do
    DANNA gives its diagnosis of the observation i
    DANNA sends the diagnosis to the IBNA
    End For
    IBNA receives the diagnosis from DANNA
    IBNA uses BN to find the variables that are out of control
    IBNA sends the report to the RA
    RA receives report about the variables involved in the fault
    RA finds the reconfiguration plan
    RA sends report to the IA
    IA receives report from RA
    End If
End

Application of the proposed model on the Tennessee Eastman process

Introduction to the Tennessee Eastman process

The Tennessee Eastman process (TEP) is proposed by Downs and Vogel (1993) to provide a simulated model and to evaluate the monitoring methods of industrial complex process. The process consists of five principal units: a condenser, a separator, a reactor, a compressor and a stripper. Four gaseous reactants (A, C, D and E) and inert B are fed to the reactor. It produces two components (G and H) and the undesired by-product F. The reaction equations are listed in equation number (1–4). All the reactions are irreversible, exothermic and approximately first order with respect to the reactant concentrations. The reaction rates are expressed as Arrhenius function of temperature. The reaction producing G has higher activation energy than that producing H, thus resulting in more sensitivity to temperature (Fig. 2).

The TEP process proposed by Downs and Vogel (1993) is open loop unstable, and it should be operated under closed loop. In this article, we use this control structure to evaluate the performance of our approach on fault diagnosis. The reactor product stream is cooled through a condenser and fed to a vapour–liquid separator. The vapour exits the separator and recycles to the reactor feed through a compressor. A portion of the recycle stream is purged to prevent the inert and by-product from accumulating. The condensed component from the separator is sent to a stripper, which is used to strip the remaining reactants. Once G and H exit the base of the stripper, they are sent to a downstream process which is not included in the diagram. The inert and by-products are finally purged as vapour from vapour–liquid separator. The process provides 41 measured and 12 manipulated variables, denoted as XMEAS(1) to XMEAS(41) and XMV(1) to XMV(12), respectively. Their brief descriptions and units are listed in

Tables 1 and 2. Fifteen preprogrammed faults IDV(1) to IDV(15) of TEP are given to represent different conditions of the process operation, as listed in Table 3.

$$A(g) + C(g) + D(g) \longrightarrow G(l) \tag{1}$$

$$A(g) + C(g) + E(g) \longrightarrow H(l) \tag{2}$$

$$A(g) + E(g) \longrightarrow F(l) \tag{3}$$

$$3D(g) \longrightarrow 2F(l) \tag{4}$$

Simulation and results analyses

The proposed approach has been implemented using the Java environment Netbeans IDE. Also, we use the agent design platform Java Agent Development framework JADE. To simplify the development of the neural network and Bayesian network with Netbeans, java offers many libraries. Moreover, we use Jess Tab which is a rule engine for the Java platform to produce our rules in the knowledge

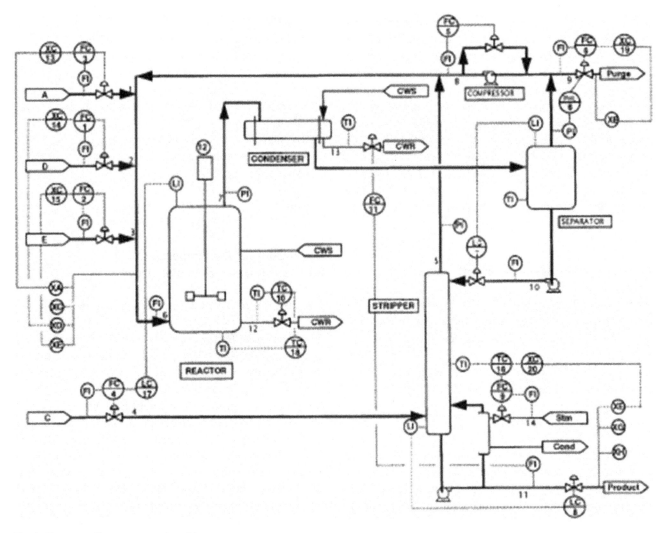

Fig. 2 Tennessee Eastman control problem

Table 1 Measurement variables in the Tennessee Eastman process

Variable	Description	Units
XMEAS(1)	A feed (stream 1)	kscmh
XMEAS(2)	E feed (stream 3)	kg/h
XMEAS(4)	Total feed (stream 4)	kg/h
XMEAS(5)	Recycle flow (stream 8)	kscmh
XMEAS(6)	Recycle flow (stream 6)	kscmh
XMEAS(7)	Reactor pressure	kPa gauge
XMEAS(8)	Reactor level	%
XMEAS(9)	Reactor temperature	C°
XMEAS(10)	Purge rate (stream 9)	kscmh
XMEAS(11)	Product sep temp	C°
XMEAS(12)	Product sep level	%
XMEAS(13)	Prod sep pressure	kPa gauge
XMEAS(14)	Prod sep underflow (stream 10)	m^3/h
XMEAS(15)	Stripper level	%
XMEAS(16)	Stripper pressure	kPa gauge
XMEAS(17)	Stripper underflow (stream 11)	m^3/h
XMEAS(18)	Stripper temperature	C°
XMEAS(19)	Stripper steam flow	kg/h
XMEAS(20)	Compressor work	kW
XMEAS(21)	Reactor cooling water outlet temp	C°
XMEAS(22)	Separator cooling water outlet temp	C°

Variable	Description	Stream
XMEAS(23)	Component A	6
XMEAS(24)	Component B	6
XMEAS(25)	Component C	6
XMEAS(26)	Component D	6
XMEAS(27)	Component E	6
XMEAS(28)	Component F	6
XMEAS(29)	Component A	9
XMEAS(30)	Component B	9
XMEAS(31)	Component C	9
XMEAS(32)	Component D	9
XMEAS(33)	Component E	9
XMEAS(34)	Component F	9
XMEAS(35)	Component G	9
XMEAS(36)	Component H	9
XMEAS(37)	Component D	11
XMEAS(38)	Component E	11
XMEAS(39)	Component F	11
XMEAS(40)	Component G	11
XMEAS(41)	Component H	11

Table 2 Manipulated variables in the Tennessee Eastman process

Variable	Description
XMV(1)	D feed flow (stream 2)
XMV(2)	E feed flow (stream 3)
XMV(3)	A feed flow (stream 1)
XMV(4)	Total feed flow (stream 4)
XMV(5)	Compressor recycle valve
XMV(6)	Purge valve (stream 9)
XMV(7)	Separator pot liquid flow (stream 10)
XMV(8)	Stripper liquid product flow (stream 11)
XMV(9)	Stripper steam valve
XMV(10)	Reactor cooling water flow
XMV(11)	Condenser cooling water flow
XMV(12)	Agitator speed

In this section, we evaluate the performances of the proposed approach on concrete example which is the TEP process. The used data represent 480 observations training for each fault and 800 tests for each faults, in addition to the normal period. The observations of training have been obtained with the simulation of each fault in a period of 24 h; moreover, the observations of the test set have been obtained in a period of 40 h. Variables are sampled every 3 min.

- The Detection

  All the persons that are worked on TEP take rate to obtain wrong alarm equal to 0.01%. In this work, we use the T^2 control chart for instability detection. A performance of detection system is evaluated by calculating its reliability (Kononenko 1991). The detection reliability is defined as: (*the number of obtained alerts in the test period/the total number of sample in the period test*).

  The MCCEA runs the T^2 control chart; if it detects an abnormal process state, it sends message to the DANNA. The detection reliability obtained in this work is the same that been obtained by Sylvain (2007). Figure 3 shows the detection reliability of MCCEA, and some faults are easily detectable [IDV (1), IDV (2), IDV (4), IDV (5), IDV (6), IDV (7), IDV (8), IDV (10), IDV (12), IDV (14)]. But other faults are difficult to detect [(IDV (3), IDV (9) and IDV (15)]. The last faults [(IDV (3), IDV (9), IDV (15)] are very identical. So the use of one chart (in this work, we use the T^2 control chart) is not sufficient. The run of many control charts simultaneously will augment the reliability of detection.

- The Diagnosis

  This task is realized by the DANNA. When it receives message from the MCCEA that the process is not stable, it creates the neural network using MLP, for the purpose

base. In this work, we use FIPA Agent Communication specifications that deal with Agent Communication Language (ACL) messages , message exchange interaction protocols and content language representations.

Table 3 The known faults of the Tennessee Eastman process

Variable	Description	Type
IDV(1)	A/C feed ratio, B composition constant (stream 4)	Step
IDV(2)	B composition, A/C ratio constant (Stream 4)	Step
IDV(3)	D feed temperature (stream 2)	Step
IDV(4)	Reactor cooling water inlet temperature	Step
IDV(5)	Condenser cooling water inlet temperature	Step
IDV(6)	A feed loss (stream 1) Step	Step
IDV(7)	C header pressure loss-reduced availability (stream 4)	Step
IDV(8)	A, B, C feed composition (stream 4)	Random variation
IDV(9)	D feed temperature (stream 2)	Random variation
IDV(10)	C feed temperature (stream 4)	Random variation
IDV(11)	Reactor cooling water inlet temperature	Random variation
IDV(12)	Condenser cooling water inlet temperature	Random variation
IDV(13)	Reaction kinetics	Slow drift
IDV(14)	Reactor cooling water valve	Sticking
IDV(15)	Condenser cooling water valve	Sticking

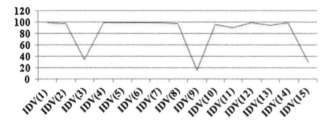

Fig. 3 The detection reliability

to find the fault that appeared in the process. In the next paragraph, we will show the diagnosis obtained by the DANNA and we will evaluate the acquired results to the result of other classifiers proposed in the literature.

Diagnosis of the known faults in the Tennessee Eastman problem

We have done the diagnosis of all the faults, i.e. IDV (1) to IDV (15) in TEP, as shown in Fig. 4. The used neural network is a MLP of three layers:

– *The input layer* contains 53 neurons that represent the process parameters,
– *The hidden layer* contains 34 neurons (*number of neurons in input layer + number neurons in output layer/2*),
– *The output layer* contains 15 neurons that represent the process faults.

Table 4 represents a comparison between the diagnosis realized by DANNA, and some other approaches proposed to the TEP faults diagnosis. Sylvain (2007) used Bayesian network for classification; however, the PC1DARMF (Li and Xiao 2011) is a supervised pattern classification method which uses one-dimensional adaptive rank-order morphological filter.

Diagnosis of IDV (4), IDV(9), IDV(11) in TEP

The most difficult faults to be classified in the TEP are: IDV (4), IDV (9) and IDV (15). The created neural network composed by 53 neurons (TEP parameters) in the input layer, 28 neurons in the hidden layer and 3 neurons in the output layer. Table 5 presents the rate of correct classification of the faults IDV (4), IDV (9) and IDV (15) of the TEP. It is a comparison between the DANNA diagnosis and the approach which proposed by El-Ferchichi (2013).

– The Identification

The IBNA is the responsible on the realization of the identification task using Bayesian net. It receives a report about the fault that appeared in the process from DANNA. To develop the Bayesian network, Sylvain (2007) used the causal decomposition of T^2. Figure 4 presents the Bayesian network that is created in the normal functionality of process. We take rate of false alarm = 0.005. The IBNA takes the observation that represents the fault, and then, it finds the variables that are involved in the fault. The variables involved in the fault have probability value under 0.995. We take the case of the observation 240 of IDV (5) that is classified as an IDV (4). The IBNA detects two variables that have a probability value under 0.995. The two variables are (XMV11) and (XMEAS21). The IBNA sends the variable identification to the RA.

– The Reconfiguration

The RA receives report from IBNA which contains the identification of the variables that cause the process instability. In our example, the identified variables are (XMV11) and (XMEAS21). The RA finds that: the variable (XMV11) represents the liquid cooling flow to the condenser, whereas the variable (XMEAS21)

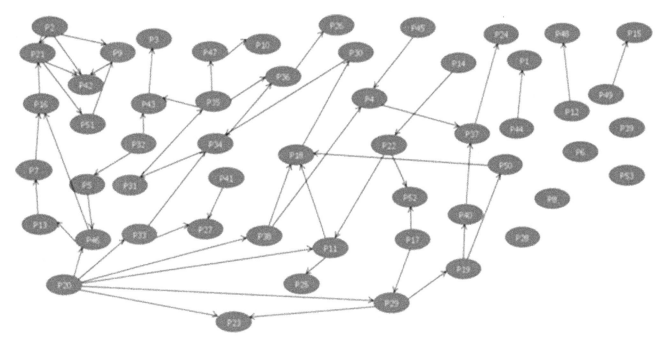

Fig. 4 The Bayesian network that used in the development of IBNA

Table 4 Classification rate of the known 15 faults in TEP

Faults	DANNA (%)	Sylvain (2007) (%)	PC1DARMF Li and Xiao (2011) (%)
IDV(1)	97.01	97.5	30
IDV(2)	95.34	98.125	95
IDV(3)	82.10	22	0.00
IDV(4)	97.34	82.375	25
IDV(5)	96.67	98	100
IDV(6)	100	100	65
IDV(7)	97.67	100	0.00
IDV(8)	100	97	5
IDV(9)	79.06	22.625	0.00
IDV(10)	71.42	86.875	15
IDV(11)	69.1	75.5	0.00
IDV(12)	96.67	98.25	5
IDV(13)	100	76.125	5
IDV(14)	93.02	98.75	5
IDV(15)	92.69	23.5	0.00

Table 5 Classification rate of IDV (4), IDV (9) and IDV (15) in TEP

Faults	DANNA (%)	El-Ferchichi (2013) (%)
IDV(4)	97.34	67.37
IDV(9)	100	66.25
IDV(15)	100	33.75

represents the cooling liquid temperature at the reactor outlet. In conclusion, these two variables involved in the fault IDV (5), so the fault that appeared in the process is the IDV (5) and not the IDV (4). It proposes the reconfiguration plan to the operator. The development of this agent requires knowledge of an expert human, which we will use to find the ideal reconfiguration plan.

Conclusion

An approach with several intelligences has been proposed in this paper for multivariate process monitoring. In this approach, we use the perfect tool for the realization of each task in a complex process monitoring. We use the multivariate control charts in the detection task. We utilize the artificial neural network classifier with MLP algorithm in the diagnosis task. For the identification task, we exploit the Bayesian network that has been proposed by Sylvain (2007). Moreover, to help the operators that are not specializing in realm, to realize the correction actions of the process, we suggest developing an expert system for reconfiguration task. To facilitate the use of the proposed approach with high efficiency, we integrate the different proposed subsystem (detection, diagnosis, identification and reconfiguration) in one system that is multi-agent system. The proposed model has been evaluated on a multivariate process (Tennessee Eastman process).

From the simulation results, we find that the proposed classifier gives a good result compared with some works applied on Tennessee Eastman process. In addition, the proposed approach gives good results for each task in the process monitoring. In the case study, we have seen that some faults are difficult for detecting; our future works will concentrate on the development of the detection task. The developed reconfiguration agent realizes the reconfiguration tasks for known faults, and we will focus also on adding the reconfiguration plan in case when a new fault appear in the process.

Acknowledgments The authors would like to express their sincere appreciation for all support provided.

References

Alt FB, Kotz NL, Johnson C (1985) R, Read multivariate quality control. Encycl Stat Sci 6:111–122

Assareh H, Noorossana R, Mengersen K-L (2013) Bayesian change point estimation in poisson-based control charts. J Ind Eng Int 9:32

Dhillon B (2005) Reliability, quality, and safety for engineers. CRC Press, Boca Raton

Downs JJ, Vogel EF (1993) A plant-wide industrial process control problem. Comput Chem Eng 17(3):245–255

Ehsan B, Sadigh R (2014) Economic design of Hotellings T^2 control chart on the presence of fixed sampling rate and exponentially assignable causes. J Ind Eng Int 10:229–238

El-Ferchichi S (2013) Ph.d. Thesis: selection and extraction of attributes for classification problem, National School of engineering of Tunis

Friedman N (2000) Using Bayesian networks to analyze expression data. J Comput Biol 7(3–4):601–620

Hotelling H (1947) Multivariate quality control. In: Eisenhart C, Hastay MW, Wallis WA (eds) Techniques of statistical analysis. McGraw-Hill, New York, pp 111–184

Jing L, Jionghua J, Jianjun S (2008) Causation-based T^2 decomposition for multivariate process monitoring and diagnosis. J Qual Technol 40(1):1–13

Kononenko I (1991) Semi-naive bayesian classier, proceeding of the European working session on learning on machine learning, pp 206–219

Li H, Xiao D (2011) Fault diagnosis of Tennessee Eastman process using signal geometry matching technique. J Adv Signal Proc (83). doi:10.1186/1687-6180-2011-83

Li J, Shi J, Satz D (2006) Modelling and analysis of disease and risk factors through learning Bayesian network from observational data, Technical report

Li J, Shi J (2007) Knowledge discovery from observational data for process control using causal Bayesian networks. IIE Trans 39(6):681–690

Lowry CA, Woodall WH, Champ CW, Rigdon SE (1992) A multivariate exponentially weighted moving average control chart. Technometrics 34(1):46–53

Mani S, Cooper GF (1999) A study in causal discovery from population-based infant birth and death records. In: Proceeding of the AMIA annual fall symposium. Philadelphia, pp 315–319

Page ES (1954) Continuous inspection schemes. Biometrika 41:100115

Pignatiello J, Runger G (1990) Comparisons of multivariate cusum charts. J Qual Technol 22(3):173186

Roberts SW (1959) Control chart tests based on geometric moving averages. Technometrics 1(3):239250

Stamatis DH (2003) Failure mode and effect analysis: FMEA from theory to execution. ASQ Quality Press, Milwaukee

Sylvain V (2007) Diagnostic et surveillance des processus complexe par rseaux baysiens.(Diagnosis and monitoring of complex process using bayesian networks), doctoral thesis, University of Angers, French

Venkatasubramanian V, Rengaswamy R, Yin K, Kavuri S (2003) A review of process fault detection and diagnosis, part I: quantitative model-based methods. Comput Chem Eng 27(3):293–311

Vijayababu V, Rukmini V-K (2014) Economic design of x-bar control charts considering process shift distributions. J Ind Eng Int 10:163171

Xia P (2015) Horizontal cumulative variance chart: A quality control scheme monitoring shifts in process variation. Int J Ind Syst Eng

Yu-Chang L, Chao-Yu C, Chung-Ho C (2015) Robustness of the EWMA median control chart to non-normality. Int J Ind Syst Eng

3

Process-based tolerance assessment of connecting rod machining process

G. V. S. S. Sharma[1] ⓘ · P. Srinivasa Rao[2] · B. Surendra Babu[3]

Abstract Process tolerancing based on the process capability studies is the optimistic and pragmatic approach of determining the manufacturing process tolerances. On adopting the define–measure–analyze–improve–control approach, the process potential capability index (C_p) and the process performance capability index (C_{pk}) values of identified process characteristics of connecting rod machining process are achieved to be greater than the industry benchmark of 1.33, i.e., four sigma level. The tolerance chain diagram methodology is applied to the connecting rod in order to verify the manufacturing process tolerances at various operations of the connecting rod manufacturing process. This paper bridges the gap between the existing dimensional tolerances obtained via tolerance charting and process capability studies of the connecting rod component. Finally, the process tolerancing comparison has been done by adopting a tolerance capability expert software.

Keywords Process tolerancing · Tolerance chart · DMAIC · Process capability · Dimensioning and tolerancing · Dimensional mapping

Introduction

The vital governing factor influencing the machining excellence is the geometric and dimensional tolerance embedded into the product as well as into the process. The two main facets of tolerancing include the arithmetic and statistical tolerancing. In arithmetic tolerancing it is assumed that the detail part dimension can have any value but within the tolerance range; whereas, in the statistical tolerancing scheme, it is assumed that detail part dimensions vary randomly according to a normal distribution, centered at the mid-point of the tolerance range and with its $\pm3\sigma$ spread covering the tolerance interval.

The main disadvantage of arithmetic tolerancing or worst-case tolerancing is that it does not follow any trend or pattern within the tolerance zone and part dimensions resulting from the machining process can possess any value within the tolerance zone. This results in checking up of each individual dimension for its correctness within the tolerance zone, which is impractical in mass production. The statistical tolerance overcomes this drawback of arithmetic tolerance and facilitates the machining to yield dimensions according to a normal distribution. Also statistical tolerancing allows some cancellation of variation from normal distribution. Hence, this paper in essence reflects the theme of statistical tolerancing.

The structure of paper is as follows. The first introduction part of the paper discusses the required introductory theoretical domain on tolerancing methods. This is followed by a literature review on the process tolerancing. A

✉ G. V. S. S. Sharma
sarma.gvss@gmail.com

P. Srinivasa Rao
psrao89@gmail.com

B. Surendra Babu
sudeepbs@gmail.com

[1] Department of Mechanical Engineering, GMR Institute of Technology, GMR Nagar, Rajam 532127, Andhra Pradesh, India

[2] Department of Mechanical Engineering, Centurion University, Parlakhemundi 761211, Odisha, India

[3] Department of Industrial Engineering, GITAM Institute of Technology, GITAM University, Visakhapatnam 530045, Andhra Pradesh, India

tolerance stack analysis with tolerance chain of the connecting rod machining is then presented. The tolerances obtained from tolerance stack are put to test for process capability studies (Sharma and Rao 2013). Then the

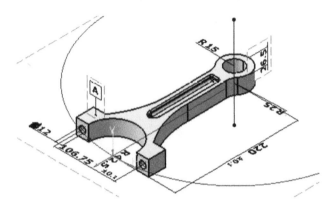

Fig. 1 Dimensioned part model of rod-end of connecting rod in CATIA V5 R14 software

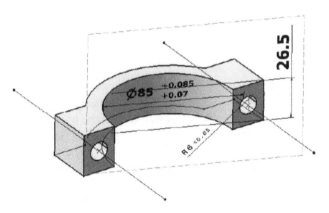

Fig. 2 Dimensioned part model of cap-end of connecting rod in CATIA V5 R14 software

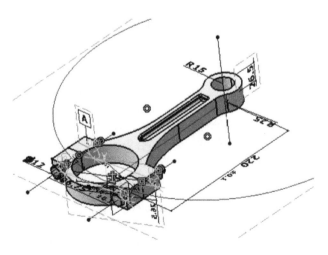

Fig. 3 Dimensioned part model of assembled view with tolerance annotations of connecting rod in CATIA V5 R14 software

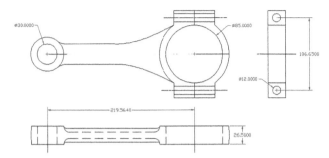

Fig. 4 Product drawing

improved C_p and C_{pk} values obtained are compared for optimum tolerance value using a tolerance capability expert software. Finally, the paper is concluded with a discussion on the results obtained.

Figures 1 and 2 show the part model rod and cap portions of the connecting rod, respectively. Figure 3 shows the assembled view of the connecting rod and Fig. 4 depicts the orthographic projection of the connecting rod product drawing.

Literature review

Primitive studies on process tolerancing were introduced through graphical representation of machining tolerance charting (Irani et al. 1989). The graphical approach and rooted tree diagram were adopted for tolerance charting (Whybrew et al. 1990). A tree theoretical representation for a tolerance chart was presented from the part blue print dimensions, stock removals and working dimensions (Ji 1993). The manufacturing process sequence was determined by using a profile representation method which incorporates a two-dimensional matrix containing a number coding system to represent the part profile (Ngoi and Ong 1993). A mathematical rooted tree model incorporating the linkage between the capability of manufacturing process and tolerance chart balancing was developed (Wei and Lee 1995). Geometrical control requirements were expressed as equivalent linear dimensions and then applied to a tolerance chart (Ngoi and Tan 1995). A backward derivation approach was traced for determining the machining tolerances starting from the last operation and computing machining allowances backwardly till the first machining operation (Ji 1996). A graphical method for presenting the process link and for obtaining the necessary working dimensions and tolerances was introduced (Ngoi and Tan 1997).

Process capability of machinery was taken into consideration for standardization of tolerances, through a nonlinear programming model (Lee and Wei 1998). This minimized the total manufacturing loss occurring due to

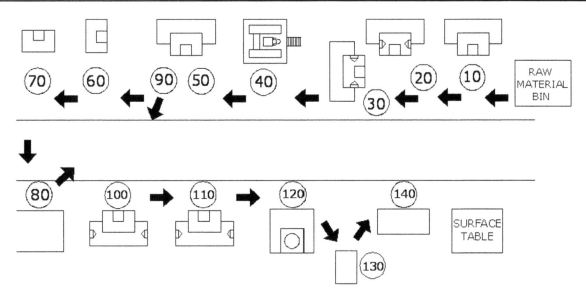

Fig. 5 Process flow chart of connecting rod manufacturing cell. Refer to Table 1 for corresponding description of connecting rod machining operations and their dimensional values

non-conforming parts. A continuous, multi-level approach to design tolerancing of electro-mechanical assemblies was outlined, wherein the assembly models for tolerancing, best practices for tolerancing, and the design process are integrated (Narahari et al. 1999). Manufacturing tolerances were allocated from forward dimensional chains, while the reverse dimensional chains were used to determine the nominal dimensions directly (Ji 1999).

Xue and Ji (2001) proposed a methodology for dealing with angular features in tolerance charting. Ji and Xue (2002) obtained the mean working dimensions from the reverse chain matrix containing reverse tolerance chains. Huang et al. (2005) devised a procedure for determining the process tolerances directly from multiple correlated critical tolerances in an assembly. Process-oriented tolerancing was focused upon, by considering all the variations arising due to tool wear, measurement device fluctuations, tolerance stack-up propagation (Ding et al. 2005). A process optimization model was introduced which considers process means and process tolerances simultaneously, with sequential operation adjustment to reduce process variability, and with part compensation to offset process shifting (Jeang et al. 2007). Peng et al. (2008) derived quality loss function of interrelated critical-to-quality dimensions. Through this quality loss function, the design-tolerances of the component are determined for achieving an improved product as well as process quality. The tolerance chart balancing was mathematically modeled for minimizing the manufacturing cost and quality loss (Jeang 2011). Concurrent tolerancing was identified as an optimization problem and a feasible solution for systematically distributing the process tolerances within the design

constraints was proposed (Sivakumar et al. 2012). Contreras (2013) proposed simplification of tolerance chains through a surface position tolerance (SPT) method for tolerance chart balancing. Chen et al. (2013) optimized the process parameters for the plastic injection molding. An improvement in the process potential capability index (C_p) and process performance capability index (C_{pk}) was registered through process capability improvement studies on thrust face thickness characteristic of connecting rod (Sharma and Rao 2013).

Recent works on tolerancing include tolerance analysis simulation during the initial design phase by a computer-aided tolerancing software (Barbero et al. 2015). The design tolerances estimated through this simulation subsequently determine the manufacturing tolerances. In another approach, complex workpiece with intricate shapes are classified based on its overall discrete geometry and tolerance analysis is performed on this overall part geometry (Schleich and Wartzack 2014). This simplifies the tolerance analysis for non-ideal complex workpiece shapes. Louhichi et al. (2015) performed realistic part tolerancing taking CAD part geometrical discrepancies into consideration. They identified the future research work as tolerance allocation by taking the manufacturing variations into consideration, which is also addressed in this paper. Considering this literature review, it can be summarized that the works on process tolerancing concentrated on the aspects of tolerance synthesis through tolerance chain and tolerance charting. In the pursuit for striking the balance between the conflicting issues of quality and cost, part tolerancing is optimized keeping the manufacturing process into consideration.

Table 1 Description of connecting rod machining operations

Machining operation number	Machining operation description	Dimensional value of machining characteristic (in mm)
10	Thrust face width rough grinding	27.250 ± 0.250
20	Gudgeon pin diameter rough boring	∅25.000 ± 0.200
30	Crank pin diameter rough boring	∅80.000 ± 0.200
40	Side face width broaching	128.300 ± 0.500
50	Thrust face width finish grinding on separate rod- and cap-end parts	26.800 ± 0.200
60	Bolt hole diameter drilling and reaming	∅6.000 ± 0.200
70	Key way slot milling	
80	Assembly of rod- and cap-end parts	
90	Thrust face width finish grinding of rod and cap connecting rod assembly	26.500 ± 0.050
100	Finish boring of gudgeon pin bore diameter	∅30.000 ± 0.200
110	Finish boring of crank pin bore diameter	∅84.090 ± 0.050
120	Crank pin bore diameter honing	∅85.077 ± 0.015
130	Magnetic crack detection	
140	Final quality check, set making and dispatch to engine assembly line	

The process capability studies on thrust face thickness, bolt hole center distance and crank pin bore diameter critical-to-quality characteristics of connecting rod were performed. After making the process capable through DMAIC approach, the end results of these process capability studies in the form of process capability values and tolerances obtained from tolerance charting of connecting rod machining process are compared with a tolerance capability expert software (Tec-ease.com 2014) and the results are documented.

Process sequence

The connecting rod manufacturing process sequence is depicted in the process flow diagram as shown in Fig. 5. The raw material from the raw material bin is the starting point of the connecting rod manufacturing process. The first roughing operation is operation no. 10 followed by a sequence of operations. The final operation is operation no. 140 where final quality check, set making and dispatch to engine assembly line are carried out. Table 1 gives the corresponding description of connecting rod machining operations.

Tolerance stack analysis of the connecting rod machining

Before proceeding to the process capability-based tolerancing study it is necessary to thoroughly examine the tolerance stack-up of the various machining processes involved in the manufacture of connecting rod. Figure 6

shows graphical representation of the tolerance chain associated with the machining of connecting rod.

The tolerance chain in the tolerance chart depicts the sequel of machining operations and their working dimensions. The tolerance stack-up and selection of reference surfaces for the subsequent machining operations is inferred from the diagram. Subsequently, the tolerances over the dimensions and the stock removal on the machining operation are also derived.

Process capability tolerancing of connecting rod

Based on the process capability improvement studies, the identified critical-to-quality characteristics in the machining of connecting rod and their initial and improved process capability values are tabulated in Table 2.

Comparison using tolerance capability expert software

The dimensional tolerances and C_{pk} values of the quality characteristic from Table 2 are the inputs into the database of the tolerance capability expert (TCE) software. In the TCE software, the worst case of manufacturing is considered, i.e., manufacturing machinery is not modern and not in good condition.

The following are the assumptions considered while using tolerance capability expert software.

1. The component or tooling used has repetitive features over a multiple references.

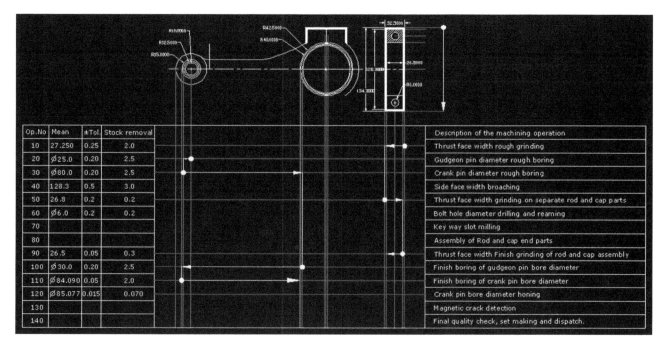

Fig. 6 Tolerance chain diagram of connecting rod in AutoCAD version 2005 software

Table 2 Process capability values of critical-to-quality characteristics of connecting rod machining process

S. no.	Quality characteristic	Dimension	Initial value			Final value		
			σ	C_p	C_{pk}	σ	C_P	C_{pk}
1	Thrust face thickness after thrust face width rough grinding	27.250 ± 0.250	0.48	0.12	0.12	0.048	1.72	1.37
2	Bolt hole center distance after bolt hole diameter drilling and reaming	106.750 ± 0.100	0.017	0.97	0.57	0.009	1.77	1.49
3	Gudgeon pin bore diameter after finish boring operation	30.000 ± 0.200	0.004	1.28	0.33	0.002	2.03	1.45
4	Crank pin bore diameter after honing operation	85.077 ± 0.015	0.005	0.5	0.34	0.002	1.52	1.45

Table 3 Graphical output from tolerance capability expert software

Characteristic	Graphical plot 1 with predicted C_{pk} for predetermined tolerance	Graphical plot 2 with predicted tolerance for predetermined C_{pk}
Thrust face thickness after thrust face width rough grinding with process dimension as 27.250 mm	See Fig. 7	See Fig. 8
Gudgeon pin bore diameter after finish boring operation with process dimension as 30.000 mm	See Fig. 9	See Fig. 10
Crank pin bore diameter after honing operation with process dimension as 85.08 mm	See Fig. 11	See Fig. 12

2. The characteristic is not along the die/mould parting line.
3. For producing this tolerance, simultaneous grinding of two parallel planes are involved.
4. The manufacturing machinery is not modern and in good condition.

5. The component size, weight, geometry and material impose additional limitations to the machine capability.
6. The feature geometry does not enable the process to be operated under good conditions of practice.
7. The process involves additional setups (for producing diesel as well as petrol variants of connecting rod).

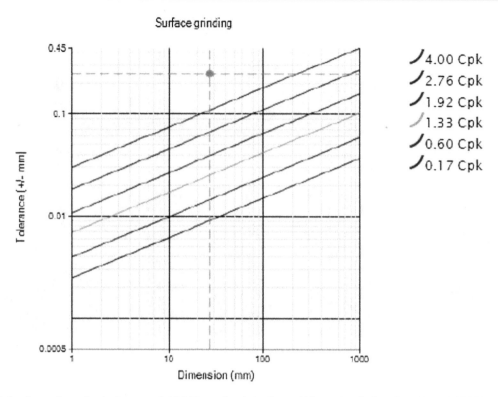

Fig. 7 Predicted C_{pk} for predetermined tolerance of ±0.250 mm for thrust face width rough grinding dimension of 27.250 mm

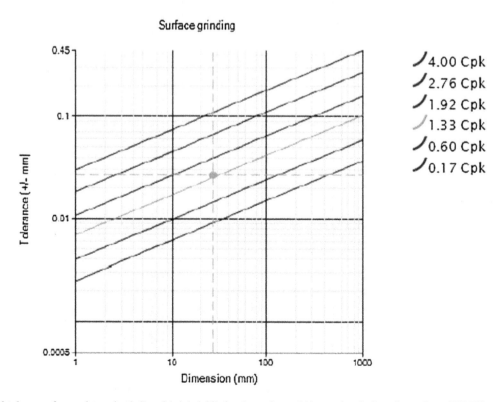

Fig. 8 Predicted tolerance for predetermined C_{pk} of 1.4 (>1.33) for thrust face width rough grinding dimension of 27.250 mm

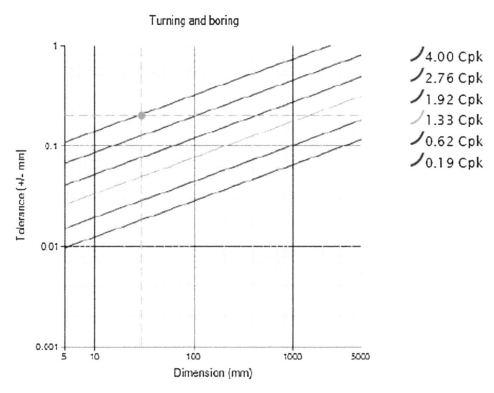

Fig. 9 Predicted C_{pk} for predetermined tolerance of ± 0.200 mm for gudgeon pin bore diameter after finish boring dimension of 30.000 mm

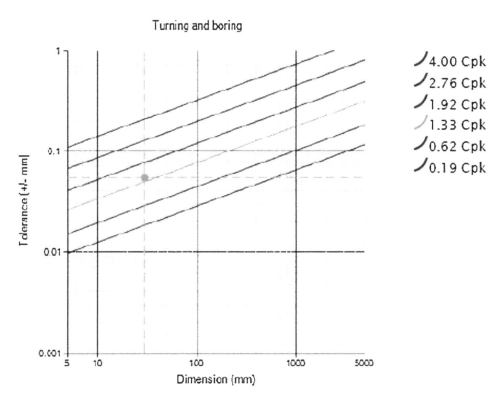

Fig. 10 Predicted tolerance for predetermined C_{pk} of 1.4 (>1.33) for gudgeon pin bore diameter after finish boring dimension of 30.000 mm

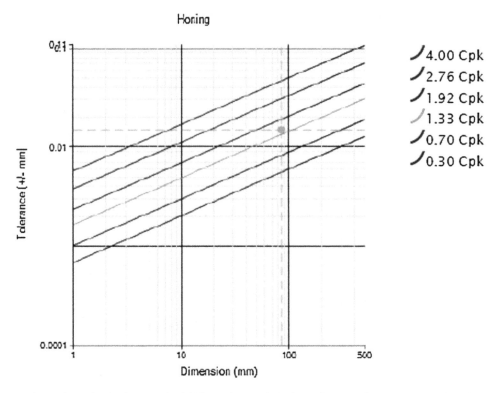

Fig. 11 Predicted C_{pk} for predetermined tolerance of ± 0.015 mm for crank pin bore diameter after honing operation dimension of 85.08 mm

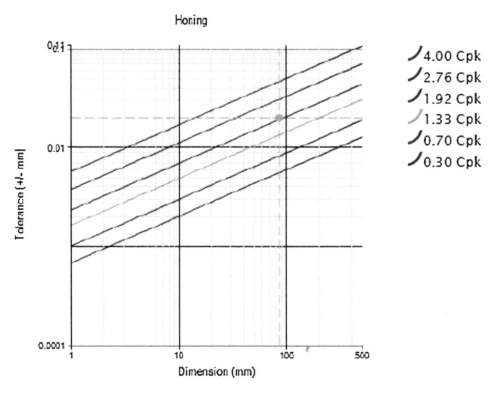

Fig. 12 Predicted tolerance for predetermined C_{pk} of 1.4 ($>$1.33) for crank pin bore diameter after honing operation dimension of 85.08 mm

Since the TCE software considers worst case of manufacturing, hence the results obtained are reliable with a certain factor of safety tolerance being inherent. The output obtained from the TCE software is tabulated in graphical form in Table 3.

Results and discussion

Table 3 depicts the various graphical outputs from the tolerance capability expert software. The dimension of the thrust face thickness after thrust face width rough grinding is 27.250 mm. Figure 7 shows that for thrust face thickness with a target tolerance of ±0.250 mm, the predicted C_{pk} lies in the region above 4.00. On the other hand, Fig. 8 shows that for a target C_{pk} of 1.4 (>1.33), the tolerance predicted is 0.027 mm, i.e., about ten times less than that in Fig. 7. The next critical-to-quality characteristic under consideration is the gudgeon pin bore diameter after finish boring operation as 30.000 mm. Figure 9 gives that for a target tolerance of ±0.200 mm, the predicted C_{pk} is 3.89; whereas, Fig. 10 gives that for a target C_{pk} of 1.45 (>1.33), the tolerance is ±0.054 mm for gudgeon pin bore diameter, leaving a large scope for improvement in this quality characteristic. The third critical-to-quality characteristic is the crank pin bore diameter after honing operation with process dimension as 85.08 mm concerning Figs. 11 and 12. In Fig. 11 it can be seen that for a target tolerance of ±0.015 mm, the predicted C_{pk} is 1.5 and from Fig. 12 it can be deciphered that for a target C_{pk} of 2.0, the predicted tolerance is ±0.020 mm. Figures 11 and 12 show close resemblance to each other and it can be deduced that the values of tolerances and C_{pk} obtained from tolerance sheet and process capability studies are in-phase with the values obtained from the tolerance capability expert software for the crank pin bore diameter after the honing operation.

Conclusion

Optimal values of the dimensional tolerance bandwidth are determined from the statistical process control charts. The process is made capable with the capability indices more than 1.33, i.e., more than a moderate level of 4σ, which is the industrial benchmark. After having made the process capable, the upper and lower tolerance bounds are shrunk to the calculated control limits inherent in the process. With the newly obtained tolerance values, the process is again calculated for its capability to be more than 4σ level. This iterative procedure of process improvement is carried out till the convergence is reached and no further noticeable process improvement is seen. Thus, the dimensional tolerances are optimized in accordance with the statistical process control improvement studies.

This paper witnesses an application of process tolerancing which proves to be a better way of finding the optimal tolerancing of the part, leading to fewer process rejections and improved quality levels. The end results of the process capability values and tolerances obtained from tolerance charting of connecting rod machining process are compared with a tolerance capability expert software. The results showed further scope of improvement for the thrust face thickness and gudgeon pin bore diameter, whereas the crank pin bore diameter after honing operation showed close resemblance between the values obtained through process capability and tolerance capability expert software.

References

Barbero BR, Aragón AC, Pedrosa CM (2015) Validation of a tolerance analysis simulation procedure in assemblies. Int J Adv Manuf Technol 76:1297–1310

Chen W-L, Huang C-Y, Huang C-Y (2013) Finding efficient frontier of process parameters for plastic injection molding. J Ind Eng Int 9:1–11

Contreras FG (2013) Maximization of process tolerances using an analysis of setup capability. Int J Adv Manuf Technol 67:2171–2181

Ding Y, Jin J, Ceglarek D, Shi J (2005) Process-oriented tolerancing for multi-station assembly systems. iiE Trans 37:493–508

Huang M-F, Zhong Y-R, Xu Z-G (2005) Concurrent process tolerance design based on minimum product manufacturing cost and quality loss. Int J Adv Manuf Technol 25:714–722

Irani S, Mittal R, Lehtihet E (1989) Tolerance chart optimization. Int J Prod Res 27:1531–1552

Jeang A (2011) Tolerance chart balancing with a complete inspection plan taking account of manufacturing and quality costs. Int J Adv Manuf Technol 55:675–687

Jeang A, Chen T, Li H-C, Liang F (2007) Simultaneous process mean and process tolerance determination with adjustment and compensation for precision manufacturing process. Int J Adv Manuf Technol 33:1159–1172

Ji P (1993) A tree approach for tolerance charting. Int J Prod Res 31:1023–1033

Ji P (1996) Determining dimensions for process planning: a backward derivation approach. Int J Adv Manuf Technol 11:52–58

Ji P (1999) An algebraic approach for dimensional chain identification in process planning. Int J Prod Res 37:99–110

Ji P, Xue J (2002) Process tolerance control in a 2D angular tolerance chart. Int J Adv Manuf Technol 20:649–654

Lee Y-C, Wei C-C (1998) Process capability-based tolerance design to minimise manufacturing loss. Int J Adv Manuf Technol 14:33–37

Louhichi B, Tlija M, Benamara A, Tahan A (2015) An algorithm for CAD tolerancing integration: generation of assembly configurations according to dimensional and geometrical tolerances. Comput-Aided Des 62:259–274

Narahari Y, Sudarsan R, Lyons KW, Duffey MR, Sriram RD (1999) Design for tolerance of electro-mechanical assemblies: an integrated approach. IEEE Trans Robot Autom Mag 15:1062–1079

Ngoi B, Ong C (1993) Process sequence determination for tolerance charting. Int J Prod Res 31:2387–2401

Ngoi B, Tan C (1995) Geometries in computer-aided tolerance charting. Int J Prod Res 33:835–868

Ngoi B, Tan C (1997) Graphical approach to tolerance charting—a "maze chart" method. Int J Adv Manuf Technol 13:282–289

Peng H, Jiang X, Liu X (2008) Concurrent optimal allocation of design and process tolerances for mechanical assemblies with interrelated dimension chains. Int J Prod Res 46:6963–6979

Schleich B, Wartzack S (2014) A discrete geometry approach for tolerance analysis of mechanism. Mech Mach Theory 77:148–163

Sharma G, Rao PS (2013) Process capability improvement of an engine connecting rod machining process. J Ind Eng Int 9:1–9

Sivakumar K, Balamurugan C, Ramabalan S (2012) Evolutionary multi-objective concurrent maximisation of process tolerances. Int J Prod Res 50:3172–3191

Tec-ease.com (2014) TCE: tolerance capability expert. [online] https://www.tec-ease.com/tce.php. Accessed 23 March 2014

Wei C-C, Lee Y-C (1995) Determining the process tolerances based on the manufacturing process capability. Int J Adv Manuf Technol 10:416–421

Whybrew K, Britton G, Robinson D, Sermsuti-Anuwat Y (1990) A graph-theoretic approach to tolerance charting. Int J Adv Manuf Technol 5:175–183

Xue J, Ji P (2001) A 2D tolerance chart for machining angular features. Int J Adv Manuf Technol 17:523–530

4

Developing a bi-objective optimization model for solving the availability allocation problem in repairable series–parallel systems by NSGA II

Maghsoud Amiri[1] · Mostafa Khajeh[2]

Abstract Bi-objective optimization of the availability allocation problem in a series–parallel system with repairable components is aimed in this paper. The two objectives of the problem are the availability of the system and the total cost of the system. Regarding the previous studies in series–parallel systems, the main contribution of this study is to expand the redundancy allocation problems to systems that have repairable components. Therefore, the considered systems in this paper are the systems that have repairable components in their configurations and subsystems. Due to the complexity of the model, a meta-heuristic method called as non-dominated sorting genetic algorithm is applied to find Pareto front. After finding the Pareto front, a procedure is used to select the best solution from the Pareto front.

Keywords Availability allocation · Series–parallel system · Repairable components · NSGA II

Introduction

In today's world with rapid technological developments and the increasing complexity of system structure, any failure in any component can lead to malfunction or serious failure to the system. Availability of the system is a suitable scale for measuring the reliability of a repairable system. Repairable system represents a system that can be repaired to operate normally in the event of any failure (Juang et al. 2008). The importance of designing reliable systems, which normally present high availability, is increasing, due to the engineering requirements of products with better quality and a higher safety level (Castro and Cavalca 2003).

Availability is the most important terminology used for evaluation on the effectiveness of any industrial plant, where most of the machines are repairable systems (Murty and Naikan 1995). It is therefore important to keep the equipments/systems always available and to lay emphasis on system availability at the highest order. System availability represents the percentage of time the system is available to users (Yusuf 2014).

A series–parallel system consists of a few subsystems connected in series, whereas each subsystem consists of a few components connected in parallel. A subsystem is failed if all the components in the subsystem are failed. Failure of any subsystem causes the failure of the whole system (Hu et al. 2012). The common structure of a parallel–series system is illustrated in Fig. 1.

On the subject of evaluating the availabilities of a system and its components, there are commonly two kinds of procedures. First, the aim of availability modeling is to develop an availability model to appraise system availability. Second, availability allocation, allocates the availability for each component based on the system's requirements or objectives (Chiang and Chen 2007).

Due to limitation in technology, the second way is better. Redundancy in a system means that the components are structured in parallel. The Redundancy allocation problem (RAP) is the most common method to meet the

✉ Mostafa Khajeh
mkhajeh2232@gmail.com

Maghsoud Amiri
mg_amiri@yahoo.com

[1] Department of Industrial Management, Allameh Tabataba'i University, Tehran, Iran

[2] Department of Industrial Management, Qom Branch, Islamic Azad University, P.O. Box 3749113191, Qom, Iran

Fig. 1 The structure of series–
parallel system

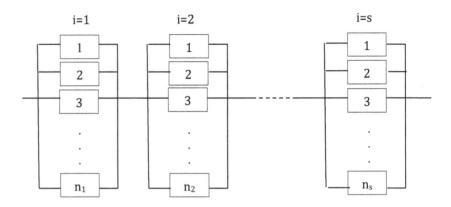

optimization of reliability and availability subject to the realistic constraints such as cost, weight, volume, etc. (Yahyatabar Arabi et al. 2014). Since the first paper on the redundancy allocation problem in a series–parallel system by (Fyffe et al. 1968) many researchers have tried to develop this knowledge. Two main approaches in the development of the RAP literature could be seen. First, proposing a fresh method to solve the previous optimization models on redundancy allocation problems. Second, develop the new optimization models for redundancy allocation problems (Amiri et al. 2014).

By investigation of literature reveals that many researchers study on the RAP in a series–parallel system for reliability optimization (Khalili-Damghani et al. 2014; Dolatshai-Zand and Khalili-Damghani 2015; Coit and Smith 1996; Wang et al. 2009; Yeh 2014; Hsieh and Yeh 2012; Azizmohammadi et al. 2013). Different heuristic and meta-heuristic methods such as genetic algorithm (GA), simulated annealing (SA), and particle swarm optimization (PSO) were proposed in this area (Khalili-Damghani and Amiri 2012; Chambari et al. 2012; Khalili-Damghani et al. 2013). A few of researchers (Elegbede and Adjallah 2003; Galikowski et al. 1996; Srivasvata and Fahim 1998; Varvarigou and Ahuja 1997) have studied on availability allocation and availability optimization. Busacca et al. (2001) presented Multi-objective optimization to maximize net profit with respect to certain availability. Elegbede and Adjallah (2003) proposed multi-objective availability allocation model and solved through Genetic Algorithm (GA). Chiang and Chen (2007) resolved the availability problem via simulated annealing (SA) based multi-objective genetic algorithm to determine the optimal solution of failure rates, repair rates, and the number of components in each subsystem, according to multi-objectives, such as system availability, system cost and system net profit. Castro and Cavalca (2003) presented an availability optimization problem of an engineering system assembled in a series configuration which has the redundancy of units and

teams of maintenance as optimization parameters. They used GA for maximized availability and considered installation and maintenance costs, weight, volume and available maintenance teams as constraints.

Yahyatabar Arabi et al. (2014)modeled availability optimization of series–parallel system using Markovian process by which the number of maintenance resources is located into the objective model under constraints such as cost, weight, and volume. They proposed meta-heuristic SA algorithm to find good results in an efficient time. Tewari et al. (2012) used genetic algorithm for calculation of the steady-state availability and performance optimization for the crystallization unit of a sugar plant. Amiri et al. (2014) investigate a multi-objective optimization model for series–parallel system with repairable components. The suggested optimization model has two objectives: maximizing the system mean time to first failure (MTTFF) and minimizing the total cost of the system. Finally a multi-objective approach of Imperialist Competitive Algorithm (ICA) is proposed to solve the model.

Tsarouhas (2015) developed analytical probability models for an automated serial production, which consists of n-machines in series. Both failure and repair rates are assumed to follow exponential distribution. In this study mathematical models of the production line have been developed using Markov process. Chandna and Ram (2014) applied fuzzy time series to forecast the availability of a standby system incorporating waiting time to repair. Faghih-Roohi et al. (2014) developed a dynamic model for availability assessment of multi-state weighted k-out-of-n systems and optimized by the genetic algorithm. For availability assessment, universal generating function and Markov process are adopted. Aggarwal et al. (2015) applied Markov modeling and reliability analysis for urea synthesis system. Lin and Droguett (2009) paired Multi-objective GA with Monte Carlo simulation to solve a bi-objective optimization of availability and cost in repairable systems.

Jiansheng et al. (2014) considered decision variables as vague factors and developed uncertain multi-objective RAP of repairable systems. They suggested artificial bee colony (ABC) algorithm to search the Pareto efficient set and showed this algorithm outperforms Non-dominated Soring Genetic Algorithm II (NSGA-II) greatly and can solve the multi-objective RAP efficiently. Srinivasa Rao and Naikan (2014) presented a hybrid approach called as Markov system dynamics (MSD) which combined the Markov approach with system dynamics simulation for reliability analysis of repairable systems.

In this paper, the RAP in repairable series–parallel systems is considered, with two objectives (1) maximizing the system asymptotic availability (2) minimizing the total cost. Furthermore, in each subsystem only one component type is allowed to be used. Each choice has different levels of failure rate, repair rate, weight and cost. The decision variables are to select the component choice and the level of redundancy. Since the considered optimization problem was proven NP-hard (Chern 1992) and Heuristic algorithms do not provide an assurance for optimization of the problem (Bashiri and Karimi 2012), therefore, meta-heuristic algorithms used to generate near optimal solutions. In this paper, proposed a Pareto-based meta-heuristic algorithm called NSGA-II to solve the problem.

The remainder of the paper is organized as follows. The mathematical formulation of the problem is introduced in "Problem description" section. The solution algorithm is presented in "Solution method" section. The numerical example is introduced in "Numerical example" section. Finally, conclusion and recommendations for future research are in "Conclusion" section.

Problem description

In this study, the mathematical model of the series–parallel system with k subsystem and repairable components is illustrated. The suggested optimization model has two objectives: maximizing the system availability and minimizing the total cost of the system. The notations and assumptions of the model are presented in the following.

Notation

k　　Total number of subsystems;
m_i　　The set of components in the i-th subsystem;
x_{ij}　　Number of type j component in subsystem i;
n_i　　Total number of component in subsystem i;
λ_{ij}　　Failure rate of component j in subsystem i;
μ_{ij}　　Repair rate of component j in subsystem i;
c_{ij}　　Cost of component j in subsystem i;
w_{ij}　　Weight of component j in subsystem i;

W　　Total weight of system;
A_s　　Availability of system;
C_s　　Cost of system.

Assumptions

- The state of each component at any point of time is one of the "good" or "failed" states.
- The state of each component is independent of the other components.
- For each subsystem, there are m_i functionally equivalent component choices that can be selected. In each subsystem only one component type is allowed to be used.
- The system conducts its function perfectly when each subsystem has at least one operable component. Therefore, for each subsystem at least one component should be selected.
- The failure and repair rate of each alternative component available for each subsystem has exponential distribution with failure rate λ_{ij} and repair rate μ_{ij}.

Mathematical model

$$\text{Maximize} A_s = \prod_{i=1}^{k}\left(1 - \left(\frac{\lambda_i}{\lambda_i + \mu_i}\right)^{n_i}\right); \tag{1}$$

$$\text{Minimize} C_s = \sum_{i=1}^{k}\sum_{j=1}^{m_i} c_{ij}x_{ij} \tag{2}$$

Subject to the following constraints:

$$\sum_{i=1}^{k}\sum_{j=1}^{m_i} w_{ij}x_{ij} \leq W \tag{3}$$

$$\lambda_i = \sum_{j=1}^{m_i} \lambda_{ij}y_{ij}; \tag{4}$$

$$\mu_i = \sum_{j=1}^{m_i} \mu_{ij}y_{ij}; \tag{5}$$

$$\sum_{j=1}^{m_i} y_{ij} = 1; \tag{6}$$

$$0 \leq x_{ij} \leq My_{ij}; \tag{7}$$

$$n_i = \sum_{j=1}^{m_i} x_{ij}; \tag{8}$$

$$1 \leq n_i \leq n_{\max,i} \tag{9}$$

The objective functions (1) and (2) maximizes the availability of system and minimizes total cost of system, respectively. The formulation of system availability is presented by Elegbede and Adjallah (2003). Constraint (3) represents the total weight of the system. The Constraints (4)–(7) make it possible to select only one type of components for each subsystem. The constraints (8) and (9) imply minimum and maximum number of components selected for each subsystem.

Solution method

There are two general approaches to multiple-objective optimization. One is to combine the individual objective functions into a single composite function or move all, but one objective to the constraint set. Determination of a single objective is possible with methods such as utility theory, weighted sum method, epsilon constraint, etc., but the problem lies in the proper selection of the weights or utility functions to characterize the decision-maker's preferences.

In the second approach, a Pareto optimal set is determined. A Pareto optimum set is a set of solutions that are non-dominated with respect to each other. Pareto optimal solution sets are often preferred to single solutions because they can be practical when considering real-life problems since the final solution of the decision-maker is always a trade-off (Konak et al. 2006). Multi-objective evolutionary algorithms (MOEA) are employed to solve the multi-objective problems and generate Pareto frontiers. Among MOEAs, NSGA-II proposed by Deb et al. (2002) is elitist and fast multi-objective genetic algorithm. NSGA-II was one of the best methods because it carried out an elite-preserving strategy and explicit diversity preserving mechanism (Li et al. 2015).

In an evolutionary cycle of the NSGA-II, a mating pool is first created and filled using binary tournament selection. Then, crossover and mutation operators apply to the members of the mating pool. Next, the old set of solutions and newly created solutions are merged to create a larger population. This new population is sorted based on two criteria: (1) rank and (2) crowding distance. Finally, a certain amount of individuals in the sorted population is selected and others are deleted. These steps are repeated until a stopping condition is met. After NSGA-II terminates, non-dominated solutions of the final population are the approximate Pareto frontier of multi-objective optimization problem (Pasandideh et al. 2013). The procedure of evolution cycle in NSGA II is shown in Fig. 2.

Selection algorithm is the most important part of NSGA-II that specifies the direction of search for finding optimal solutions. Those of solutions with better ranking are transferred to the next step. If two solutions are same rank, the solution with the larger crowding distance is selected. Figure 3 illustrates the ranking and crowding distance used

in NSGA-II. In the following subsection, the steps of this algorithm are described.

Solution representation

A series of genes that arrange sequentially is called a chromosome. The number of genes in a chromosome is equal to the number of decision variables. Chromosome description is one of the most significant parts of the algorithm that is taken into account as the code form. In this paper, the solution encoding for this problem is a $2 \times s$ matrix. The elements of the first row illustrate the type of component, selected for the related subsystem. The element in the second row of each subsystem column, verifies the number of selected components for related subsystem. An example of the solution representation is illustrated in Fig. 4.

Initial population

The generation of an initial population is necessary to start solving the optimization problem with a GA. The size of any population is given and remains the same in each generation. The main difficulty in the initial population is that the individuals may not satisfy all or part of the constraints of the problem (Elegbede and Adjallah 2003). In this paper, initial population size is considered 100. This population size has been used for a lot of researches like safari (2012), zoulfaghari et al. (2014) and Deb et al. (2002). As mentioned safari (2012), in problems with very large solutions paces, the population size must be selected no <100.

Crossover

The crossover operator explores a new solution space and provides the possibility of generating new solutions called offspring through mating pairs of chromosomes (Pasandideh et al. 2015). The most common crossover techniques are: (1) One-point crossover (2) Two-point crossover (3) Uniform crossover. At a single crossover point, two parents selected and all data beyond that point with certain probability are swapped between two parents. The resulting chromosomes are the children. In this paper, one-point crossover is used. Figure 5 depicts the crossover performed in the NSGA- II.

Mutation

Mutation operator because of its ability to enter new genes into the chromosomes has extraordinary importance. The mutation operator is also used at a certain rate less than the crossover rate. The main purpose of applying the mutation operator is to increase diversity and avoid being trapping

Fig. 2 An evolution cycle in NSGA II (Galikowski et al. 1996)

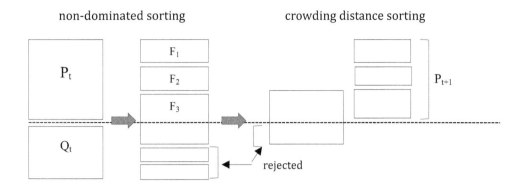

Fig. 3 a Non-dominated ranking and **b** the crowding distance calculation (Kumar et al. 2009)

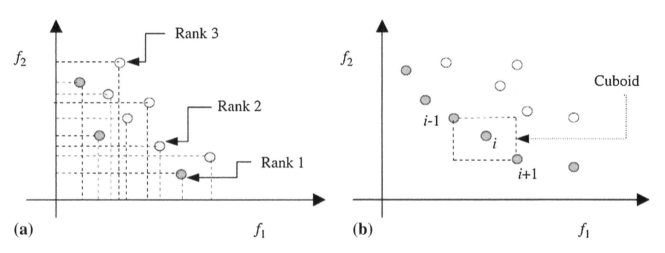

Fig. 4 Structure of the solution representation

Subsystem index	1	2	3	4
Type of Component	3	2	3	1
Number of components	2	1	2	2

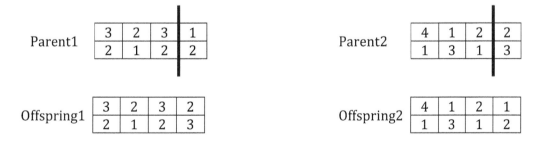

Fig. 5 Example of one-point crossover

into local optimization (Zoulfaghari et al. 2014). Since, in reality, the mutation rarely happened, probability of mutation is considered very low. In this paper, one subsystem is selected. Then, type and number of components in this subsystem are replaced with each other. Figure 6 illustrated the mutation operator that applied in this paper.

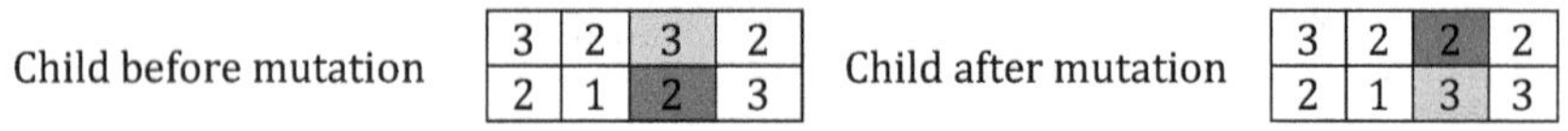

Fig. 6 Example of mutation

Child before mutation

3	2	3	2
2	1	2	3

Child after mutation

3	2	2	2
2	1	3	3

Table 1 Subsystems alternative components

Subsystem	Maximum number of components in subsystem	Minimum number of components in subsystem	Component types	Failure rate	Repair rate	Weight	Cost
1	5	1	1	2	10	100	120
			2	5	18	80	100
			3	4	25	85	140
			4	4	2	90	110
2	6	1	1	5	40	250	400
			2	6	42	200	380
			3	10	100	200	500
3	4	1	1	4	22	450	800
			2	4	28	550	800
			3	7	20	250	800
			4	7	18	300	800
4	4	1	1	5	30	500	1200
			2	7	35	500	1500
			3	3	25	500	1500

Maximum weight of the system = 4500

Table 2 The parameters for NSGA II

Parameter	Value
Population size	100
Mutation rate	0.08
Crossover rate	0.9
Number of iterations	100

Stopping criteria

The algorithm terminates after certain iterations. Number of iterations in this problem considered 100 iterations.

Numerical example

In this section, to evaluate the performance of the proposed NSGA-II, the example that the data it is presented in (Amiri et al. 2014) has been used. In this paper, a series–parallel system with four parallel subsystems is considered, and each subsystem has three or four repairable components of choice. Failure and repair rates of all components are negative exponential. The maximum total weight of the system is 4500. Table 1 includes details of the problem. The objective is to maximize the system availability and minimize the system cost. The decision variables are to select the component choice and the level of redundancy in each subsystem.

To solve the problem, the proposed NSGA-II was used. The NSGA-II was implemented using MATLAB software and was run on a computer with 2G of RAM. The parameters of NSGA-II approach are shown in Table 2. After solving the problem, like other multi-objective optimization models, the Pareto optimal solutions were obtained. The Pareto optimal solutions contain the solutions that were not dominated by other solutions. Table 3 showed the non-dominated solutions obtained with NSGA-II.

Although determination of Pareto optimal solutions can be considered as one of strengths of multi-objective optimization algorithms, but the decision maker will be confused in choosing the best solution. There are some methods for determining the best solution in a Pareto set. The most widely used method that described in (Eschenauer et al. 1990) is the L_P-norm. This technique minimizes the normalized distance from the Pareto set to an ideal solution (i.e., utopia point) to find the optimal solution according to the following formula (Kasprzak and Lewis 2000):

$$\text{Minimize} \left(\sum_{i=1}^{m} \left(\frac{f_i(x) - f_i^{\min}}{f_i^{\max} - f_i^{\min}} \right)^p \right)^{\frac{1}{p}}, \quad p = 1, 2, \ldots, \infty$$

(10)

Table 3 The non-dominated solutions resulted from using the NSGA-II

Answer number	System characteristic									L_2
	Availability	Cost	Weight	Decision variable						
				Subsystem number		1	2	3	4	
1	0.5135869	2480	1330	z_i		2	2	2	1	1
				n_i		1	1	1	1	
2	0.5657327	2520	1335	z_i		3	2	2	1	0.89228
				n_i		1	1	1	1	
3	0.6252362	2580	1410	z_i		2	2	2	1	0.76945
				n_i		2	1	1	1	
4	0.6380208	2620	1450	z_i		1	2	2	1	0.74316
				n_i		2	1	1	1	
5	0.6688466	2780	1420	z_i		3	3	2	1	0.68047
				n_i		2	1	1	1	
6	0.7033908	2960	1610	z_i		2	2	2	1	0.61140
				n_i		2	2	1	1	
7	0.7242354	3040	1610	z_i		3	2	2	1	0.56998
				n_i		2	2	1	1	
8	0.7712303	3600	1950	z_i		2	1	2	3	0.49222
				n_i		5	2	1	1	
9	0.7913146	3760	2160	z_i		2	2	2	1	0.46086
				n_i		2	2	2	1	
10	**0.864036**	**4260**	**2400**	**z_i**		**1**	**1**	**2**	**3**	**0.36825**
				n_i		**3**	**2**	**2**	**1**	
11	0.8757403	4700	2670	z_i		2	1	2	3	0.39481
				n_i		4	3	2	1	
12	0.8765127	4720	2600	z_i		1	2	2	3	0.39591
				n_i		4	3	2	1	
13	0.8775877	4900	2850	z_i		1	1	2	3	0.41398
				n_i		5	3	2	1	
14	0.9043596	4960	2560	z_i		2	2	2	1	0.39055
				n_i		2	2	2	2	
15	0.9228515	5000	2700	z_i		1	2	2	1	0.37818
				n_i		2	2	2	2	
16	0.9311598	5040	2670	z_i		3	2	2	1	0.37660
				n_i		2	2	2	2	
17	0.9957989	8820	4500	z_i		2	2	1	1	0.86850
				n_i		5	4	4	3	
18	0.9965262	9120	4375	z_i		3	2	2	3	0.90959
				n_i		5	4	3	3	
19	0.9966230	9500	4450	z_i		1	3	2	3	0.96165
				n_i		5	4	3	3	
20	0.9967914	9600	4475	z_i		3	1	1	3	0.97534
				n_i		5	3	4	3	

The bold row is the best non-dominated solution

z_i component type, n_i numbers of component

where f_i^{min} and f_i^{max} are the minimum and maximum value for the i-th objective function in the Pareto optimal set. In this formula all objective functions must be minimized.

In this paper, we apply L_2-norm. For using this method, first objective function (maximize system availability) must be transformed to minimization. For this purpose, system unavailability is calculated. The best non-dominated solution is shown at row 10 in Table 3.

Conclusion

In this paper, we have developed a bi-objective model for solving availability allocation problem in series–parallel systems with repairable components. The considered system in this study has components with constant failure and repair rate, therefore considering systems comprising of components without exponential distribution for their repair and failure times could be a good challenge for future studies.

In this study, the designed optimization model is solved by a meta-heuristic algorithm, NSGA-II; the main goal of the paper was to propose an optimization model and a solving algorithm to attain the optimal structure of a repairable series–parallel system. Using other algorithms to solve the proposed optimization model and comparing the results of the solutions resulted in this paper could be the goal for future works.

Acknowledgments The authors would like to thank the Editor in Chief and referees for providing very helpful comments and suggestions.

References

Aggarwal AK, Kumar S, Singh V, Grab TK (2015) Markov modeling and reliability analysis of urea synthesis system of a fertilizer plant. J Ind Eng Int 11:1–14

Amiri M, Sadeghi MR, Khatami Firoozabadi A, Mikaeili F (2014) A multi objective optimization model for redundancy allocation problems in series–parallel systems with repairable components. Int J Ind Eng Prod Res 25(1):71–81

Azizmohammadi M, Amiri M, Tavakkoli Moghaddam R, Mohammadi M (2013) Solving a redundancy allocation problem by a hybrid multi-objective imperialist competitive algorithm. Int J Eng 26(9):1031–1042

Bashiri M, Karimi H (2012) Effective heuristics and meta-heuristics for the quadratic assignment problem with tuned parameters and analytical comparisons. J Ind Eng Int 8(1):1–9

Busacca PG, Marseguerra M, Zio E (2001) Multi-objective optimization by genetic algorithms: application to safety systems. Reliab Eng Syst Saf 72:59–74

Castro HF, Cavalca KL (2003) Availability optimization with genetic algorithm. Int J Qual Reliab Manag 20(7):847–863

Chambari A, Rahmati SHA, Najafi AA, Karimi A (2012) A bi-objective model to optimize reliability and cost of system with a choice of redundancy strategies. Comput Ind Eng 63:109–119

Chandna R, Ram M (2014) Forecasting availability of a standby system using fuzzy time series. J Reliab Stat Stud 7:01–08

Chern MS (1992) on the computational complexity of reliability redundancy allocation in a series system. Oper Res Lett 11:309–315

Chiang CH, Chen LH (2007) Availability allocation and multi-objective optimization for parallel–series systems. Eur J Oper Res 180:1231–1244

Coit DW, Smith AE (1996) Reliability optimization of series–parallel systems using a genetic algorithm. IEEE Trans Reliab 45(2):254–260

Deb K, Pratap A, Agerwal S, Meyarivan T (2002) A fast and elitist multi objective genetic algorithm: NSGA-II. IEEE Trans Evol Comput 6:182–197

Dolatshai-Zand A, Khalili-Damghani K (2015) Design of SCADA water resource management control center by a bi-objective redundancy allocation problem and particle swarm optimization. Reliab Eng Syst Saf 133:11–21

Elegbede C, Adjallah K (2003) Availability allocation to repairable systems with genetic algorithms: a multi-objective formulation. Reliab Eng Syst Saf 82:319–330

Eschenauer H, Koski J, Osyczka AE (1990) Multicriteria design optimization: procedures and applications. Springer, New York

Faghih-Roohi S, Xie M, Ng KM, Yam RCM (2014) Dynamic availability assessment and optimal component design of multi-state weighted k-out-of-n systems. Reliab Eng Syst Saf 123:57–62

Fyffe DE, Hines WW, Lee NK (1968) System reliability allocation and a computational algorithm. IEEE Trans Reliab R-17(2):64–69

Galikowski C, Sivazlian BD, Chaovalitwongse P (1996) Optimal redundancies for reliability and availability of series systems". Microelectron Reliab 36(10):1537–1546

Hsieh TJ, Yeh WC (2012) Penalty guided bees search for redundancy allocation problem with a mix of components in series–parallel systems. Comput Oper Res 39:2688–2704

Hu L, Yue D, Li J (2012) Availability analysis and design optimization for a repairable series–parallel system with failure dependencies. Int J Innov Comput Inf Control 8(10):6693–6705

Jiansheng G, Zutong W, Mingfa Z, Ying W (2014) Uncertain multi-objective redundancy allocation problem of repairable systems based on artificial bee colony algorithm. Chin J Aeronaut 27(6):1477–1487

Juang YS, Lin SS, Kao HP (2008) A knowledge management system for series–parallel availability optimization and design. Expert Syst Appl 34:181–193

Kasprzak EM, Lewis KE (2000) An approach to facilitate decision trade-offs in Pareto solution sets. J Eng Valuat Cost Anal 3(1):173–187

Khalili-Damghani K, Amiri M (2012) Solving binary-state multi-objective reliability redundancy allocation series–parallel problem using efficient epsilon-constraint, multi-start partial bound enumeration algorithm, and DEA. Reliab Eng Syst Saf 103:35–44

Khalili-Damghani K, Abtahi AR, Tavana M (2013) A new multi-objective particle swarm optimization method for solving reliability redundancy allocation problems. Reliab Eng Syst Saf 111:58–75

Khalili-Damghani K, Abtahi AR, Tavana M (2014) Decision support system for solving multi-objective redundancy allocation problems. Qual Reliab Eng Int 30(8):1249–1262

Konak A, Coit DW, Smith AE (2006) Multi-objective optimization using genetic algorithms: a tutorial. Reliab Engi Syst Saf 91:992–1007

Kumar R, Izui K, Yoshimura M, Nishiwaki S (2009) Multi-objective hierarchical genetic algorithms for multilevel redundancy allocation optimization. Reliab Eng Syst Saf 94:891–904

Li Y, Liao S, Liu G (2015) Thermo-economic multi-objective optimization for a solar-dish Brayton system using NSGA-II and decision making. Electr Power Energy Syst 64:167–175

Lin ID, Droguett EL (2009) Multiobjective optimization of availability and cost in repairable systems design via genetic algorithms and discrete event simulation. Pesqui Oper 29(1):43–66

Murty ASR, Naikan VNA (1995) Availability and maintenance cost optimization of a production plant. Int J Qual Reliab Manag 12(2):28–35

Pasandideh SHR, Akhavan Niaki ST, Sharafzadeh S (2013) Optimizing a bi-objective multi-product EPQ model with defective items, rework and limited orders: NSGA-II and MOPSO algorithms. J Manuf Syst 32:764–770

Pasandideh SHR, Akhavan Niaki ST, Asadi K (2015) Bi-objective optimization of a multi-product multi-period three-echelon supply chain problem under uncertain environments: NSGA-II and NRGA. Inf Sci 292:57–74

Safari J (2012) Multi-objective reliability optimization of series–parallel systems with a choice of redundancy strategies. Reliab Eng Syst Saf 180:10–20

Srinivasa Rao M, Naikan VNA (2014) Reliability analysis of repairable systems using system dynamics modeling and simulation. J Ind Eng Int 10:69

Srivasvata VK, Fahim A (1998) K-out-of-m system availability with minimum-cost allocation of spears. IEEE Trans Reliab 37(3): 287–292

Tewari PC, Khanduja R, Gupta M (2012) Performance enhancement for crystallization unit of a sugar plant using genetic algorithm. J Ind Eng Int 8:1

Tsarouhas PH (2015) Performance evaluation of the croissant production line with repairable machines. J Ind Eng Int 11:101–110

Varvarigou TA, Ahuja S (1997) MOFA: a model for fault and availability in complex services. IEEE Trans Reliab 46(2):222–232

Wang Z, Chen T, Tang K, Yao X (2009) A multi-objective approach to redundancy allocation problem in parallel–series systems. In: Proceedings of IEEE Congress on Evolutionary Computation, Trondheim, pp 582–589

Yahyatabar Arabi A, Eshraghniaye Jahromi A, Shabannataj M (2014) Developing a new model for availability optimization applied to a series–parallel system. Int J Ind Eng Prod Res 24(2):101–106

Yeh WC (2014) Orthogonal simplified swarm optimization for the series–parallel redundancy allocation problem with a mix of components. Knowl Based Syst 64:1–12

Yusuf I (2014) Comparative analysis of profit between three dissimilar repairable redundant systems using supporting external device for operation. J Ind Eng Int 10:199–207

Zoulfaghari H, Zeinal Hamadani A, Abouei Ardakan M (2014) Bi-objective redundancy allocation problem for a system with mixed repairable and non-repairable components. ISA Trans 53:17–24

5

Two models of inventory control with supplier selection in case of multiple sourcing: a case of Isfahan Steel Company

Masood Rabieh[1] · Mohammad Ali Soukhakian[2] · Ali Naghi Mosleh Shirazi[2]

Abstract Selecting the best suppliers is crucial for a company's success. Since competition is a determining factor nowadays, reducing cost and increasing quality of products are two key criteria for appropriate supplier selection. In the study, first the inventories of agglomeration plant of Isfahan Steel Company were categorized through VED and ABC methods. Then the models to supply two important kinds of raw materials (inventories) were developed, considering the following items: (1) the optimal consumption composite of the materials, (2) the total cost of logistics, (3) each supplier's terms and conditions, (4) the buyer's limitations and (5) the consumption behavior of the buyers. Among diverse developed and tested models—using the company's actual data within three pervious years—the two new innovative models of mixed-integer non-linear programming type were found to be most suitable. The results of solving two models by lingo software (based on company's data in this particular case) were equaled. Comparing the results of the new models to the actual performance of the company revealed 10.9 and 7.1 % reduction in total procurement costs of the company in two consecutive years.

Keywords Inventory control · Supplier selection · Multiple sourcing · Mathematical models

✉ Masood Rabieh
  M_Rabieh@sbu.ac.ir

[1] Department of Industrial Management, Shahid Beheshti University, Tehran, Iran

[2] Department of Management, Shiraz University, Shiraz, Iran

Introduction

Supplier selection is turning to become one of the crucial decisions in operations management area for many companies. Nowadays that competition plays a major role in business, two factors, namely, cost reduction and increase in quality of products, are keys to success of a company. Attaining these two factors is heavily dependent on having appropriate suppliers. Therefore, selecting appropriate suppliers can increase the competitiveness of a business.

The main cost of a product is mostly dependent on the cost of raw material and component parts in most industries (Ghodsypour and O'Brien 2001). Under such a condition the raw material supply and its inventory control can play a key role in the efficiency and effectiveness of a business and have a direct impact on cost reduction, profitability and its flexibility. Regarding supplier selection, there are two general situations:

Single sourcing A situation in which there is no constraint and a single supplier of an item is able to satisfy all requirements of the buyer.

Multiple sourcing In this situation there are many suppliers of a required item, but no single suitable supplier can satisfy all requirements of the buyer. Thus, the buyer must choose "an appropriate set of suitable supplies" to work with (Ghodsypour and O'Brien 1998).

Considering many factors such as variations in price, terms and conditions, quality, quantity, transportation costs and distances, etc. of each supplier, the multiple sourcing situations usually involves taking complex decisions.

While there is a paucity of research that takes into account different aspects of this complex decision situation, only a limited number of mathematical models have been proposed for such decisions. Many of the proposed models consider "net price" as the main factor, a few of them consider "the total costs of logistics".

The present study investigates the issue of multiple sourcing and proposes mathematical models based on considering factors such as net price, transportation costs, inventory costs and shrinkage problems.

The rest of this paper is organized as follows. In "Background" section literature review is presented. In "The situation" section, the case study is described. The mathematical formulating of problem is presented in "Formulating the models" section. Data collection and parameters are described in "Parameters of model" section. Computational result is presented in "Model runs and results" section and finally, some concluding remarks are given in "Discussion and conclusion" section.

Background

Supplier selection literature may generally be divided into two areas: First, descriptive, survey type approaches and, second, quantitative modeling methods. In the first area, the researches of Dickson (1966) and Weber et al. (1991) should be mentioned as the most comprehensive ones. Dickson has identified and summarized a number of criteria that purchasing managers consider for supplier selection. In his view, the most important criteria are quality, delivery, and the performance history of the supplier. Weber et al. (1991) in a review of 74 articles on supplier selection criteria, found that the most important factor is net price, yet, they suggested that supplier selection is dependent on a multitude of factors with different priorities, depending on the particular purchasing situation.

In the second area, which is more relevant to this article, a few number of fine research attempts should be mentioned here.

Benton (1991) applied Lagrange relaxation to develop a non-linear program for supplier selection under various conditions including multiple suppliers, multiple items, resource limitations and quantity discount. Ghodsypour and O'Brien (1997) suggested integrated analytical hierarchy process (AHP) with mixed integer programming to develop a decision support system (DSS). Their objective was to reduce the number of suppliers. Ghodsypour and O'Brien (1998) also developed a model to take into account both qualitative and quantitative factors. This approach was based on the integration of AHP and linear programming model. In a further development Ghodsypour and O'Brien (2001) presented a mixed integer non-linear programming model to solve the multiple sourcing problems. Their model takes the total cost of logistics into consideration. Kumar et al. (2004) advised a fuzzy goal programming approach to solve the vendor selection problem in case of multiple

objectives. Chen et al. (2006) presented a fuzzy decision making approach to solve the supplier selection problem. They proposed linguistic values to evaluate the ratings for a number of quantitative and qualitative factors including quality, price, flexibility, and delivery performance. Their model shown to be very good tool for supplier selection decision making situation. Basnet and Leung (2005) investigated the problem of supplier selection considering the lot-sizing. Amid et al. (2006) represented multi objective linear programming model to supplier selection. Lin and Chang (2008) propose mixed-integer programming and fuzzy TOPSIS approach to solve the supplier selection problem. Aissaouia et al. (2007) have extended previous survey papers by presenting a literature review that covers the entire purchasing process and covers internet-based procurement environments. In the mentioned work they have focused especially on the final selection stage that consists of determining the best mixture of vendors and allocating orders among them so as to satisfy different purchasing requirements. Also, they have concentrate mainly on works that employ operations research and computational models. Farzipoor saen (2007) has considered widespread application of manufacturing philosophies such as just-in-time (JIT), emphasis has shifted to the simultaneous consideration of cardinal and ordinal data in supplier selection process and proposed an innovative method, which is based on imprecise data envelopment analysis (IDEA) to selected the best suppliers in the presence of both cardinal and ordinal data. Ustun and Aktar Demirtas (2008) have recommended an integrated approach of analytic network process (ANP) and multi-objective mixed integer linear programming (MOMILP) for supplier selection problem. Their approach considers both tangible and intangible factors in choosing the best suppliers and defines the optimum quantities among selected suppliers to maximize the total value of purchasing (TVP), and to minimize the total cost and total defect rate and to balance the total cost among periods. Soukhakian et al. (2007) developed a model based on the Ghodsypour and O'Brien (2001) model. The contribution of the developed model is compared with basic the basic model which consider limitations such as integer number of orders and minimum assigned order quantity to each supplier. Due to the complexity of model and its non-linearity, the model is solved by genetic algorithm. Rabieh et al. (2008) developed a new model based on the Ghodsypour and O'Brien (2001) model for a real case of the agglomeration unit of Isfahan Steel Company. In this model assume that some suppliers of iron concentrate in have to cover the inventory in turn during each ordering cycle (T), while other suppliers of iron ore deliver their shipments simultaneously. In end, the non-linear model is solved by LINGO 8 software. Jafarnezhad et al. (2009) introduced a fuzzy decision making approach for supplier selection problem in case of single sourcing. In this research, the fuzzy TOPSIS method was developed for ranking and selecting suppliers. At the end, a numerical example was introduced for showing performance of the developed method. Wu and Blackhurst (2009) presented a supplier selection and evaluation method based on an extension of data envelopment analysis (DEA) that can efficiently evaluate suppliers. Kuo and Lin (2012) introduced an integrated approach of analytic network process (ANP) and data envelopment analysis (DEA) in solving supplier selection problem. Their model also considered green indicators due to environmental protections issues. Finally, Mendoza and Ventura (2012) presented two mixed integer nonlinear programming models to select the best suppliers and determine order quantities. Their research integrated the issues of inventory management and supplier selection. Rao et al. (2013) developed a new approach to design a multi-echelon, multi-facility, and multi-product supply chain in uncertain environment in fuzzy form. In this research, a mixed integer programming was formulated at strategic level and a non-linear programming model was presented in tactical level. In the tactical level, inventory control of raw material of suppliers was considered (Table 1).

The situation

The agglomeration unit of Isfahan Steel works—one of the largest steel manufacturing firms in the ME region located in central Iran—is the case studied in this research. The main task of this unit is to agglomerate different kinds of raw materials in specific proportions. Most of the raw materials come from different quarries and plants scattered all over the country. The materials are bought and transported to the works mainly via railroads and sometimes by trucks in distances even up to 1300 km. The functional and financial importance of each required raw materials for agglomeration unit found to be different in nature. So, as the first step, a classification of inventory items should have been curried out prior to actual modeling. The following three popular classification methods, so called selective inventory control techniques, are usually applied for grouping inventory items:

ABC analysis, classifies items in terms of annual financial requirement.
VED analysis, classifies items in terms of their functional importance (Vital, Essential, Desirable).
FNS analysis, classifies items in terms of their movement speed (Fast, Normal, Slow; Nair 2002).

Using ABC and VED methods, the inventory items of agglomeration plant were analyzed, and iron ore and iron concentrate were found to be the most important raw materials respectively. Thus, modeling in this study was focused around the purchasing and supply of these two items.

Iron concentrate is a supplementary material which is very similar to iron ore in appearance and should be mixed with iron ore in agglomeration process. Since it contains more Fe; its price is much higher than iron ore. However, to obtain a desired and consistent quality of the agglomeration process output, a right percentage of these two materials should be mixed together each time. The needed iron ore and concentrate for agglomeration plant is purchased from five different suppliers, none of which has the sufficient capacity to supply the whole annual requirements. Furthermore, there are some quality variations in their products and each supplier has its own supply characteristics.

The developed models in this study take into account such variations, and are formulated in a way to obtain a right combination of the raw materials in one hand, and minimize the total inventory costs in the other.

Formulating the models

Defining model parameters and variables

Before describing the model, the pertaining parameters and variables are defined as follows:

Decision variables

Q: Ordered quantity to all suppliers in each period.
Q_i: Ordered quantity to ith supplier in each period.
X_i: Percentage of Q assigned to ith supplier.

$$Y_i = \begin{cases} 1 & \text{if } X_i > 0 \\ 0 & \text{if } X_i = 0 \end{cases}$$

Parameters

D: Annual iron ore and concentrate demand (in term of tons).
T: The length of each period.
T_i: Part of the period in which the lot of ith supplier (Q_i) is used.
n: Number of suppliers
C_i: Annual capacity of the ith supplier to supply raw material.
C_{ti}: Transportation cost for ith supplier per unit of raw material.

Table 1 The features of other reviewed quantitative researches and our research

Researches	Criterion of comparison							
	Mathematical model	Multi attribute decision making (MADM)	Mathematical model and MADM approach	Deterministic model	Uncertain model	Non-linearity	Supplier selection and inventory control	Real case
Our research	✓			✓		✓	✓	✓
Benton (1991)	✓			✓		✓	✓	
Ghodsypour and O'Brien (1997)	✓		✓	✓				
Ghodsypour and O'Brien (1998)	✓		✓	✓				
Ghodsypour and O'Brien (2001)	✓			✓		✓	✓	
Kumar et al. (2004)	✓				✓		✓	✓
Basnet and Leung (2005)	✓			✓				
Chen et al. (2006)	✓	✓			✓			
Amid et al. (2006)	✓				✓			
Soukhakian et al. (2007)	✓			✓		✓	✓	✓
Rabieh et al. (2008)	✓	✓		✓		✓	✓	
Jafarnezhad et al. (2009)	✓				✓			
Farzipoor saen (2007)	✓				✓			
Ustun and Aktar Demirtas (2008)	✓		✓	✓			✓	✓
Wu and Blackhurst (2009)	✓			✓				✓
Kuo and Lin (2012)	✓		✓	✓			✓	✓
Mendoza and Ventura (2012)	✓			✓		✓		
Rao et al. (2013)	✓				✓	✓		

The research gap: compared with the basic model and other researches, the study combined supplier selection and inventory control to be applied in a real case (two designed model matched with condition of real case)

r: Inventory holding cost rate.

A_i: Ordering cost of ith supplier's raw material.

P_i: Selling price of ith supplier's raw material.

h_i: Percentage of moisture in the item of the ith supplier.

D': Speed of material consumption.

P': Speed of receiving materials.

SS: Safety stock.

SS_i: Safety stock of the ith supplier's item.

$$\beta_i = (P_i + C_{ti}) \quad \text{and} \quad \alpha_i = \left(1 - \frac{D_i'}{P_i'}\right)$$

Other parameters and variables will be described later.

The basic assumptions

- Constant annual demand (D)
- Infinite raw materials storage space
- Stable prices over the year
- Gradual receiving and consumption of raw materials
- Stable safety-stock levels
- Stock-out is not allowed.

Graphical explanation of models

The basic model

The basic model is built following the approach and assumptions of Ghodsypour and O'Brien (2001) model. This model assumes instant, in-simultaneous order receives from different suppliers and gradual consumption of the materials. Figure 1 shows the behavior of inventory levels of an item under the assumptions of this model.

In Fig. 1, total order cycle (T) is equal to the sum of order cycles of every supplier (T_i) and at the time one supplier's inventory is used up, the next supplier's shipment would arrive in.

In general, this model is applicable to situations where, the quality specification of receiving items from different sources is identical. And no mixing of different items is required.

The new models

Because of the need for mixing iron ore and concentrating on the agglomeration process, the basic model couldn't be applied for the current situation. Furthermore, the inventory supply is not instantaneous, but placed orders are shipped gradually. So, the basic model had to be manipulated to fit the situation correctly. Two slightly different possibilities were considered as shown in Figs. 2 and 3.

Model A The two supplier of iron concentrate in Model A (Q_4 and Q_5) have to cover the inventory in turn during each ordering cycle (T), while the three suppliers of iron ore deliver their shipments simultaneously.

Model B In Model B all suppliers send their shipments simultaneously, therefore, they are under less pressure to keep up with a tight delivery schedule.

Comparing with The basic model, at the first glance, one may expect a rise in average inventory in Models A and B, which leads to an increase in total annual carrying costs as a result. But, as we will see later, the inherent flexibility of the new models paves the way for formulating more effective ordering policies. This would prevent such increase in costs to materialize in practice.

Obviously, Models A and B were formulated for different purchasing behaviors. The formulation process of both models is very similar to each other. The slight differences actually are in formulating the carrying cost in objective function and in the quality constraints of the models. So, we skip from presenting such details here, and continue our model formulation only for Model A.

Formulating the objective function

Because of the objective function of this model is formed from inventory related costs such as the purchasing price,

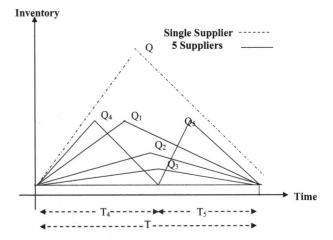

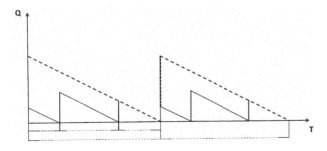

Fig. 1 Inventory behavior in the basic model

Fig. 2 Inventory behavior in Model A

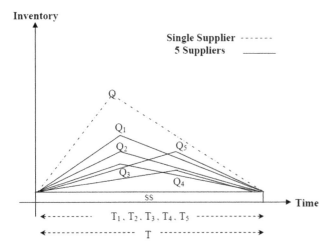

Fig. 3 Inventory behavior in Model B

transportation costs, carrying and ordering costs, shrinkage cost, it is a minimizing type objective function. The shrinkage cost is mainly related to the evaporation of raw materials moisture during the agglomeration process. Since the iron ore quarries are located in both dray and wet areas of the country, the water content of their stones differ significantly, and should be taken into account as a part of the total annual purchasing cost.

Annual purchasing cost (APC)

Since ordering quantity (Q) should be shared by $n = 5$ suppliers, we have the following:

$$Q = \sum_{i=1}^{n} Q_i \quad Q_i = X_i Q \quad T_i = X_i T \quad 0 \leq X_i \leq 1$$

$$\sum_{i=1}^{n} X_i = 1 \quad i = 1, 2, \ldots, n$$

Since annual purchasing from ith supplier is $X_i D$ and its price is P_i, APC is:

$$APC = \sum_{i=1}^{n} X_i P_i D \quad i = 1, 2, \ldots, n$$

Annual transportation cost (ATC)

ATC is computed by multiplying annual purchasing quantity, and transportation tariff, for ith supplier, thus:

$$ATC = \sum_{i=1}^{n} X_i C_{ti} D \quad i = 1, 2, \ldots, n$$

Annual weight reduction cost (AWRC)

As we mentioned earlier, the moisture content of receiving shipments from each supplier is significantly different.

Therefore, the weight reduction of materials due to evaporation in agglomeration process should be taken into account. The data for this is obtainable from Isfahan Steel works daily Lab Reports. Let h_i be the average moisture fraction of ith supplier, then, AWRC is computed as:

$$AWRC = \sum_{i=1}^{n} X_i D(h_i)(P_i + C_{ti}) = \sum_{i=1}^{n} X_i D h_i (P_i + C_{ti})$$

The above formula considers the fact that the evaporated moisture is actually bought and paid for its transportation.

To avoid unnecessary repetition, let β_i be equal to $(P_i + C_{ti})1$. Hence:

$$AWRC = \sum_{i=1}^{n} X_i D h_i \beta_i \quad i = 1, 2, \ldots, 5$$

And, the sum of forgoing three costs is:

$$APC + ATIC + AWRC = \sum_{i=1}^{n} X_i D(1 + h_i)(P_i + C_{ti})$$
$$= \sum_{i=1}^{n} X_i D(1 + h_i)\beta_i$$

Therefore, it is inferred that the unit cost of ith supplier's material in agglomeration process is equal to $(1 + h_i)\beta_i$. We will apply this formula to compute annual holding (carrying) cost.

Annual holding cost (AHC)

Referring to different behavior of inventory levels in Models A and B, especially in regard with iron concentrate, obviously, the formulation of AHC differs slightly. To save us time, we proceed with formulating AHC for Model A only.

Average inventory in gradual receipt of shipments is equal to: $\frac{Q}{2}\left(1 - \frac{D'}{P'}\right) + SS$, or in general is $\frac{Q}{2}\left(1 - \frac{D'_i}{P'_i}\right) + SS_i$, so, the average holding cost in T_i would be:

$$\left(X_i \frac{Q}{2}\left(1 - \frac{D'_i}{P'_i}\right) + SS_i\right) r p_i$$

In order to avoid unnecessary repetitions, let $\left(1 - \frac{D'_i}{P'_i}\right)$ be equal to α_i. Since SS is a constant value, it will be omitted in differentiation process any way. Thus, Total Holding Cost per Period (THCP) is formulated as follows:

$$THCP = X_1 \frac{Q}{2}\alpha_1 r(1 + h_1)\beta_1 T_1 + X_2 \frac{Q}{2}\alpha_2 r(1 + h_2)\beta_2 T_2$$
$$+ \cdots + X_n \frac{Q}{2}\alpha_n r(1 + h_n)\beta_n T_n$$

In Model A, for those suppliers that we have on hand inventory during the whole order cycle, $T_i = T$, $i = 1, 2, \ldots, m$, and for those vendors that we have on hand inventory just during a part of cycle, $T_i = \frac{X_i}{\sum_{i=m+1}^{n} X_i} T$,

$i = m + 1, \ldots, n$. Now, we know that $T = \frac{Q}{D}$, and considering the shape of the model, and the fact that $n = 5$, we have:

$$T_1 = T_2 = T_3 = T = \frac{Q}{D}, \quad T_4 = \frac{X_4}{X_4 + X_5}\frac{Q}{D},$$

$$T_5 = \frac{X_5}{X_4 + X_5}\frac{Q}{D}$$

(Here the flexibility and adoptability of the new models become clearer, as when the model, for any reason, does not allow purchase from supplier 4, for instance, then, we have: $X_4 = 0$, $T_4 = 0$, and $T_5 = T$).

However, the detailed computation of THCP is:

$$THCP = X_1\frac{Q}{2}\alpha_1 r(1+h_1)\beta_1\frac{Q}{D} + X_2\frac{Q}{2}\alpha_2 r(1+h_2)\beta_2\frac{Q}{D}$$

$$+ X_3\frac{Q}{2}\alpha_3 r(1+h_3)\beta_3\frac{Q}{D}$$

$$+ X_4\frac{Q}{2}\alpha_4 r(1+h_4)\beta_4\frac{X_4}{X_4 + X_5}\frac{Q}{D}$$

$$+ X_5\frac{Q}{2}\alpha_5 r(1+h_5)\beta_5\frac{X_5}{X_4 + X_5}\frac{Q}{D}$$

or

$$THCP = \sum_{i=1}^{3} X_i Q^2\frac{\alpha_i r(1+h_i)\beta_i}{2D}$$

$$+ \frac{X_4^2}{X_4 + X_5}Q^2\frac{\alpha_4 r(1+h_4)\beta_4}{2D}$$

$$+ \frac{X_5^2}{X_4 + X_5}Q^2\frac{\alpha_5 r(1+h_5)\beta_5}{2D}$$

Still, the shorter form for THCP is

$$THCP = \frac{rQ^2}{2D}\left(\sum_{i=1}^{3} X_i\alpha_i(1+h_i)\beta_i\right)$$

$$+ \frac{rQ^2}{2D}\frac{1}{\sum_{i=4}^{5} x_i}\left(\sum_{i=4}^{5} X_i^2\alpha_i(1+h_i)\beta_i\right)$$

The Annual Holding Cost (AHC) is computed by multiplying THCP and the number of order cycles per year, or

$$AHC = (THCP)\times\frac{1}{T} = \frac{(THCP)D}{Q}$$

And with suitable substitutions, we have

$$AHC = \frac{rQ^2}{2D}\left(\sum_{i=1}^{3} X_i\alpha_i(1+h_i)\beta_i\right)\frac{D}{Q}$$

$$+ \frac{rQ^2}{2D}\frac{1}{\sum_{i=4}^{5} x_i}\left(\sum_{i=4}^{5} X_i^2\alpha_i(1+h_i)\beta_i\right)\frac{D}{Q}$$

or, simply

$$AHC = \frac{rQ}{2}\left(\sum_{i=1}^{3} X_i\alpha_i(1+h_i)\beta_i\right)$$

$$+ \frac{rQ}{2}\frac{1}{\sum_{i=4}^{5} x_i}\left(\sum_{i=4}^{5} X_i^2\alpha_i(1+h_i)\beta_i\right)$$

Also holding cost of Model B is:

$$AHC = \frac{rQ}{2}\left(\sum_{i=1}^{5} X_i\alpha_i(1+h_i)\beta_i\right)$$

Annual ordering cost (AOC)

Due to the fact that the required raw materials are ordered and purchased from n suppliers, the Ordering Cost each Period (OCP) is:

$$OCP = \sum_{i=1}^{n} A_i Y_i, \quad i = 1, 2, \ldots, 5$$

where $Y_i = \begin{cases} 1 & \text{if } X_i > 0 \\ 0 & \text{if } X_i = 0 \end{cases} \quad i = 1, 2, \ldots, 5$

AOC is obtained from multiplication of OCP by the number of periods per year:

$$AOC = (OCP)\times\frac{1}{T} \Rightarrow AOC = \left(\sum_{i=1}^{n} A_i Y_i\right)\frac{1}{T}$$

$$= \left(\sum_{i=1}^{n} A_i Y_i\right)\frac{D}{Q}$$

Having formulated the annual costs of purchased materials and AHC and AOC, the Total Annual Costs (TAC) is simply computed by adding up all these costs:

$$TAC = APC + ATC + AWRC + AHC + AOC$$

$$TAC = \sum_{i=1}^{5} X_i P_i D + \sum_{i=1}^{5} X_i C_{ti} D + \frac{rQ}{2}\left(\sum_{i=1}^{3} X_i\alpha_i(1+h_i)\beta_i\right)$$

$$+ \frac{1}{\sum_{i=4}^{5} x_i}\times\frac{rQ}{2}\left(\sum_{i=4}^{5} X_i^2\alpha_i(1+h_i)\beta_i\right) + \frac{D}{Q}\left(\sum_{i=1}^{5} A_i Y_i\right)$$

$$+ \sum_{i=1}^{n} h_i\beta_i X_i D$$

Manipulating the above equation a bit and considering that $\beta_i = P_i + C_{ti}$, a simpler from of TAC will be:

$$\mathrm{TAC} = \sum_{i=1}^{5} X_i D (1+h_i)\beta_i + \frac{rQ}{2}\left(\sum_{i=1}^{3} X_i \alpha_i (1+h_i)\beta_i\right) + \frac{1}{\sum_{i=4}^{5} x_i}$$

$$\times \frac{rQ}{2}\left(\sum_{i=4}^{5} X_i^2 \alpha_i (1+h_i)\beta_i\right) + \frac{D}{Q}\left(\sum_{i=1}^{5} A_i Y_i\right)$$

As usual, we differentiate the above equation in respect to Q to obtain the optimum order quantities. Doing so, we have:

$$(\mathrm{TAC})' = \frac{r}{2}\left(\sum_{i=1}^{3} X_i \alpha_i (1+h_i)\beta_i\right) + \frac{r}{2}$$

$$\times \frac{1}{\sum_{i=4}^{5} X_i}\left(\sum_{i=4}^{5} X_i^2 \alpha_i (1+h_i)\beta_i\right)$$

$$- \frac{D}{Q^2}\left(\sum_{i=1}^{5} A_i Y_i\right)$$

Finally, omitting Q, the objective function for Model A is:

The model constraints

The constraints of this model actually pertains to the buyer's annual demand and quality of receiving materials, on one hand, and the suppliers' allocable capacity, on the other. In the following section, we present formulation of these constraints, as they were introduced to the model:

Demand constraint

Assuring D is the Isfahan works, annual demand for iron ore and iron concentrate, as we mentioned earlier, $n = 5$ vendors can satisfy D at the present time. Therefore, we have:

$$\sum_{i=1}^{n} X_i D = D \quad i = 1, 2, \ldots, 5$$

Omitting D from both sides of equation, then we have:

$$\mathrm{TAC} = \sum_{i=1}^{5} X_i D (1+h_i)\beta_i + \sqrt{2Dr\left(\sum_{i=1}^{3} X_i \alpha_i (1+h_i)\beta_i\right) + \frac{1}{\sum_{i=4}^{5} X_i}\left(\sum_{i=4}^{5} X_i^2 \alpha_i (1+h_i)\beta_i\right)} \times \sqrt{\sum_{i=1}^{5} A_i Y_i}$$

And the optimum order quantity, Q^*, is:

$$Q^* = \sqrt{\frac{2D\left(\sum_{i=1}^{n} A_i Y_i\right)}{r\left(\sum_{i=1}^{m} X_i \alpha_i (1+h_i)\beta_i\right) + \frac{1}{\sum_{i=m+1}^{n} X_i}\left(\sum_{i=1}^{n} X_i^2 \alpha_i (1+h_i)\beta_i\right)}}$$

the objective function for Model B after omitting Q is:

$$\mathrm{Min\,TAC} = \sum_{i=1}^{5} X_i D (1+h_i)\beta_i$$

$$+ \sqrt{2Dr\left(\sum_{i=1}^{5} X_i \alpha_i (1+h_i)\beta_i\right)\left(\sum_{i=1}^{5} A_i Y_i\right)}$$

And the optimum order quantity, Q^*, is:

$$Q^* = \sqrt{\frac{2D\left(\sum_{i=1}^{n} A_i Y_i\right)}{r\left(\sum_{i=1}^{n} X_i \alpha_i (1+h_i)\beta_i\right)}}$$

$$\sum_{i=1}^{n} X_i = 1 \quad i = 1, 2, \ldots, 5$$

Fe quality constraint

In practice, to get a quality agglomeration process output with a prescribed Fe content, a calculated mix of input materials, based on their Fe content is used. In this case we have:

$$\sum_{i=1}^{5} X_i D q_{\mathrm{Fe}i} \geq q_{\mathrm{aFe}} D$$

And, omitting D from both sides, we have:

$$\sum_{i=1}^{5} X_i q_{\mathrm{Fe}i} \geq q_{\mathrm{aFe}}$$

where q_{aFe} is the minimum acceptable percent of Fe in the input mix, and q_{Fei} is the percent of Fe content in the ith supplier's material.

Due to the assumptions of Model A, which requires breaking the order cycle into two parts, we should divide the Fe quality constraint into two parts as well, and introduce it to the model as follows:

$$(X_1 Dq_{Fe1} + X_2 Dq_{Fe2} + X_3 Dq_{Fe3}$$
$$+ (1 - X_1 + X_2 + X_3)) Dq_{Fe4}) \geq Dq_{aFe}$$
$$(X_1 Dq_{Fe1} + X_2 Dq_{Fe2} + X_3 Dq_{Fe3}$$
$$+ (1 - X_1 + X_2 + X_3)) Dq_{Fe5}) \geq Dq_{aFe}$$

$$\sum_{i=1}^{5} DX_i q_{Fei} - 10^{20} Y_4 D - 10^{20} Y_5 D \geq Dq_{afe}$$

Knowing that:

$$X_4 \leq (1 - (X_1 + X_2 + X_3))$$
$$X_5 \leq (1 - (X_1 + X_2 + X_3))$$

And omitting D from both sides, we have:

$$(X_1 q_{Fe1} + X_2 q_{Fe2} + X_3 q_{Fe3} + (1 - X_1 + X_2 + X_3)) q_{Fe4}) \geq q_{afe}$$
$$(X_1 q_{Fe1} + X_2 q_{Fe2} + X_3 q_{Fe3} + (1 - X_1 + X_2 + X_3)) q_{Fe5}) \geq q_{afe}$$

$$\sum_{i=1}^{5} X_i q_{Fei} - 10^{20} Y_4 - 10^{20} Y_5 \geq q_{afe}$$

$$X_4 \leq (1 - (X_1 + X_2 + X_3))$$
$$X_5 \leq (1 - (X_1 + X_2 + X_3))$$

Additionally the number 10^{20} that represented is a very large number in the model.

Quality Constraint for Model B is:

$$\sum_{i=1}^{5} X_i q_{Fei} \geq q_{aFe}$$

Capacity constraint

This constraint stems from the fact that the ith supplier can satisfy only a fraction of the annual buyer's needs, C_i, each year. Thus: $X_i D \leq C_i$.

Finally, we have to make sure that Y_i has an integer value of 0 or 1. To introduce this constraint to model, and knowing that X_i is always equal or less than 1, then we have:

$$X_i \leq Y_i \quad i = 1, 2, \ldots, n$$
$$X_i \geq \varepsilon Y_i$$

where, ε is a little bit greater than 0.

Instead, in above constraint formulas, where ever we have X_i, we can multiply it by Y_i.

Fe quality constraint

Model A formulation

Now, we summarize *Model A* formulation as follows:

$$\mathrm{Min\,TAC} = \sum_{i=1}^{n} X_i D(1+h_i)\beta_i + \frac{rQ}{2}\left(\sum_{i=1}^{m} X_i \alpha_i (1+h_i)\beta_i\right)$$
$$+ \frac{1}{\sum_{i=m+1}^{n} X_i} \times \frac{rQ}{2}\left(\sum_{i=m+1}^{n} X_i^2 \alpha_i (1+h_i)\beta_i\right) + \frac{D}{Q}\left(\sum_{i=1}^{n} A_i Y_i\right)$$

$$\sum_{i=1}^{n} X_i = 1 \quad i = 1, 2, \ldots, m, \ldots m+1, \ldots n$$

$$\left(\sum_{i=1}^{m} X_i q_{Fei} + \left(1 - \left(\sum_{i=1}^{m} X_i\right)\right) q_{Fe(m+1)}\right) \geq q_{aFe} Y_{m+1}$$

$$\left(\sum_{i=1}^{m} X_i q_{Fei} + \left(1 - \left(\sum_{i=1}^{m} X_i\right)\right) q_{Fe(m+2)}\right) \geq q_{aFe} Y_{m+2}$$

$$\vdots \qquad \vdots \qquad \vdots$$

$$\left(\sum_{i=1}^{m} X_i q_{Fei} + \left(1 - \left(\sum_{i=1}^{m} X_i\right)\right) q_{Fe(n)}\right) \geq q_{aFe} y_n$$

$$\sum_{i=1}^{n} X_i q_{Fei} - 10^{20} \sum_{i=m+1}^{n} Y_i \geq q_{aFe}$$

$$0 \leq X_{m+1} \leq \left(1 - \left(\sum_{i=1}^{m} X_i\right)\right)$$

$$0 \leq X_{m+2} \leq \left(1 - \left(\sum_{i=1}^{m} X_i\right)\right)$$

$$\vdots \qquad \vdots$$

$$0 \leq X_n \leq \left(1 - \left(\sum_{i=1}^{m} X_i\right)\right)$$

$$X_i D \leq C_i \quad i = 1, 2, \ldots, n$$
$$Q_i = X_i Q$$
$$X_i \leq Y_i$$
$$X_i \geq \varepsilon Y_i$$

$$Q = \sum_{i=1}^{n} Q_i$$

$$X_i \geq 0, Y_i = 0, 1, \quad i = 1, 2, \ldots, m, \ldots, n$$

Model B formulation

We summarize *Model B* formulation as follows:

$$\text{Min TAC} = \sum_{i=1}^{n} X_i D(1+h_i)\beta_i + \frac{rQ}{2}\left(\sum_{i=1}^{n} X_i \alpha_i (1+h_i)\beta_i\right)$$

$$+ \frac{D}{Q}\left(\sum_{i=1}^{n} A_i Y_i\right)$$

$$\sum_{i=1}^{n} X_i = 1 \quad i = 1, 2, \ldots, n$$

$$\sum_{i=1}^{n} X_i q_{\text{Fe}i} \geq q_{\text{aFe}} \quad i = 1, 2, \ldots, n$$

$$X_i D \leq C_i \quad i = 1, 2, \ldots, n$$

$$Q_i = X_i Q$$

$$X_i \leq Y_i$$

$$X_i \geq \varepsilon Y_i$$

$$Q = \sum_{i=1}^{n} Q_i$$

$$X_i \geq 0, Y_i = 0, 1, \quad i = 1, 2, \ldots, m \ldots, n$$

Parameters of model

It should be considered that the model was tested for two time periods (two successive years). Some of the parameters are the same for two time period and others are different thus they are represented separately (Tables 2, 3).

Table 2 Common parameters for two successive years

i	r	α_i	$q_{\text{Fe}i}$	h_i	C_i
1	0.16	0.685	60.19	0.0189	2,021,000
2	0.16	0.594	61.56	0.0434	142,000
3	0.16	0.837	60.41	0.0773	84,000
4	0.16	0.282	67.26	0.0915	3,000,000
5	0.16	0.084	68.06	0.0832	4,000,000

Table 3 Different parameters for two successive years

i	First year		Second year	
	A_i	β_i	A_i	β_i
1	10,485,422	157,682	10,083,210	122,000
2	10,485,422	170,000	10,083,210	128,000
3	10,485,422	198,000	10,083,210	167,500
4	10,485,422	254,000	10,083,210	254,000
5	10,485,422	215,000	10,083,210	215,000

$q_{\text{a first year}} = 61.17$, $q_{\text{a second year}} = 60.23$, $D_{\text{first year}} = 2,082,368$, $D_{\text{second year}} = 1,861,518$

Table 4 Model A results

	First year	Second year
Objective function (Rials)	286,884,300,000	300,929,000,000
X_1	0.8191556	0.9949174
X_2	0.0681916	0
X_3	0	0
X_4	0	0
X_5	0.1126528	0.005082592
Y_1	1	1
Y_2	1	0
Y_3	0	0
Y_4	0	0
Y_5	1	1
Q	100,844.4	66,727.33
Q_1	82,640.02	66,388.18
Q_2	6879.469	0
Q_3	0	0
Q_4	0	0
Q_5	11,364.91	339.1478

Model runs and results

In order to run the formulated models and compare its results with actual performance, we had to compute model parameters from Company's records. This was done carefully using data of two consecutive financial years. We employed the global option of Lingo Version 8 for running the models. The results are summarized in Table 4.

Take notice of the fact that the model has chosen three of the suppliers for the first year and only two of them in the second year. Exactly, the same results obtained when running Model B too. Obtaining the same results from Models A and B is rather exceptional, and relates only to this studied situation, and stems from the fact that both models rejected buying iron concentrate from a particular vendor. Models are based on important criteria such as cost, quality and capacity.

Actually, Isfahan Steel Co. on a regular basis, has been buying raw materials form all five suppliers during those 2 years. A comparison of the model results with the actual cost performance of the Company is made in Table 5.

Discussion and conclusion

Examining the results presented in Tables 4 and 5, reducing cost by 10.9 and 7.1 %, increasing company's annual profit, attract any top manager's attention. One might argue that real world mangers of large processing firms like Isfahan Steel Company keep purchasing from different sources to ensure a continuous and reliable stream of

Table 5 Comparing the model results with actual data

Results	First year	Second year
Total cost of the Model A	286,884,300,000	300,292,000,000
Total actual cost (Rials)	321,929,098,600	323,290,000,000
Cost reduction (Rials)	35,044,798,600	22,998,000,000
% of reduction	10.9	7.1

supplies. However, it turned out to be a very expensive way to get such assurances. The responsible managers of the said company had not been aware of the tremendous differences that a wrong suppliers selection decisions could create. Furthermore, one of the real potential values of mathematical modeling is reminding the mangers of alternative ways of doing their daily affairs.

However, some points should be mentioned about the new models and the present study:

1. The models determine the percentages of iron ore and iron concentrate to be mixed in agglomeration process. This is done by considering the minimum Fe contents required for specified output quality of the process, q_{afe}, and other constraints, and objective function. The models recommended a mix of 88.73472 % iron ore and 11.26528 %, concentrate for the first year and 99.49174 % iron ore, 0.005082592 % concentrate for the second year.

2. A sensitivity analysis of the models was assumed by changing capacity and quality constraints' parameters, the number of placed orders in a year, D/Q. The results of the sensitivity analysis revealed that even in the most pessimistic conditions, the models would result in total costs reduction.

   One of the interesting findings of the sensitivity analysis of the model was that if the second iron ore supplier had no capacity limitations, and could supply the Company's whole annual needs, a substantial costs reduction could happened. Based on this finding, we recommended the management helping that particular supplier to invest in increasing its capacity through a joint venture project.

3. The non-linear assumption of these models makes them closer to the real world managerial problems. Most of the relationships in socio-economic systems are non-linear in nature. Yet, Model B is flexible enough to be changed to a linear model simply by assuming D/Q as constant. As a result, the new models are applicable to a fairly large area of operations management special problems.

4. The new models are not just a simple inventory model. As we have seen, the models can handle a multi-criteria situation comprising cost and quality and help us select appropriate suppliers. They also can present a

purchasing schedule to tell us when and how much to buy from each vendor. At the same time, the models can suggest an optimum consumption mix of the materials.

In summary, the new models presented in this article have notable advantages and improvements over the previously introduced ones, such as Ghodsypour and O'Brien model (2001), with a considerable number of real world special applications.

References

Aissaouia N, Haouaria M, Hassinib E (2007) Supplier selection and order lot sizing modeling: a review. Comput Oper Res 34:3516–3540

Amid A, Ghodsypour SH, O' Brein CO (2006) Fuzzy multiobjective linear model for supplier selection in a supply chain. Int J Prod Econ 104:394–407

Basnet C, Leung JMY (2005) Inventory lot-sizing with supplier selection. Comput Oper Res 32:1–14

Benton WC (1991) Quantity discount decision under conditions of multiple items, multiple suppliers and resource limitation. Int J Prod Econ 27:1953–1961

Chen C-T, Lin C-T, Huang S-F (2006) A fuzzy approach for supplier evaluation and selection in supply chain management. Int J Prod Econ 102(2):289–301

Dickson GW (1966) An analysis of vendor selection systems and management. J Purch 2(1):5–17

Farzipoor Saen R (2007) Suppliers selection in the presence of both cardinal and ordinal data. Eur J Oper Res 183:741–747

Ghodsypour SH, O'Brien C (1997) A decision support system for reducing the number of suppliers and managing the supplier partnership in a JIT/TQM environment. In: Proceedings of 3rd international symposium on logistics, University of Padua, Italy

Ghodsypour SH, O'Brien C (1998) A decision support system for supplier selection using an integrated analytic hierarchy process and linear programing. Int J Prod Econ 56–57:199–212

Ghodsypour SH, O'Brien C (2001) The total cost of logistics in supplier selection, under conditions of multiple souring, multiple criteria and capacity constraint. Int J Prod Econ 73:15–27

Jafarnezhad A, Esmaelian M, Rabieh M (2009) Evaluation and selection of supplier in supply chain in case of single sourcing with fuzzy approach. Manag Res Iran 12(4):127–153

Kumar M, Vrat P, Shankar R (2004) A fuzzy goal programming approach for vendor selection problem in a supply chain. Comput Ind Eng 46(69):58

Kuo RJ, Lin Y (2012) Supplier selection using analytic network process and data envelopment analysis. Int J Prod Res 50(11):2852–2863

Lin HT, Chang WL (2008) Order selection and pricing methods using flexible quantity and fuzzy approach for buyer evaluation. Eur J Oper Res 187:415–428

Mendoza A, Ventura JA (2012) Analytical models for supplier selection and order quantity allocation. Appl Math Model 36:3826–3835

Nair NG (2002) Resource management. Vikas Publishing House, New Delhi

Rabieh M, Soukhakian MA, Jafarnezhad A (2008) Designing a nonlinear compatible model of supplier selection in case of multiple sourcing: a case study of Isfahan Steel Company. Iran J Trade Stud 12(45):57–83

Rao KN, Subbaiaha KV, Singh GVP (2013) Design of supply chain in fuzzy environment. J Ind Eng Int. 9:9. doi:10.1186/2251-712X-9-9

Soukhakian MA, Rabieh M, Afsar A (2007) Designing a inventory control model with a consideration of total cost of logistics in case of multiple sourcing. J Soc Sci Humanit Shiraz Univ 26(150):75–94

Ustun O, Aktar Demirtas E (2008) An integrated multi-objective decision-making process for multi-period lot-sizing with supplier selection. Omega 36:509–521

Weber CA, Current JR, Benton WE (1991) Vendor selection criteria and methods. Eur J Oper Res 50:2–18

Wu T, Blackhurst J (2009) Supplier evaluation and selection: an augmented DEA approach. Int J Prod Res 47(16):4593–4608

6

Efficacy of fuzzy MADM approach in Six Sigma analysis phase in automotive sector

Rajeev Rathi[1] · Dinesh Khanduja[1] · S. K. Sharma[1]

Abstract Six Sigma is a strategy for achieving process improvement and operational excellence within an organization. Decisions on critical parameter selection in analysis phase are always very crucial; it plays a primary role in successful execution of Six Sigma project and for productivity improvement in manufacturing environment and involves the imprecise, vague and uncertain information. Using a case study approach; the paper demonstrates a tactical approach for selection of critical factors of machine breakdown in center less grinding (CLG) section at an automotive industry using fuzzy logic based multi attribute decision making approach. In this context, we have considered six crucial attributes for selection of critical factors for breakdown. Mean time between failure is found to be the pivotal selection criterion in CLG section. Having calculated the weights pertinent to criteria through two methods (fuzzy VIKOR and fuzzy TOPSIS) critical factors for breakdown are prioritized. Our results are in strong agreement with the perceptions of production and maintenance department of the company.

Keywords Six Sigma · Analytical hierarchy process · Fuzzy logic · MADM · Center less grinding · Automotive industry

✉ Rajeev Rathi
rajeevrathi_1443@nitkkr.ac.in

Dinesh Khanduja
dineshkhanduja@yahoo.com

S. K. Sharma
sksharma49nitk@yahoo.com

[1] Department of Mechanical Engineering, National Institute of Technology, Kurukshetra, Haryana 136119, India

Introduction

Companies are continuously facing the resistance to settle into the ever changing technological environment. Six Sigma has been recognized for many years as an efficient strategy and has helped several companies to rise to this challenge. This is one of the most important and popular developments in the field of process improvement. It has saved large amounts of money and improved the processes for a large number of manufacturing organizations worldwide (Neuman and Cavanagh 2000; Snee and Hoerl 2003; Harry and Schroeder 2005). Six Sigma has gone through a considerable evolution since the early exposition. Initially it was a quality improvement methodology based on statistical concepts. Then it transformed to a disciplined process improvement technique. In its current existence; it is commonly presented as 'a breakthrough strategy' of best in class. It is accepted that in current scenario Six Sigma is applicable to various environments such as service, manufacturing, process, software industry regardless of the size of the business, and, if successfully implemented, it may lead to nearly perfect solutions and services (Banuelas et al. 2005; Antony et al. 2006; Chakrabarty and Tan 2007). Six Sigma has enormous potential to reduce breakdown costs, improve performance, grow revenue, strengthen focus, and empower resources (Snee and Hoerl 2004). It is a commanding strategy that employs a regimented approach to undertake process variability using the application of statistical and non-statistical tools and techniques in an accurate way (Jiju 2004). This teaches everyone in the organization to become more effective and efficient (Eckes 2003). Business leaders must be aware that successful implementation of Six Sigma requires not only technical understanding, but also behavioral awareness (Linderman et al. 2003). Most of the SMEs (Small and medium-sized enterprises) are not aware of Six Sigma and many not have the proper resources to execute Six Sigma projects (Jiju et al. 2005). In comparison with conventional approaches of quality and process improvement, Six Sigma is the most effective approach because of the interrelation between planning, organizational structures, measures, tools and techniques (Tilo et al. 2004; Zu et al. 2008). Six Sigma is a process improvement strategy which includes various phases logically related with each other acronym DMAIC methodology (define, measure, analyze, improve and control) is used for continuous improvement in any system or processes (Amer et al. 2008). It is the strategy of achieving key improvements in the process by applying DMAIC methodology through elimination of causes. Manufacturing units can put into action such strategies to enhance productivity of their manufacturing processes (Singh and Singh 2014).

In this case study, we are focusing on the analysis phase of DMAIC methodology through which all Six Sigma projects are executed. In the analysis phase, all measurements will be analyzed by understanding them and to make basic problem easier. The idea is to search for the factors having the biggest impact on process performance and determine the roots causes. In this case we have identified various factors for breakdown/failure in center less grinding section of machine shop in an automotive industry. The aim of study is to prioritize the critical breakdown factors in CLG section for further improvement. In this context, the key attributes/impacts were identified that depended on the views of various decision makers (such as machine operators, maintenance experts, production manager, technical and financial experts, etc.) and there are no crystalline themes among the views of these decision makers. Therefore it turned out to be necessary to forecast the excellent solution in terms of selecting critical factors for such problems using a decision-making technique. Such problems can be attempted with multiple attribute decision making (MADM) approach. MADM models are used to select best alternative from the large number of alternatives for a set of selection criteria. This approach has been effectively applied in broad range of decision-making problems in engineering and scientific fields (Perego and Rangone 1998; Pahlavani 2010). A variety of methods are reported under MADM category in literature (Chen et al. 1992; Tönshoff et al. 2007). MADM approach includes analytic hierarchy process (AHP) (Saaty 2014), graph theory and matrix approach (GTMA) (Rabbani et al. 2014), VlseKriterijumska Optimisacija I Kompromisno Resenje (VIKOR) (Liu et al. 2014; Singh and Kumar 2014), technique for order preference by similarity to ideal solution (TOPSIS) (Chu 2002; Chu and Lin 2003; Khanna et al. 2011; Dey et al. 2014), simple additive weighting (SAW) (Afshari et al. 2010) multiplicative analytical hierarchy

process (MAHP) (Cheng and Mon 1994), weighted product method (WPM), Group decision making (GDM) (Chen 2000) and many others. These techniques have been successfully applied to various fields of engineering and among these, VIKOR and TOPSIS are outstanding multiple attribute decision making (MADM) approaches. These have been applied to various problems ranging from advanced manufacturing (Kulak and Kahraman 2005), production planning (Chen and Liao 2003), supplier selection (Azar et al. 2011), decision making (Sanayei et al. 2010), machine tool selection (Nguyen et al. 2014), supply chain management (Wei et al. 2007) and many more (Vats and Vaish 2013; Ding and Kamaruddin 2014; Tahriri et al. 2014; Tiwary et al. 2014; Vats and Vaish 2014a, b). These approaches work on crisp value of attributes/impacts. The aim of present study is to select critical factors for breakdown/failure in CLG section under fuzzy environment using fuzzy VIKOR and fuzzy TOPSIS methodology using AHP weights. The present study is one of the first efforts to evaluate failure parameters using fuzzy MADM approach in Six Sigma analysis phase in Indian automotive sector.

Evaluation criteria

Six attributes have been identified for evaluation of the critical breakdown factors in center less grinding section of the selected automotive industry. These are based on the discussion with various technical experts, machine operators, production manager, maintenance manager and studies conducted by various researchers (Ayağ and Özdemir 2011; Nguyen et al. 2014).

Attributes/ impact	Symbol	Depiction
Ease of maintenance	C_1	It describes the ease with which a machine can be maintained in order to correct defects or their causes. Ease of Maintenance is the means whereby the Project Team confirms whether equipment can be maintained in-service and meets the maintainability and ease of maintenance criteria within the maintenance strategy
Safety	C_2	There are common hazards associated with the use of machine shop equipment and tools. Working safely is the first thing because the safe way is the correct way. The costs of accidents and ill health to engineering machine shops may be disproportionately high. Many employees are 'key' workers whose losses through injury or ill health severely disrupt production and lowers productivity and profitability

continued

Attributes/ impact	Symbol	Depiction
MTBF	C_3	It is the prime factor for selecting critical reasons of breakdown in machine shops. MTBF is stated as the average time between system failures of the entire machine shop. It defines of how reliable a component is. It shows the failure rate of each parameter responsible for breakdown in CLG section
Cost	C_4	It is also a key factor for investigating critical reasons of breakdown. It includes the cost of breakdown, maintenance, repair and all activities necessary to meet all its functional requirements throughout the service life. This becomes a critical to estimate such costs
Green effect	C_5	Green Effects go beyond just energy efficiency and attempt to rate an effort with regard to the total environmental stewardship of a machine shop. It includes minimum wastage, low energy consumption and user friendly environment. In this regard green effects are significantly more encompassing than just energy. An energy efficient shop floor may not be a green shop floor, but a true shop floor will be energy efficient
Repair time	C_6	It is the Portion of breakdown time during which one or more experts are working on a system to effect a repair. Repair time includes preparation time, fault detection time, fault correction time and final bind up time

Methods

As discussed in previous section, the present study emphasizes on finding out critical factors responsible for breakdown time in CLG's to improve their availability and to enhance company profit. This is done by first optimizing the parameters using AHP and then using VIKOR and TOPSIS with fuzzy logic to sum up the result.

Analytical hierarchy process (AHP)

Analytic hierarchy process is a decision making model that aids us in making decision in our complex world, developed by Satty (1980, 1988). AHP provides a framework to cope with multiple criteria situations involving intuitive, rational, qualitative and quantitative aspects. It has been one of the most widely used techniques for complex decision found especially suitable for planning at strategic level. It is a three part process which includes, identifying and organizing decision objectives, criteria constraints and alternatives into a

hierarchy. The process requires the decision-maker to develop a hierarchical structure of the factors in the given problem and to provide judgments about the relative importance of each of these factors and ultimately to specify a preference for each decision alternative with respect to each factor (Bhutta and Huq 2002). The elements of the hierarchy are related to an aspect of the decision problem which can be carefully measured or roughly estimated anything at all that applies to the decision making. Generally hierarchy has three levels; the goal, the criteria, the alternatives. The levels of hierarchy describe a system from the lowest level (sets of alternatives), through the intermediate levels (sub criteria and criteria), to the highest level (general object) (Liu and Hai 2005). It is the essence of the AHP that human judgments, and not just the underlying information, can be used in performing the evaluations. In order to compare distinct attributes, numeric priority values are assigned to the attributes on the scale of 1–9 (Saaty 1990). AHP is used as a framework to formulize the evaluation of trade-offs between the conflicting selections criteria associated with the various suppliers' offers (Nydick and Hill 1992; Radcliffe and Schniederjans 2003). The comparison is based on expert opinion, some inconsistency may occur in the system. The consistency of system can be checked by the consistency ratio (CR):

$$CR = \frac{CI}{RI} \tag{1}$$

where CI is the consistency index which can be written as:

$$CI = \frac{\lambda_{max} - m}{m - 1} \tag{2}$$

The random consistency index (R.I.) is the predefined value (Satty 1994).

Fuzzy logic

Fuzzy approach was introduced to undertake the problem where there are no clear edges between the two parameters (Azar et al. 2011). It deals with the problems where it is hard to differentiate between members and non-member objects of a set. Fuzzy approach was used for multiple criteria decision making where the stress is on likelihood rather than probability (Wei et al. 2007). Fuzzy logic is based on a set theory and contains a membership function within the interval (0, 1) which depicts the extent of significance of an element for being the member of the set (Bevilacqua et al. 2006). Linguistic variables are used for all the assessments, in which numerical values are assigned without any riddle. A linguistic variable is a variable whose value is denoted in words or sentences in a natural or artificial language (Zadeh 1975). For example, if the values of quality are presumed to be the fuzzy variables marked as good, bad and worst in place of actual numbers, then

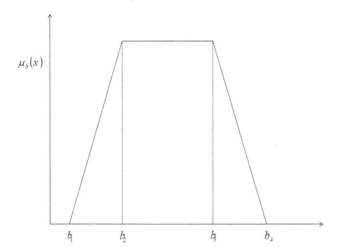

Fig. 1 Trapezoidal fuzzy number

quality is a linguistic variable. Fields of artificial intelligence, linguistics, pattern recognition, human decision processes, psychology, economics etc. have origin in the linguistic approach (Bellman and Zadeh 1970). Different fuzzy numbers are used depending on their situation. In present case we use trapezoidal fuzzy numbers (b_1, b_2, b_3, b_4) for $\{b_1, b_2, b_3, b_4 \in R; \ b_1 \leq b_2 \leq b_3 \leq b_4\}$ as in Fig. 1. Because of its simplicity and information processing in a fuzzy environment; it is often suitable to work with trapezoidal fuzzy numbers. The membership function $\mu_b(x)$ of trapezoidal fuzzy number is defined as

$$\mu_b(x) = \begin{cases} \dfrac{x - b_1}{b_2 - b_1}, & x \in [b_1, b_2] \\ 1, & x \in [b_2, b_3] \\ \dfrac{b_4 - x}{b_4 - b_3}, & x \in [b_3, b_4] \\ 0, & \text{otherwise} \end{cases} \tag{3}$$

VIKOR

Opricovic (2011) developed VIKOR, the Serbian name: VlseKriterijumska Optimizacija I Kompromisno Resenje; method to determine the compromise solution for a set of alternatives. Compromise solution is a feasible solution closest to the ideal solution for a MADM problem. The compromise solutions could be the basis for agreements, involving the decision maker's preferences by criteria weight. This method focuses on ranking and selecting from a set of alternatives, and determines compromise solutions for a problem with conflicting criteria, which can help the decision makers to reach a final decision (Sanayei et al.

2010). VIKOR algorithm determines the weight stability intervals for the obtained compromise solution with the input weights given by the experts.

TOPSIS

TOPSIS (Technique for order preference by similarity to an ideal solution) method was presented by Hwang and Yoon (Yoon and Hwang 1995). TOPSIS uses different weighting schemes and distance metrics to compares results of different sets of weights applied to set of multiple criteria data (Olson 2004; Önüt and Soner 2008). The basic principle is that the chosen alternative should have the shortest distance from the ideal solution and the farthest distance from the negative ideal solution. The ideal solution is a solution that maximizes the benefit criteria and minimizes the cost criteria, whereas the negative ideal solution maximizes the cost criteria and minimizes the benefit criteria. Benefit criteria is for maximization, while the cost criteria is for minimization. The best alternative is the one, which is closest to the ideal solution and farthest from the negative ideal solution (Wang and Elhag 2006).

Methodology used

This section explains the steps involved in the proposed subjective fuzzy VIKOR and Fuzzy TOPSIS approach for calculation of critical factors responsible for breakdown in CLG's. The approach utilizes AHP weights for inter-comparison among all criteria followed by fuzzy logic approach with VIKOR and TOPSIS methods. Figure 2 shows the flow chart of proposed methodology used in present study and make clears how the views of the decision makers are quantitatively compiled. It includes following steps:

Step 1 Calculation of AHP weights.
As discussed in "Analytical hierarchy process (AHP)" section, AHP weights (W_j) are calculated for all breakdown parameters. This provides the weights of different criteria.

Step 2 Define linguistic terms, relevant membership function and corresponding fuzzy numbers.
A set of fuzzy rates is required in order to compare all the alternatives for each criterion. These fuzzy terms are assigned by the decision makers and responsible for intra criterion comparisons of the alternatives.

Step 3   Decision matrix formation.

Let p be the parameters and q be the alternative. For k number of decision makers in the proposed model for the aggregated fuzzy rating for C_j criterion is represented as $x_{ijk} = \{x_{ijk1}, x_{ijk2}, x_{ijk3}, x_{ijk4}\}$. For $i = 1, 2,\dots p; j = 1, 2,\dots q; k = 1, 2,\dots k$, x_{ijk} is calculated as (Kahraman et al. 2003; Kwong and Bai 2003):

$$\begin{cases} x_{ij1} = \min_k\{b_{ijk1}\} \\ x_{ij2} = \dfrac{1}{k}\sum b_{ijk2} \\ x_{ij3} = \dfrac{1}{k}\sum b_{ijk3} \\ x_{ij4} = \max_k\{b_{ijk4}\} \end{cases} \tag{4}$$

Thus the obtained decision matrix (M) is shown as:

$$M = \begin{bmatrix} x_{11} & x_{12} & \cdots & x_{1p} \\ x_{21} & x_{22} & \cdots & x_{2p} \\ \vdots & \vdots & \ddots & \vdots \\ \vdots & \vdots & \ddots & \vdots \\ x_{q1} & x_{q2} & \cdots & x_{qp} \end{bmatrix}$$

Step 4   Defuzzification.

Defuzzification is performed to obtain the crisp values for each criterion corresponding to each alternative. This provides a quantitative value for the linguistic variables and fuzzy numbers assigned based on the verbal reasoning of the decision makers. Following equation lead to the crisp values:

The crisp values, thus obtained are integrated with AHP weights to calculate final ranking using- VIKOR and TOPSIS approach as discussed below.

VIKOR approach steps

Step 5   Determination of ideal and negative ideal solutions;

The ideal solution f^* and negative ideal solution f^- are determined as

$$f^* = \{\max f_{ij}\} \tag{6}$$

$$f^- = \{\min f_{ij}\} \tag{7}$$

Step 6   Calculation of utility and regret measures

$$S_i = \sum_{j=1}^{n} W_j \frac{\left(f_j^* - f_{ij}\right)}{\left(f_j^* - f_j^-\right)}; \quad \forall i \tag{8}$$

$$R_i = \max_j \left[W_j \frac{\left(f_j^* - f_{ij}\right)}{\left(f_j^* - f_j^-\right)} \right]; \quad \forall i \tag{9}$$

where S_i and R_i represent the utility and regret measures, respectively and W_j is the relative weight assigned to the jth parameter using AHP.

$$\begin{aligned} f_{ij} = \text{Defuzz}\left(x_{ij}\right) &= \frac{\int \mu(x) \cdot x\,dx}{\int \mu(x) \cdot dx} \\ &= \frac{\int_{x_{ij1}}^{x_{ij2}} \{(x - x_{ij1})/(x_{ij2} - x_{ij1})\} \cdot x\,dx + \int_{x_{ij2}}^{x_{ij3}} x\,dx + \int_{x_{ij3}}^{x_{ij4}} \{(x_{ij4} - x)/(x_{ij4} - x_{ij3})\} \cdot x\,dx}{\int_{x_{ij1}}^{x_{ij2}} \{(x - x_{ij1})/(x_{ij2} - x_{ij1})\}\,dx + \int_{x_{ij2}}^{x_{ij3}} dx + \int_{x_{ij3}}^{x_{ij4}} \{(x_{ij4} - x)/(x_{ij4} - x_{ij3})\} \cdot x\,dx} \\ &= \frac{-x_{ij1}x_{ij2} + x_{ij3}x_{ij4} + (1/3)(x_{ij4} - x_{ij3})^2 + (1/3)(x_{ij2} - x_{ij1})^2}{-x_{ij1} - x_{ij2} - x_{ij3} + x_{ij4}} \end{aligned} \tag{5}$$

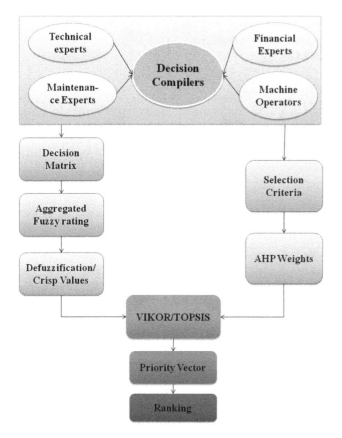

Fig. 2 Flow chart for methodology used

Step 7 Calculation of VIKOR index

$$Q_i = v\left[\frac{S_i - S^*}{S^- - S^*}\right] + (1-v)\left[\frac{R_i - R^*}{R^- - R^*}\right]; \quad \forall i \quad (10)$$

where Q_i represents ith alternatives VIKOR value, v is the group utility weight, it is generally considered as 0.5 (unsupervised) and;

$$S^* = \min_i(S_i); \quad (11)$$

$$S^- = \max_i(S_i); \quad (12)$$

$$R^* = \min_i(R_i); \quad (13)$$

$$R^- = \max_i(R_i); \quad (14)$$

Breakdown factor with least value of VIKOR index Q_i is preferred.

TOPSIS approach steps

Step 5 Normalized the matrix as given below:

$$r_{ij} = \frac{f_{ij}}{\sqrt{\sum_{i=1}^{m}(f_{ij})^2}}; \quad \forall j \quad (15)$$

Step 6 Calculate the weighted normalized decision matrix as given:

$$V_{ij} = [r_{ij}]_{m \times n} \times [W_j]_{n \times m}^{\text{diagonal}} \quad (16)$$

Step 7 Calculate the positive ideal and negative ideal solution:
The positive ideal solution V_j^+ and negative ideal solution V_j^- are as given below:

$$V_j^+ = \{(\max V_{ij}, j \in J_1), (\min V_{ij}, j \in J_2), \\ i = 1, 2, 3 \dots m\}; \quad \forall j \quad (17)$$

$$V_j^- = \{(\min V_{ij}, j \in J_1), (\max V_{ij}, j \in J_2), \\ i = 1, 2, 3 \dots m\}; \quad \forall j \quad (18)$$

where J_1 and J_2 represents higher best and lower best criteria respectively.

Step 8 Calculate the distance d_i^+ and d_i^- from the positive ideal solution and negative ideal solution respectively

$$d_i^+ = \left[\sum_{j=1}^{n}\left(V_{ij} - V_j^+\right)^2\right]^{0.5}, \\ i = 1, 2, 3, \dots m \quad (19)$$

$$d_i^- = \left[\sum_{j=1}^{n}\left(V_{ij} - V_j^-\right)^2\right]^{0.5}, \\ i = 1, 2, 3, \dots m \quad (20)$$

Step 9 Calculation of TOPSIS rank index:

$$C_i^+ = \frac{d_i^-}{d_i^- + d_i^+} \quad (21)$$

Breakdown factor with highest rank index C_i^+ are preferred.

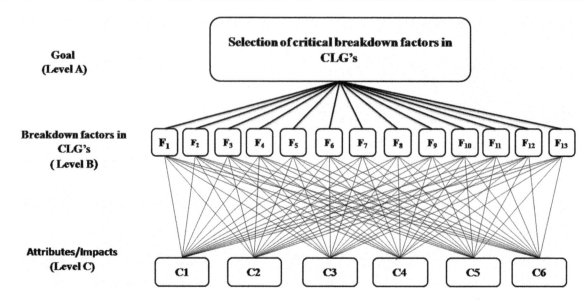

Fig. 3 The hierarchical structure for the selection of the critical factors of Breakdowns in CLG's

Table 1 Subjective weights of the evaluation criteria calculated using AHP

Attributes/impact	C_1	C_2	C_3	C_4	C_5	C_6	Weights	Rank
Ease of maintenance (C_1)	1	5	0.11	0.14	5	0.14	0.0768	4
Safety (C_2)	0.20	1	0.11	0.14	3	0.14	0.0381	5
MTBF (C_3)	9	9	1	9	9	9	0.4945	1
Cost (C_4)	7	7	0.11	1	7	7	0.2187	2
Green effect (C_5)	0.20	0.33	0.11	0.14	1	0.14	0.0239	6
Repair time (C_6)	7	7	0.11	0.14	7	1	0.1478	3

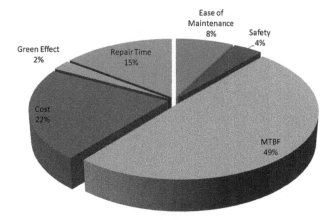

Fig. 4 Contribution of all dominating attributes about selection of critical factors for breakdown in CLG's

Table 2 Linguistic variables and corresponding fuzzy numbers

Linguistic variable	Fuzzy number
Absolutely high (AH)	(0.8, 0.9, 1.0, 1.0)
Very high (VH)	(0.7, 0.8, 0.8, 0.9)
High (H)	(0.5, 0.6, 0.7, 0.8)
Above average (AA)	(0.4, 0.5, 0.5, 0.6)
Below average (BA)	(0.2, 0.3, 0.4, 0.5)
Very poor (VP)	(0.1, 0.2, 0.2, 0.3)
Absolutely poor (AP)	(0.0, 0.0, 0.1, 0.2)

Results and discussion

The hierarchical structure for the selection of critical breakdowns factors in CLG's is demonstrated in Fig. 3. Level A specifies our goal on selection of the critical factors that have to be selected from the identified thirteen important factors of failure indicated in level B. Further, in

brainstorming session with decision makers like machine operators, maintenance experts, production manager, technical and financial experts etc.; we concluded that selection of the critical factors of breakdowns in CLG's depends on six criteria as discussed in "Evaluation criteria" section and these are illustrated in level C of the Fig. 3 as attributes/impacts. Breakdown parameters are fully interdependent on these attributes and it shows the intricacy of the problem. Moreover, this is a time consuming process and significant knowledge of both technological as well as economic aspects is needed in this case. After the

Table 3 Linguistic decision matrix of factors for breakdown in CLG's for all evaluation criteria

Evaluation Criteria (attribute/ impact)	Breakdown factors in Centre Less Grinding's (alternatives)												
	Conveyor malfunction (F_1)	Loader failure (F_2)	Gear box fault (F_3)	Coolant pump malfunction (F_4)	Hydraulic motor not working (F_5)	Hydraulic oil leakage (F_6)	Slide failure (F_7)	Spindle jam (F_8)	CWD unit fault (F_9)	Electrical faults (F_{10})	Sensor faults (F_{11})	Grinding wheel fault (F_{12})	Improper lubrication (F_{13})
C_1	AP	VP	BA	AP	BA	VP	AP	VP	VP	VP	H	BA	VH
C_2	VP	BA	VP	BA	BA	VP	VP	BA	AP	BA	H	AA	AA
C_3	AP	VP	BA	VP	AA	VP	VP	VP	VP	VP	H	AA	AA
C_4	AH	VH	H	VH	AA	VH	AH	VH	AH	H	BA	AA	VP
C_5	VH	VH	AA	VH	BA	H	VH	VH	H	VH	VP	H	VH
C_6	AH	AH	H	AH	VH	AH	AH	AH	AH	VH	VP	H	AP

attributes are identified, the next issue is to prioritize these attributes, as to which one has more impact on the identified breakdown reasons. AHP approach is used to prioritize these attributes and in order to compare these distinct attributes, numeric priority values are assigned to the attributes on a scale of 1–9 and pair-wise comparison is made. Table 1 shows the relative decision matrix formed on the basis of pair-wise comparison (AHP approach) and the weights calculated for all the considered criteria. MTBF appease as the most dominant attribute for the selection of these critical reasons of breakdown; while green effect is found to be the least dominant factor. Figure 4 shows contribution of all these dominating attributes towards selection of these critical breakdown factors. It is clearly observed that contributions of these attributes vary from shop floor of company to company.

In next step, fuzzy hypothesis analysis is performed on conclusions of the decision makers for comparison of all alternatives for each attribute. Fuzzy logic approach dealt well with such problems. Linguistic variables were used for selection of the critical factors of breakdowns. These were further converted into fuzzy numbers as shown in Table 2 for the current study. The highest range is termed absolutely high (AH) and the least is termed as absolutely poor (AP). Table 3 demonstrates the linguistic decision matrix filled during brainstorming session with decision makers. Here a single decision matrix has been formed rather than having a separate decision matrix for each decision maker. However, it is clearly known that final decision matrix can change as per the requirements and existing conditions. Further, fuzzy values are finally transformed into crisp values as shown in Eq. (5). Table 4 shows the calculated crisp values obtained from aggregated fuzzy ratings. Calculated crisp values are used with VIKOR approach as shown in Eqs. (6)–(14) and these values are used with TOPSIS approach, using Eqs. (15)–(21) to obtain the rank indices of all alternatives. Table 5 shows corresponding rank indices and ranks for the factors of breakdowns in CLG's. The ranking of alternatives obtained by VIKOR and TOPSIS approach are exactly same. This shows the robustness of the results used. Our computation shows that conveyor malfunction is the prime factor for breakdown in CLG section. Other main reasons for breakdown are slide failure, CWD unit fault, coolant pump malfunction and hydraulic oil leakage, respectively (refer Table 5). It is also observed that sensor faults are having least effect on failure of this section. Improper lubrication and grinding wheel fault and electrical faults are also rarely responsible for failure. We found that our results are in good agreement with long term perceptions of auto companies under normal working conditions. These critical causes of capacity waste need immediate monitoring, so that productivity loss could be checked for future.

Table 4 Calculated crisp values for assigned fuzzy rates

Evaluation criteria	Breakdown factors in CLG's (alternatives)												
	F_1	F_2	F_3	F_4	F_5	F_6	F_7	F_8	F_9	F_{10}	F_{11}	F_{12}	F_{13}
C_1	0.0778	0.2333	0.3667	0.0778	0.3667	0.2333	0.0778	0.2333	0.2333	0.2333	0.6667	0.3667	0.8333
C_2	0.3833	0.3667	0.2333	0.3667	0.3667	0.2333	0.2333	0.3667	0.0778	0.3667	0.6667	0.5333	0.5333
C_3	0.1778	0.2333	0.3667	0.2333	0.5333	0.2333	0.2333	0.2333	0.2333	0.2333	0.6667	0.5333	0.5333
C_4	0.9444	0.8333	0.6667	0.8333	0.5333	0.8333	0.9444	0.8333	0.9444	0.6667	0.3667	0.5333	0.2333
C_5	0.9833	0.8333	0.5333	0.8333	0.3667	0.6667	0.8333	0.8333	0.6667	0.8333	0.2333	0.6667	0.8333
C_6	0.9444	0.9444	0.6667	0.9444	0.8333	0.9444	0.9444	0.9444	0.9444	0.8333	0.2333	0.6667	0.0778

Table 5 Calculated VIKOR and TOPSIS ranking

Evaluation criteria	Breakdown factors in CLG's (alternatives)												
	F_1	F_2	F_3	F_4	F_5	F_6	F_7	F_8	F_9	F_{10}	F_{11}	F_{12}	F_{13}
VIKOR rank index	0	0.09	0.297	0.08	0.485	0.088	0.054	0.09	0.061	0.133	0.75	0.503	0.648
VIKOR ranks	1	7	9	4	10	5	2	6	3	8	13	11	12
TOPSIS ranks	1	7	9	4	10	5	2	6	3	8	13	11	12
TOPSIS rank index	0.9987	0.9806	0.7315	0.9825	0.3599	0.9811	0.9912	0.9806	0.9894	0.9333	0.0172	0.2982	0.0574

Conclusions

Fuzzy MADM method has been used for the selection of the critical factors of breakdown in CLG section of an automotive industry. Analytical Hierarchy Process (AHP) method is used to calculate weights of all persuasive attributes for selection of the failure parameters. MTBF has been found to be the most serious and green effect as least critical attribute. Further priority order of critical breakdown factors in CLG's is determined using fuzzy VIKOR and fuzzy TOPSIS approach with AHP weights. Conveyor malfunction, slide not working, and CWD unit fault, coolant pump malfunction and hydraulic oil leakage are found to be the critical factors of breakdowns in CLG section. This study explores the feasibility of fuzzy VIKOR and fuzzy TOPSIS methods in Six Sigma analysis phase for selection of the breakdown/failure parameters. Briefly, the main features of this study are summarized as follows:

(a) The study helps to highlight the importance of 'Analysis Phase' for successful implementation of Six Sigma project.
(b) The study has also helped to prove that Fuzzy-MADM approach can be effectively used to select most critical CTQs, which can be further improved to achieve better sigma rating.
(c) Within MADM, the study has successfully explored the efficacy of AHP, VIKOR and TOPSIS methods

to prioritize the CTQs which are highly important for execution of Six Sigma project.
(d) The study will help the managers and practitioners to implement Six Sigma more effectively and more scientifically, using MADM approaches instead of using conventional statistical tools.

Authors' contribution All authors made substantial contribution to conception or design of the work, data collection, data analysis and interpretation, drafting the article, critical revision of the article and final approval of the version to be published.

References

Afshari A, Mojahed M et al (2010) Simple additive weighting approach to personnel selection problem. Int J Innov Manag Technol 1(5):511–515

Amer Y, Luong L et al (2008) Optimizing order fulfillment using design for six sigma and fuzzy logic. Int J Manag Sci Eng Manag 3(2):83–99

Antony J, Kumar A et al (2006) World class applications of six sigma. Routledge, London

Ayağ Z, Özdemir RG (2011) An intelligent approach to machine tool selection through fuzzy analytic network process. J Intell Manuf 22(2):163–177

Azar A, Olfat L et al (2011) A BSC method for supplier selection strategy using TOPSIS and VIKOR: a case study of part maker industry. Manag Sci Lett 1(4):559–568

Banuelas R, Antony J et al (2005) An application of Six Sigma to reduce waste. Qual Reliab Eng Int 21(6):553–570

Bellman RE, Zadeh LA (1970) Decision-making in a fuzzy environment. Manag Sci 17(4):B-141–B-164

Bevilacqua M, Ciarapica F et al (2006) A fuzzy-QFD approach to supplier selection. J Purch Supply Manag 12(1):14–27

Bhutta KS, Huq F (2002) Supplier selection problem: a comparison of the total cost of ownership and analytic hierarchy process approaches. Supply Chain Manag Int J 7(3):126–135

Chakrabarty A, Tan KC (2007) The current state of six sigma application in services. Manag Serv Qual 17(2):194–208

Chen C-T (2000) Extensions of the TOPSIS for group decision-making under fuzzy environment. Fuzzy Sets Syst 114(1):1–9

Chen Y-K, Liao H-C (2003) An investigation on selection of simplified aggregate production planning strategies using MADM approaches. Int J Prod Res 41(14):3359–3374

Chen SJ, Hwang CL, Hwang FP (1992) Fuzzy multiple attribute decision making: methods and applications. Springer, New York, NY

Cheng C-H, Mon D-L (1994) Evaluating weapon system by analytical hierarchy process based on fuzzy scales. Fuzzy Sets Syst 63(1):1–10

Chu T-C (2002) Selecting plant location via a fuzzy TOPSIS approach. Int J Adv Manuf Technol 20(11):859–864

Chu T-C, Lin Y-C (2003) A fuzzy TOPSIS method for robot selection. Int J Adv Manuf Technol 21(4):284–290

Dey PP, Pramanik S, et al (2014) TOPSIS approach to linear fractional bi-level MODM problem based on fuzzy goal programming. J Ind Eng Int 10(4):173–184

Ding S-H, Kamaruddin S (2015) Assessment of distance-based multi-attribute group decision-making methods from a maintenance strategy perspective. J Ind Eng Int 11:73–85

Eckes G (2003) Six Sigma for everyone. Wiley, New York

Harry M, Schroeder R (2005) Six Sigma: the breakthrough management strategy revolutionizing the world's top corporations. Random House, LLC, New York

Jiju A (2004) Some pros and cons of six sigma: an academic perspective. TQM Mag 16(4):303–306

Jiju A, Maneesh K et al (2005) Six sigma in small- and medium-sized UK manufacturing enterprises. Int J Qual Reliab Manag 22(8):860–874

Kahraman C, Cebeci U et al (2003) Multi-criteria supplier selection using fuzzy AHP. Logist Inf Manag 16(6):382–394

Khanna HK, Sharma D et al (2011) Identifying and ranking critical success factors for implementation of total quality management in the Indian manufacturing industry using TOPSIS. Asian J Qual 12(1):124–138

Kulak O, Kahraman C (2005) Multi-attribute comparison of advanced manufacturing systems using fuzzy vs. crisp axiomatic design approach. Int J Prod Econ 95(3):415–424

Kwong C, Bai H (2003) Determining the importance weights for the customer requirements in QFD using a fuzzy AHP with an extent analysis approach. IIE Trans 35(7):619–626

Linderman K, Schroeder RG et al (2003) Six Sigma: a goal-theoretic perspective. J Oper Manag 21(2):193–203

Liu F-HF, Hai HL (2005) The voting analytic hierarchy process method for selecting supplier. Int J Prod Econ 97(3):308–317

Liu H-C, You J-X et al (2014) Site selection in waste management by the VIKOR method using linguistic assessment. Appl Soft Comput 21:453–461

Neuman RP, Cavanagh R (2000) The six sigma way: how GE, Motorola, and other top companies are honing their performance. McGraw-Hill, New York

Nguyen H-T, Dawal SZM et al (2014) A hybrid approach for fuzzy multi-attribute decision making in machine tool selection with consideration of the interactions of attributes. Expert Syst Appl 41(6):3078–3090

Nydick RL, Hill RP (1992) Using the analytic hierarchy process to structure the supplier selection procedure. J Supply Chain Manag 28(2):31

Olson DL (2004) Comparison of weights in TOPSIS models. Math Comput Model 40(7–8):721–727

Önüt S, Soner S (2008) Transshipment site selection using the AHP and TOPSIS approaches under fuzzy environment. Waste Manag 28(9):1552–1559

Opricovic S (2011) Fuzzy VIKOR with an application to water resources planning. Expert Syst Appl 38(10):12983–12990

Pahlavani A (2010) A new fuzzy MADM approach and its application to project selection problem. Int J Comput Intell Syst 3(1):103–114

Perego A, Rangone A (1998) A reference framework for the application of MADM fuzzy techniques to selecting AMTS. Int J Prod Res 36(2):437–458

Rabbani M, Monshi M, et al. (2014) A new AATP model with considering supply chain lead-times and resources and scheduling of the orders in flowshop production systems: a graph-theoretic view. Appl Math Model 38(24):6098–6107

Radcliffe LL, Schniederjans MJ (2003) Trust evaluation: an AHP and multi-objective programming approach. Manag Decis 41(6):587–595

Saaty TL (1988) What is the analytic hierarchy process?. Springer, New York

Saaty TL (1990) How to make a decision: the analytic hierarchy process. Eur J Oper Res 48(1):9–26

Saaty TL (2014) The analytic hierarchy process without the theory of Oskar Perron. Int J Anal Hierarchy Process 5(2). doi:10.13033/ijahp.v5i2.191

Sanayei A, Farid Mousavi S et al (2010) Group decision making process for supplier selection with VIKOR under fuzzy environment. Expert Syst Appl 37(1):24–30

Satty TL (1980) The analytic hierarchy process. McGraw-Hill, New York

Satty TL (1994) Fundamentals of decision making and priority theory with the analytic hierarchy process. RWS Publications, Pittsburgh

Singh H, Kumar R (2014). Selection of chain-material in automobile sector using multi attribute decision making approach. Paper presented at the annual meeting of the ISAHP. Grand Hyatt Hotel, Washington, D.C

Singh J, Singh H (2014) Performance enhancement of manufacturing unit using Six Sigma DMAIC approach: a case study. In: Proceedings of the international conference on research and innovations in mechanical engineering. Springer

Snee RD, Hoerl RW (2003) Leading Six Sigma: a step-by-step guide based on experience with GE and other Six Sigma companies. FT Press, Upper Saddle River

Snee R, Hoerl R (2004) Six Sigma beyond the factory floor: deployment strategies for financial services, health care, and the rest of the real economy. PH Professional Business

Tahriri F, Mousavi M et al (2014) The application of fuzzy Delphi and fuzzy inference system in supplier ranking and selection. J Ind Eng Int 10(3):1–16

Tilo P, Wolf R et al (2004) Integrating six sigma with quality management systems. The TQM Magazine 16(4):241–249

Tiwary A, Pradhan B et al. (2014) Application of multi-criteria decision making methods for selection of micro-EDM process parameters. Adv Manuf 2:251–258

Tönshoff HK, Reinsch S et al (2007) Soft-computing algorithms as a tool for the planning of cyclically interlinked production lines. Prod Eng Res Devel 1(4):389–394

Vats G, Vaish R (2013) Selection of lead-free piezoelectric ceramics. Int J Appl Ceram Technol 11(5):883–893

Vats G, Vaish R (2014a) Phase change materials selection for latent heat thermal energy storage systems (LHTESS): an industrial engineering initiative towards materials science. Adv Sci Focus 2(2):140–147

Vats G, Vaish R (2014b) Selection of optimal sintering temperature of $K_{0.5}Na_{0.5}NbO_3$ ceramics for electromechanical applications. J Asian Ceram Soc 2(1):5–10

Wang Y-M, Elhag TMS (2006) Fuzzy TOPSIS method based on alpha level sets with an application to bridge risk assessment. Expert Syst Appl 31(2):309–319

Wei C-C, Liang G-S et al (2007) A comprehensive supply chain management project selection framework under fuzzy environment. Int J Proj Manag 25(6):627–636

Yoon KP, Hwang C-L (1995) Multiple attribute decision making: an introduction. Sage Publications, Beverley Hills

Zadeh LA (1975) The concept of a linguistic variable and its application to approximate reasoning—I. Inf Sci 8(3):199–249

Zu X, Fredendall LD et al (2008) The evolving theory of quality management: the role of Six Sigma. J Oper Manag 26(5):630–650

7

Ranking efficient DMUs using minimizing distance in DEA

Shokrollah Ziari[1] · Sadigh Raissi[2]

Abstract In many applications, ranking of decision making units (DMUs) is a problematic technical task procedure to decision makers in data envelopment analysis (DEA), especially when there are extremely efficient DMUs. In such cases, many DEA models may usually get the same efficiency score for different DMUs. Hence, there is a growing interest in ranking techniques yet. The main purpose of this paper is to overcome the lack of infeasibility and unboundedness in some DEA ranking methods. The proposed method is for ranking extreme efficient DMUs in DEA based on exploiting the leave-one out and minimizing distance between DMU under evaluation and virtual DMU.

Keywords Data envelopment analysis (DEA) · Ranking · Efficiency · Extreme efficient

Introduction

Data envelopment analysis (DEA) was initiated by Charnes et al. (1978) as a method to assess relative efficiency of homogeneous decision making units with multiple inputs and multiple outputs. Then, Banker et al. (1984) extended basic DEA models under returns to scale. As regards, the most models of DEA are introduced the more than one efficient DMU in evaluating the relative efficiency DMUs, thus the investigating rank of efficient DMUs is an interesting research topic. A DMU is called extremely efficient if it cannot be represented as a linear combination (with nonnegative coefficients) of the remaining DMUs (Cooper et al. 2007). In data envelopment analysis, there are several methods for ranking of the extreme efficient DMUs, e.g. AP (Andersen and Petersen 1993) method, MAJ (Mehrabian et al. 1999) method. Andersen and Petersen proposed a new procedure to rank efficient DMUs. The AP method exhibits the rank of a given DMU by removing it from the reference set and by computing its super efficiency score. However, the AP model may be infeasible in some cases. It is proved that super efficient DEA models are infeasible (see Thrall 1996, Cooper et al. 2007, Seiford and Zhu 1999, Charnes et al. 1989). Mehrabian et al. (Charnes et al. 1978) suggested as MAJ model for complete ranking efficient DMUs, but their approach lacks infeasibility in some cases, too. To overcome the drawbacks of the AP (Andersen and Petersen 1993) and MAJ (Mehrabian et al. 1999) models, Jahanshahloo et al. (2004a) presented a method to rank the extremely efficient DMUs in DEA models with constant and variable returns to scale using L_1-norm. The proposed model is a nonlinear programming form which has the computational complexity in solving. A complex treatment was applied in Jahanshahloo et al. (2004a) to convert the nonlinear model into a linear one which provides an approximately optimal solution. Wu and Yan (2010) have also used an effective transformation to convert the nonlinear model in Jahanshahloo et al. (2004a) into a linear model. Also Jahanshahloo et al. (2004b) have applied gradient line for ranking efficient units. Rezai Balf et al. (2012) applied Tchebycheff norm (L_1-norm) introduced in (Briec 1998; Tavares et al. 2001) for complete ranking

✉ Shokrollah Ziari
 shok_ziari@yahoo.com

 Sadigh Raissi
 Raissi@azad.ac.ir

[1] Department of Mathematics, Firoozkooh Branch, Islamic Azad University, Firoozkooh, Iran

[2] School of Industrial Engineering, Islamic Azad University, South Tehran Branch, Tehran, Iran

efficient units. Amirteimoori et al. (2005) introduced a method for ranking of extreme efficient DMUs, based on distance. Hashimoto (1999) proposed a super efficiency DEA model with assurance region in order to rank the DMUs completely. Torgesen et al. (1996) suggested a method for ranking efficient units, by their importance as benchmarks for the inefficient units. Sexton et al. (1986) investigated a ranking method for DMUs based on a cross-efficiency ratio matrix. The cross-efficiency ranking method computes the efficiency score of each DMU that determines a set of optimal weights using linear programs corresponding to each DMU. Then by taking the average of scores of given DMU is obtained the rank of that DMU. Liu and Peng (2008) determined one common set of weights for ranking efficient DMUs, that DMUs are ranked according to the efficiency score weighted by the common set of weights. Bal et al. (2008) suggested a DEA model for ranking of DMUs based on defining the coefficient of variation for input–output weights. Khodabakhshi and Aryavash (2012) proposed a method to rank the efficient DMUs. According to their method, first the minimum and maximum efficiency values of each DMU are computed under the assumption that the sum of efficiency values of all DMUs is equal to unity. Then, the rank of each DMU is determined in proportion to a combination of its minimum and maximum efficiency values. Shetty and Pakkala (2010) suggested a method for ranking efficient units, which is created the average of the corresponding inputs and outputs of all DMUs. Early, Jahanshahloo and Firoozi Shahmirzadi (2013) modified the model which was proposed by Bal et al. (2008). They introduced two new models for ranking efficient DMUs based on L_1-norm and using mean of input–output weights. For our new method it does not need any additional constraints.

In this paper, we suggest a new method for ranking extreme efficient DMUs. The rest of the paper is organized as follows. In "DEA models and ranking models review", we review the concept of DEA framework. We review some ranking methods in "The proposed ranking model for efficient DMUs", "Extension to variable returns to scale" proposes the new model for ranking efficient units. "Illustrated examples" includes some numerical examples. The last section concludes the study.

DEA model and ranking model review

DEA model review

DEA is a methodology for assessing the relative efficiency of decision making units (DMUs) where each DMU has multiple inputs used to secure multiple outputs.

It is assumed in DEA that there are n DMUs and for each DMU_j $(j = 1, \ldots, n)$ is considered a column vector of inputs (X_j) to produce a column vector of outputs (Y_j), where $X_j = (x_{1j}, x_{2j}, \ldots, x_{mj})^T$ and $Y_j = (y_{1j}, y_{2j}, \ldots, y_{sj})^T$. Here, the superscript (T) indicates a vector transpose. It is also assumed that $X_j \geq 0, Y_j \geq 0, X_j \neq 0,$ and $Y_j \neq 0$ for every $j = 1, \ldots, n$.

The following input-oriented CCR model [see (Cooper et al. 2007)] in the envelopment form with constant Returns to Scale measures the level of DEA efficiency (θ) of the kth DMU (X_k, Y_k):

$$\theta^* = \min \theta$$

$$\text{s.t.} \sum_{j=1}^{n} \lambda_j x_{ij} \leq \theta x_{ik}, \quad i = 1, \ldots, m$$

$$\sum_{j=1}^{n} \lambda_j y_{rj} \geq y_{rk}, \quad r = 1, \ldots, s \tag{1}$$

$$\lambda_j \geq 0, \quad j = 1, \ldots, n$$

Here, $\lambda = (\lambda_1, \ldots, \lambda_n)^T$ is a column vector of unknown variables used for components of the input and output vectors by a combination. θ^* represents the efficiency score of DMU_k in (1), where the superscript (*) indicates optimality.

DMU_k is relatively efficient if and only if on optimality, the objective of (1) equals to one and all the slacks are zero.

Similarly, the output-oriented CCR model, corresponding to (1), is formulated as follows:

$$\phi^* = \max \phi$$

$$\text{s.t.} \sum_{j=1}^{n} \lambda_j x_{ij} \leq x_{ik}, \quad i = 1, \ldots, m$$

$$\sum_{j=1}^{n} \lambda_j y_{rj} \geq \phi y_{rk}, \quad r = 1, \ldots, s \tag{2}$$

$$\lambda_j \geq 0, \quad j = 1, \ldots, n$$

Here, $1/\phi^*$ intends the DEA efficiency score in the output-oriented model.

Also, the following input-oriented BCC model [see Banker et al. (1984)] in the envelopment form with variable Returns to Scale measures the level of DEA efficiency (θ) of the kth DMU (X_k, Y_k):

$$\theta^* = \min \theta$$

$$\text{s.t.} \sum_{j=1}^{n} \lambda_j x_{ij} \leq \theta x_{ik}, \quad i = 1, \ldots, m$$

$$\sum_{j=1}^{n} \lambda_j y_{rj} \geq y_{rk}, \quad r = 1, \ldots, s \tag{3}$$

$$\sum_{j=1}^{n} \lambda_j = 1,$$

$$\lambda_j \geq 0, \quad j = 1, \ldots, n$$

DMU_k is relative efficient if and only if on optimality, the objective of (3) equals to one and all the slacks are zero.

Similarly, the output-oriented BCC model, corresponding to (3) which obtains from (2) by adding constraint,

$$\sum_{j=1}^{n} \lambda_j = 1.$$

Moreover, the following additive model is based on input and output slacks which accounts the possible input decreases as well as output increases simultaneously.

$$\max \sum_{j=1}^{n} s_i^- + \sum_{j=1}^{n} s_r^+$$

$$\text{s.t.} \sum_{j=1}^{n} \lambda_j x_{ij} + s_i^- \leq x_{ik}, \quad i = 1,\ldots,m \qquad (4)$$

$$\sum_{j=1}^{n} \lambda_j y_{rj} - s_r^+ \geq \phi y_{rk}, \quad r = 1,\ldots,s$$

$$\lambda_j, s_i^-, s_r^+, \geq 0,$$

DMU_k is relative efficient if and only if on optimality, the objective of (4) equals to zero.

Ranking models

In this subsection we review the some ranking models in data envelopment analysis. The first ranking model proposed by Anderson and Peterson (1993) which is the supper efficiency model. In the AP model DMU under evaluation is excluded from reference set and by using other units, the rank of given DMU is obtained.

The AP model using the CRS super-efficiency model is as follows:

$$AP: \min \theta$$

$$\text{s.t.} \sum_{j=1j\neq k}^{n} \lambda_j x_{ij} \leq \theta x_{ik}, \quad i = 1,\ldots,m \qquad (5)$$

$$\sum_{j=1j\neq k}^{n} \lambda_j y_{rj} \geq y_{rk}, \quad r = 1,\ldots,s$$

$$\lambda_j \geq 0, \quad j = 1,\ldots,n, j \neq k$$

The main drawbacks of this model are infeasibility and instability for some DMUs. It is said that a model is stable if a DMU under evaluation is efficient, it is remains efficient after perturbation on data.

The second ranking model under investigation proposed by Mehrabian et al. (1999) to solve infeasibility of AP models in some cases. The following model is MAJ model:

$$MAJ: \min 1 + w$$

$$\text{s.t} \sum_{j=1j\neq k}^{n} \lambda_j x_{ij} \leq x_{ik} + w, \quad i = 1,\ldots,m$$

$$\sum_{j=1j\neq k}^{n} \lambda_j y_{rj} \geq y_{rk}, \quad r = 1,\ldots,s$$

$$\lambda_j \geq 0, \quad j = 1,\ldots,n, j \neq k \qquad (6)$$

The third ranking model proposed by Jahanshahloo et al. (2004a), that their proposed method to rank the extremely efficient DMUs in DEA models with constant and variable Returns to Scale using the omitted DMU under evaluation from production possibility set and applying L_1-norm. It is shown that the proposed method is able to overcome the existing difficulties in the AP (Andersen and Petersen 1993) and MAJ (Mehrabian et al. 1999) models. On the other hand, the proposed model is the form of nonlinear programming which is difficult to be solved. The model of Jahanshahloo et al. (2004a) is presented as follows:

$$L_1 - norm: \min \sum_{i=1}^{m} |x_i - x_{ik}| + \sum_{r=1}^{s} |y_r - y_{rk}|$$

$$\text{s.t.} \sum_{j=1j\neq k}^{n} \lambda_j x_{ij} \leq x_i, \quad i = 1,\ldots,m$$

$$\sum_{j=1j\neq k}^{n} \lambda_j y_{rj} \geq y_r, \quad r = 1,\ldots,s$$

$$x_i \geq 0, y_r \geq 0 \quad i = 1,\ldots,m, r = 1,\ldots,s$$

$$\lambda_j \geq 0, \quad j = 1,\ldots,n, j \neq k \qquad (7)$$

The fourth ranking model proposed by Rezai Balf et al. (2012) which applies for ranking extreme efficient units using the leave-one-out idea and L_∞-norm. The proposed model is always feasible and so, it is able to remove the existing difficulties in some methods, such as Andersen and Petersen (1993). The model of Rezai Balf et al. (2012) is formulated as follows:

$$L_\infty - norm : \min v_k$$

$$\text{s.t.} v_k \geq \sum_{j=1j\neq k}^{n} \lambda_j x_{ij} - x_{ik}, \quad i = 1,\ldots,m$$

$$v_k \geq y_{rk} - \sum_{j=1j\neq k}^{n} \lambda_j y_{rj}, \quad r = 1,\ldots,s \qquad (8)$$

$$\lambda_j \geq 0, \quad j = 1,\ldots,n, j \neq k$$

$$v_k \geq 0$$

The proposed ranking model for efficient DMUs

In this section, we suppose that the DMUk is extreme efficient. By excluding the DMUk from the CCR production possibly set, it is obtained a new efficiency frontier. In order to gain the ranking score of DMUk by exploiting the new efficiency frontier, we suggest a new model by by using the leave-one out idea and minimizing distance between DMU under evaluation and virtual DMU. The proposed model is as follows:

$$\min \sum_{i=1}^{m} \alpha_i + \sum_{r=1}^{s} \beta_r$$

$$\text{s.t.} \sum_{j=1 j \neq k}^{n} \lambda_j x_{ij} \leq x_{ik} - \alpha_i, \quad i = 1, \ldots, m$$

$$\sum_{j=1 j \neq k}^{n} \lambda_j y_{rj} \geq y_{rk} + \beta_r, \quad r = 1, \ldots, s$$

$$\lambda_j \geq 0, \qquad j = 1, \ldots, n, j \neq k,$$

$$\alpha_i \geq 0, \quad \beta_r \geq 0 \qquad i = 1, \ldots, m \quad r = 1, \ldots, s,$$

$$(9)$$

where $\alpha = (\alpha_1, \ldots, \alpha_m)$, $\beta = (\beta_1, \ldots, \beta_s)$ and $\lambda = (\lambda_1, \ldots, \lambda_{k-1}, \lambda_{k+1}, \ldots, \lambda_n)$ are the variables of the model (9).

Theorem 1 *The model (9) is feasible and bounded.*

Proof For $p \neq k$ we set $\lambda_p = 1, \lambda_j = 0, j = 1, \ldots, n, j \neq k, p; \alpha_i = \min\{x_{ik} - x_{ip}\}, i = 1, \ldots, m; \beta_r = \min\{y_{rp} - y_{rk}\}, r = 1, \ldots, s.$

Obviously, it can be seen that (λ, α, β) according to above selection is a feasible solution of the model (9). Moreover, the objective function of model (9) is bounded below zero, because the variables of model are nonnegative. Also, the target function is zero when $\alpha_i = 0$ and $\beta_r = 0$ for all i, r $\square$.

Extension to variable returns to scale

In this section, the proposed model in previous section is extended to variable Returns to Scale model. For this purpose, the model (9) is reformulated by adjoining the following convexity constraint to the model:

$$\sum_{\substack{j=1 \\ j \neq k}}^{n} \lambda_j = 1, \quad \lambda_j \geq 0.$$

So, in order to get the ranking score under variable returns to Scale assumption is solved the following model:

Table 1 Input and output data for Example 1

DMU	Input 1	Input 2	Output 1	Output 2
1	81	87.6	5191	205
2	85	12.8	3629	0
3	56.7	55.2	3302	0
4	91	78.8	3379	8
5	216	72	5368	639
6	58	25.6	1674	0
7	112.2	8.8	2350	0
8	293.2	52	6315	414
9	186.6	0	2865	0
10	143.4	105.2	7689	66
11	108.7	127	2165	266
12	105.7	134.4	3963	315
13	235	236.8	6643	236
14	146.3	124	4611	128
15	57	203	4869	540
16	118.7	48.2	3313	16
17	58	47.4	1853	230
18	14	650.8	4578	217
19	0	91.3	0	508

$$\min \sum_{i=1}^{m} \alpha_i + \sum_{r=1}^{s} \beta_r$$

$$\text{s.t.} \sum_{j=1 j \neq k}^{n} \lambda_j x_{ij} \leq x_{ik} - \alpha_i, \quad i = 1, \ldots, m$$

$$\sum_{j=1 j \neq k}^{n} \lambda_j y_{rj} \geq y_{rk} + \beta_r, \quad r = 1, \ldots, s$$

$$\sum_{j=1 j \neq k}^{n} \lambda_j = 1,$$

$$\lambda_j \geq 0, \qquad j = 1, \ldots, n, \quad j \neq k,$$

$$\alpha_i \geq 0, \quad \beta_r \geq 0 \qquad i = 1, \ldots, m, \quad r = 1, \ldots, s,$$

$$(10)$$

Theorem 2 *The model (10) is feasible and bounded.*

Proof The proof of this theorem is similar to the proof of Theorem 1.

Table 2 Results of ranking by different models

DMU	1	2	5	9	15	19
AP ranking results	4	1	3	–	2	–
MAJ ranking results	5	3	2	6	4	1
L_1-norm ranking results	4	3	2	6	5	1
L_∞-norm ranking results	5	2	3	6	4	1
Proposed model ranking results	5	3	4	2	6	1

Table 3 The value of inputs and outputs

DMU# cities/zones	Input 1	Input 2	Output 1	Output 2	Output 3
Dalian	2874.8	16,738	160.89	80,800	5092
Qinhuangdao	946.3	691	21.14	18,172	6563
Tianjin	6854.0	43,024	375.25	44,530	2437
Qingdao	2305.1	10,815	176.68	70,318	3145
Yantai	1010.3	2099	102.12	55,419	1225
Weihai	282.3	757	59.17	27,422	246
Shanghai	17,478.6	116,900	1029.09	351,390	14,604
Lianyungang	661.8	2024	30.07	23,550	1126
Ningbo	1544.2	3218	160.58	59,406	2230
Wenzhou	428.4	574	53.69	47,504	430
Guangzhou	6228.1	29,842	258.09	151,356	4649
Zhanjiang	697.7	3394	38.02	45,336	1555
Beihai	106.4	367	7.07	8236	121
Shenzhen	4539.3	45,809	116.46	56,135	956
Zhuhai	957.8	16,947	29.20	17,554	231
Shantou	1209.2	15,741	65.36	62,341	618
Xiamen	972.4	23,822	54.52	25,203	513
Hainan	2192.0	10,943	25.24	40,267	895

Table 4 Results for several models ranking

DMU	1	2	5	6	7	9	10	11	13	16
AP	9	1	8	4	6	3	2	7	5	10
MAJ	1	8	3	4	9	7	10	6	2	5
L_1-norm	4	8	3	6	9	1	10	7	2	5
L_∞-norm	3	8	4	6	9	1	10	7	2	5
Proposed model	1	7	3	2	8	9	10	5	6	4

Illustrated examples

In this section, we employ the above DEA model (6) and (7) on the two data sets which they are introduced here, with the assumption of constant returns to scale.

Example 1 As can be seen from Table 1, the data set consists of 19 DMUs with 2 inputs and 2 outputs. The data originally are used by Rezai Balf et al. (2012). Table 2 reports the results of ranking for 6 extremely efficient DMUs $(D_1, D_2, D_5, D_9, D_{15}, D_{19})$ in model (7) with constant Returns to Scale and the proposed method is compared with Ap, MAJ, L_1 and L_∞. The results imply that the model proposed in this paper provides a easy tool for ranking extremely efficient DMUs. The value of inputs and outputs.

Example 2 (Empirical example). We employ DEA model (10) on the empirical example used in Zhu (1998), with the assumption of variable Returns to Scale. The data set in Table 3 provides 13 open coastal Chinese cities and five Chinese special economic zones in 1989. Two inputs and three outputs were chosen to characterize the technology of those cities/zones. Two inputs include Investment in fixed assets by state-owned enterprises, Foreign funds actually used. Three outputs include Total industrial output value, Total value of retail sales and Handling capacity of coastal ports. Table 4 reports the results of ranking for 10 extremely efficient DMUs $(D_1, D_2, D_5, D_6, D_7, D_9, D_{10}, D_{11}, D_{13}, D_{16})$ in model (10) with variable returns to scale and the proposed method are compared with other methods.

Conclusion

Many DEA researches are proposed on ranking of efficient decision making units, but they have a problem, e.g. the AP model may be infeasible in some cases. In the present paper, we proposed a model for ranking extreme efficient DMUs in DEA by exploiting the leave-one out and minimizing distance between DMU under evaluation and virtual DMU. The proposed model is linear form and always feasible and bounded. Therefore, it is able to rank all extreme efficient DMUs in the DEA methods with constraint and variable Returns to Scale and so, eliminate the existing difficulties in some methods. In addition, it can be easily used when the number of inputs and outputs is much larger than the number of DMUs. Illustrative examples are included to show good ranking results by the proposed method.

References

Amirteimoori A, Jahanshahloo GR, Kordrostami S (2005) Ranking of decision making units in data envelopment analysis: a distance-based approach. Appl Math Comput 171:122–135

Andersen P, Petersen NC (1993) A procedure for ranking efficient units in data envelopment analysis. Manag Sci 39:1261–1264

Bal H, Horkcu H, Celebioglu S (2008) A new method based on the dispersion of weights in data envelopment analysis. J Comput Ind Eng 54:502–512

Banker RD, Charnes A, Cooper WW (1984) Some methods for estimating technical and scale inefficiencies in data envelopment analysis. Manag Sci 30(9):1078–1092

Briec W (1998) Hölder distance function and measurement of technical efficiency. J Prod Anal 11:111–131

Charnes A, Cooper WW, Rhodes E (1978) Measuring the efficiency of decision making units. Eur J Oper Res 2(6):429–444

Charnes A, Cooper WW, Li S (1989) Using DEA to evaluate relative efficiencies in the economic performance of Chinese-key cities. Soc Econ Plan Sci 23:325–344

Cooper WW, Seiford LM, Tone K (2007) Data envelopment analysis: a comprehensive text with models, applications, references and DEA-solver Software, Second Edition. Springer

Hashimato A (1999) A ranked voting system using a DEA/AR exclusion model: a note. Eur J Oper Res 97:600–604

Jahanshahloo GR, Firoozi Shahmirzadi P (2013) New methods for ranking decision making units based on the dispersion of weights and Norm 1 in data envelopment analysis. Comput Ind Eng 65:187–193

Jahanshahloo GR, Hosseinzadeh Lotfi F, Shoja N, Tohidi G, Razavian S (2004a) Ranking by using L_1-norm in data envelopment analysis. Appl Math Comput 153:215–224

Jahanshahloo GR, Sanei M, Hosseinzadeh Lotfi F, Shoja N (2004b) Using the gradient line for ranking DMUs in DEA. Appl Math Comput 151:209–219

Khodabakhshia M, Aryavash K (2012) Ranking all units in data envelopment analysis. Appl Math Lett 25:2066–2070

Liu FF, Peng HH (2008) Ranking of units on the DEA frontier with common weights. Comput Oper Res 35:1624–1637

Mehrabian S, Alirezaee MR, Jahanshahloo GR (1999) A complete efficiency ranking of decision making units in data envelopment analysis. Comput Optim Appl 14:261–266

Rezai Balf F, Zhiani Rezai H, Jahanshahloo GR, Hosseinzadeh Lotfi F (2012) Ranking efficient DMUs using the Tchebycheff norm. Appl Math Model 36:46–56

Seiford LM, Zhu J (1999) Infeasibility of super-efficiency data envelopment analysis models. INFOR 37(2):174–187

Sexton TR, Silkman RH, Hogan AJ (1986) Data envelopment analysis: critique and extensions. In: Silkman RH (ed) Measuring efficiency: an assessment of data envelopment analysis. Jossey-Bass, San Francisco, pp 73–105

Shetty U, Pakkala TPM (2010) Ranking efficient DMUs based on single virtual DMU in DEA. Oper Res Soc 47(1):20–72

Tavares G, Antunes CH (2001) A Tchebycheff DEA Model. Rutcor Research Report

Thrall RM (1996) Duality, classification and slacks in DEA. Ann Oper Res 66:109–138

Torgersen AM, Forsund FR, Kittelsen SAC (1996) Slack-adjusted efficiency measures and ranking of efficient units. J Prod Anal 7:379–398

Wu J, Yan H (2010) An effective transformation in ranking using L_1-norm in data envelopment analysis. Appl Math Comput 217:4061–4064

Zhu J (1998) Data envelopment analysis vs. principal component analysis: an illustrative study of economic performance of Chinese cities. Eur J Oper Res 111:50–61

8

Closed loop supply chain network design with fuzzy tactical decisions

Mahtab Sherafati[1] · Mahdi Bashiri[2]

Abstract One of the most strategic and the most significant decisions in supply chain management is reconfiguration of the structure and design of the supply chain network. In this paper, a closed loop supply chain network design model is presented to select the best tactical and strategic decision levels simultaneously considering the appropriate transportation mode in activated links. The strategic decisions are made for a long term; thus, it is more satisfactory and more appropriate when the decision variables are considered uncertain and fuzzy, because it is more flexible and near to the real world. This paper is the first research which considers fuzzy decision variables in the supply chain network design model. Moreover, in this study a new fuzzy optimization approach is proposed to solve a supply chain network design problem with fuzzy tactical decision variables. Finally, the proposed approach and model are verified using several numerical examples. The comparison of the results with other existing approaches confirms efficiency of the proposed approach. Moreover the results confirms that by considering the vagueness of tactical decisions some properties of the supply chain network will be improved.

Keywords Supply chain network design · Closed loop · Transportation mode · Fuzzy mathematical programming

✉ Mahdi Bashiri
bashiri@shahed.ac.ir

Mahtab Sherafati
S_m_sherafati@azad.ac.ir

[1] Department of Industrial Engineering, South Tehran Branch, Islamic Azad University, Tehran, Iran

[2] Department of Industrial Engineering, Shahed University, Tehran, Iran

Introduction

Supply chain management (SCM) has received great attention in both industry and academia recently. SCM consists of procurement of raw materials and components, turning them into finished products, and distribution of products to customers aiming at minimization of the total cost and while satisfying the customer demands as much as possible (Singh 2014). In SCM studies, based on the time horizon, three planning levels are recognized for such researches: strategic level decisions, tactical level decisions and operational level decisions. Strategic level decisions belong to the long-term horizon and they are the most important decisions because they influence performances of all of the supply chain echelons significantly. One of these strategic decisions is supply chain network design (SCND) that it is the most vital and critical decision in supply chain. Its purpose is location of supply chain facilities and allocation of elements to each other with minimum total cost (Pishvaee and Razmi 2012; Govindan et al. 2015). Tactical level decisions are made for the medium term horizon usually once a year or once in each season. These decisions include transportation mode selection, inventory policies, production decisions, supplier selection and so on. The operational decisions are relevant to the short-term horizon and they are made once a weak or day to day such as routing and scheduling and so on.

Most papers in the supply chain area speak about strategic decisions and determine the location of facilities and the quantity of shipments between them. Considering of other tactical decisions (except the flow) with strategic level decision is studied rarely. For example, (Goetschalckx et al. 2002) surveyed and reviewed integration of the strategic and tactical models in the numerous researches. Badri et al. (2013) proposed a new SCND mathematical model considering different time resolutions for tactical and strategic decisions. Salema et al. (2010) integrated strategic and tactical decisions in a novel SCND model. Because of existing a research gap and of course suggestion of previous researchers, in this study, it is tried to regard tactical decisions during the making of strategic decisions.

Today one of the basic features of SCND is uncertainty, lack of the exact and clear around information and dynamic environment. A real supply chain should be operated in an uncertain environment and disregarding any effects of uncertainty causes to design a worthless supply chain (Xu and Zhai 2008) and also as (Nenes and Nikolaidis 2012) said coping dynamically with variations is essential for any remanufacturing enterprise. The crisp methods cannot solve problem with decision-makers' uncertainties, ambiguities and vagueness (Kulak and Kahraman 2005). Some methods are applied to overcome it, such as stochastic programming and fuzzy optimization. Stochastic programming is used when the parameters (for example costs and demands) are stochastic and they may fluctuate widely (Snyder 2006). The reader is referred to some papers which solved SCND problems using stochastic programming such as: (Santoso et al. 2005; Ramezani et al. 2013; Pishvaee et al. 2009) and so on. In the fuzzy environment, it is assumed that the related parameters, coefficient, objective goals and constraints have ambiguous. Baykasoğlu and Göçken (2008) classified fuzzy mathematical programming models according to the components which are aspiration values of the objectives (F), the right hand side value of the constraints (b), the coefficients of the objectives (c) and the coefficients of the constraints (A) and their combinations (in total, there are fifteen types of fuzzy mathematical programming models according to the fuzzy components). Using deterministic and stochastic models may not cause to fully satisfactory consequences and these drawbacks can be removed by application of fuzzy models (Aliev et al. 2007) and overcome the problem of imprecision that usually occurs in supply chain (Sarkar and Mohapatra 2006). Moreover, computational efficiency and flexibility in fuzzy mathematical programming is more than the deterministic and stochastic programming techniques (Liang 2011). In addition, one of the main reasons to use fuzzy theory in SCM problems is unexpected variations throughout the supply chain (Sadeghi et al. 2014) and as (Hasani et al. 2012) said the uncertainty issue is more important in reverse and closed loop supply chain network design, because of reverse material flows have the inherent uncertainty in comparison to forward material flows and quantity and quality of the returned products are usually uncertain.

A significant shortcoming of the previous works of the SCND in fuzzy environment is that they have ignored fuzziness of decision variables. They suppose that the parameters are fuzzy but it is assumed that decision variables are crisp. Fuzziness of decision variables in the supply chain can lead to have more flexibility in the later decisions in each element of the supply chain (Kabak and Ülengin 2011). This omission appeared anomalous considering tactical and strategic level decisions. By considering of ambiguous for decision variables of the tactical decisions, more flexible and efficient strategic decisions can be made in which is closer to the real-world situations. Moreover, quantity of shipments in SCM is always uncertain and it is hard to estimate these parameters (Qin and Ji 2010; Fazlollahtabar et al. 2012).

As authors best knowledge this study is the first research considering fuzziness of decision variables in supply chain network design problems. In the other fields the variables are handled as fuzzy numbers, for example (Fazlollahtabar et al. 2012) proposed a vehicle routing problem with fuzzy decisions and also (Kabak and Ülengin 2011) considered a mathematical model containing the resources allocation and outsourcing decisions which the two decision variables are considered to be fuzzy such as the input parameters. A resources management problem with social and environmental factors presented by (Tsakiris and Spiliotis 2004), a multi-item solid transportation problem proposed by (Giri et al. 2015) and inventory models for items with shortage backordering stated by (Mahata and Goswami 2013) are other researches which include fuzzy decision variables in the other fields.

The main contributions and advantages of this research that distinguishes this paper from the presented existing ones in the related literature are listed as follows:

- Considering suppliers and evaluation and selection of them: The supplier selection has been acknowledged an important and considerable challenge in supply chain and it has a growing effect on the success or failure of a business (Bevilacqua et al. 2006). Supplier selection problem receives attention of so many researchers; so that (Chai et al. 2012) selected and reviewed 123 journal articles and (Igarashi et al. 2013) examined 60 green supplier selection articles.
- Presenting a novel closed loop multi-echelon supply chain network design model which considers tactical and strategic decision levels which this gap is suggested as a necessary and significant future research (Pishvaee and Razmi 2012; Govindan et al. 2015).
- Regarding the life cycle assessment of a product (cradle-to-grave).
- Considering transportation mode selection: recently (Govindan et al. 2015) reviewed 328 papers about reverse logistics and closed loop supply chain and said that these decisions are distinguished as a remarkable gap and future opportunity.

- Offering an efficient programming model which can be used in a real cases such as automotive industry (Olugu and Wong 2012), computer (Kusumastuti et al. 2008), outdated products (such as IC chips and mobile phones) (Yang et al. 2010), plastics (Pohlen and Farris 1992), carpet (Biehl et al. 2007), battery (Fernandes, Gomes-Salema, and Barbosa-Povoa 2010; Kannan, Sasikumar, and Devika 2010, Sasikumar and Haq 2011), medical needle and syringe (Pishvaee and Razmi 2012), Waste electrical and electronic equipment recycling (Nagurney and Toyasaki 2005; Tsai and Hung 2009), copiers (Krikke 2011) and in a wide various of process industries including chemicals, food, rubber, and plastics (French and LaForge 2006).
- Computational analysis is provided by using a hospital industrial case study to present the significance of the presented model as well as the efficiency of the proposed solution method.
- Considering the parameters and the decision variables as a triangular fuzzy number: This is disregarded by the previous researchers, while some authors emphasis that the variables should be assumed with ambiguous such as (Qin and Ji 2010; Kabak and Ülengin 2011; Fazlollahtabar et al. 2012).

The remainder of this paper is organized as follows. At first, a literature review of supply chain network design using fuzzy mathematical programming is presented in "Literature review of SCND using fuzzy mathematical programming". The proposed supply chain network design is described in "The proposed closed loop supply chain network design". Then in "Formulation of the model", indices, parameters, decision variables are introduced and the formulation of the proposed SCND model is stated. "The proposed fuzzy optimization approach" contains a new fuzzy optimization approach and its applications. In "Numerical examples" several numerical examples are discussed to verify the proposed approach and model. After that some comparisons are presented to show the priority of proposed approach. "Case study and managerial implications" consider and analysis a real case study and the differences of the proposed method with other approaches show the advantage of the proposed method and then some managerial insights are drawn. Finally, the concluding remarks and the future study directions are expressed in "Conclusion".

Literature review of SCND using fuzzy mathematical programming

Generally, the supply chain network design studies in the fuzzy environment can be categorized in three following classes:

- Forward supply chain network design using fuzzy mathematical programming.
- Reverse logistics network design using fuzzy mathematical programming.
- Closed loop supply chain network design using fuzzy mathematical programming.

Forward supply chain network design using fuzzy mathematical programming

In the forward supply chain network models, the components and raw materials are purchased from suppliers and the products are manufactured in the plants and then they are shipped to downstream facilities (likely warehouses, distribution center (DC) and retailers and certainly the customers) and on the other hand the information and financial flows are given back to the upstream echelons. Some of the parameters such as demands, capacities and cost are handled as fuzzy numbers. Some of the papers trying to optimize the forward supply chain network design using fuzzy mathematical programming can be seen in Table 1. In most of them, the decision variables are considered as crisp, only (Bashiri and Sherafati 2012) optimized a multi-product supply chain network design using a fuzzy optimization introduced by (Kumar et al. 2010). In that study, both of the decision variables and parameters are assumed as fuzzy values. Additionally just three of them integrate the transportation mode selection issues as tactical decisions with strategic network design decisions.

Reverse logistics supply chain network design using fuzzy mathematical programming

If the products are returned to the manufactures in order to recycle, reuse, remanufacturing, the model is named "reverse logistics". With reverse logistic activities, customer service level and competence of enterprises can be improved and the motivation and demands of customers are increased, because it provides a green image to the enterprises (Özceylan and Paksoy 2013a, b).

Thin part of literature is dedicated to this classification, these studies are presented briefly as follows.

Qin and Ji (2010) applied a fuzzy programming tool to design the product recovery network. That research is one of the first studies which regarded to logistics network design with product recovery in fuzzy environment. Three fuzzy programming models were formulated and a hybrid intelligent algorithm was designed to solve the proposed models.

Moghaddam (2015) developed a general reverse logistics network as a fuzzy multi-objective mathematical model. To find the Pareto optimal and solve the model, a

Table 1 Summary of supply chain network design literature in fuzzy environment

Study	Model elements with fuzzy values	Reverse/ closed loop	Tactical decision	Considering transportation mode selection	Echelons				
					Supplier	Plant	DC	CC	Customer
Chen and Lee (2004)	F		Inventory decision	✔	✔	✔	✔		
			Transportation plan						
			Production plan						
			Material flow decision						
Chen et al. (2007)	F		Material flow decision		✔	✔	✔		✔
Xu et al. (2009)	A, c, b		Material flow decision		✔	✔	✔		✔
Xu et al. (2008)	A, c, b		Material flow decision		✔	✔	✔		✔
Selim and Ozkarahan (2008)	F		Material flow decision		✔	✔	✔		✔
Gumus et al. (2009)	b		Material flow decision			✔	✔		
Pinto-Varela et al. (2011)	A, c, b		Capacities determination		✔	✔	✔		✔
			Process planning						
			Material flow decision						
Pishvaee et al. (2012a, b)	A, c, b		Production technology selection			✔	✔		✔
			Material flow decision						
Pishvaee et al. (2012a, b)	A, c, b		Transportation mode selection	✔		✔	✔		✔
			Production technology selection						
			Material flow decision						
Paksoy and Yapici Pehlivan (2012)	b		Transportation mode selection	✔	✔	✔			
			Material flow decision						
Paksoy et al. (2012)	b		Material flow decision		✔	✔	✔		✔
Bashiri and Sherafati (2012)	A, c, b, x		Material flow decision		✔	✔			
Jouzdani et al. (2013)	b		Material flow decision		✔	✔			✔
Tabrizi and Razmi (2013)	A, c, b		Material flow decision		✔	✔	✔		✔
Ozgen and Gulsun (2014)	A, c, b		Material flow decision		✔	✔	✔		✔
Qin and Ji (2010)	c		Material flow decision					✔	✔
Dhouib (2014)			Material flow decision						
Moghaddam (2015)			Material flow decision						
Pishvaee and Torabi (2010)	A, c, b	✔	Material flow decision			✔	✔		✔
Zarandi et al. (2011)	F	✔	Material flow decision		✔	✔	✔		✔
Pishvaee and Razmi (2012)	b	✔	Material flow decision			✔	✔		✔
Vahdani et al. (2013)	F	✔	Material flow decision		✔	✔	✔		✔
Özkır and Başlıgil (2012)	F, b	✔	Material flow decision			✔	✔		✔
Özceylan and Paksoy (2013a, b)	F, A, b	✔	Material flow decision		✔	✔			✔
Subulan et al. (2014)	F	✔	Material flow decision			✔	✔		✔
Current study	A, c, b, x (fully fuzzy model)	✔	Transportation mode selection	✔	✔	✔	✔		✔
			Material flow decision						

simulation method is proposed and applied. Finally a real case study is adopted to evaluate and validate the model and the solution method.

Closed loop supply chain network design using fuzzy mathematical programming

If both of the forward and reverse flows are integrated simultaneously, the mode is known as "closed loop supply chain network". In the following, some of the papers trying to optimize the closed loop supply chain network design using fuzzy mathematical programming are mentioned.

Pishvaee and Torabi (2010) presented a bi-objective possibilistic mixed-integer programming model to design a closed loop supply chain network. In that research, an interactive fuzzy solution method is proposed to solve the possibilistic optimization problem. Selim and Ozkarahan (2008) optimized a supply chain network design model using a new approach. In their study, objectives targets are handled as fuzzy but the parameters and right hand side of constraints are considered as crisp. The fuzzy objectives are minimization of total costs, minimization of plants and warehouses investment cost, and maximization of the total service level provided to the retailers. Zarandi et al. (2011) integrated (Selim and Ozkarahan 2008)'s model with backward flows and they assumed fuzziness for objectives targets and also crispness for parameters. In the mentioned research, the solution approach is similar to (Selim and Ozkarahan 2008)'s research and the proposed closed loop supply chain network is optimized by that approach. Pishvaee and Razmi (2012) used a multi-objective fuzzy mathematical programming to design a green supply chain network with forward and reverse flows. The proposed model minimizes the total cost and also minimizes the total environmental impact. The fuzzy model is transformed to the crisp equivalent auxiliary model using (Jiménez et al. 2007). Then the modified ε-constraint method and an interactive fuzzy solution approach are applied to solve the crisp problem. Finally, the proposed model and fuzzy optimization approach are applied in a real industrial case. Vahdani et al. (2012) presented a reliable closed loop supply chain as a bi-objective mathematical programming. A new hybrid solution approach (which is combination of robust optimization approach, queuing theory and fuzzy multi-objective programming) is applied to design the proposed reliable closed logistics network under uncertainty. Özkır and Başlıgil (2012) suggested a fuzzy multiple objective optimization model to a closed loop supply chain network design problem and various recovery processes. The considered objective functions are: maximizing the satisfaction levels of the closed loop supply chain stakeholders, sourced by sales and purchasing price, maximizing the fill rate of customer demands, maximizing the

total profit. The proposed model is solved using Baron Solver and it also is validated. Özceylan and Paksoy (2013a, b) proposed a fuzzy multi-objective model to optimize a general closed loop supply chain network. In that research two objectives and the capacities, demands and reverse rates are uncertain and are assumed as fuzzy number to incorporate the logistics manager's imprecise aspiration levels. They converted the model to the crisp equivalent and solved it. Subulan et al. (2014) developed a multi-objective and multi-product closed loop supply chain model for a real case study, i.e., the lead/acid battery. Fuzzy goal programming approach is applied to solve the presented model. One of the contributions of the mentioned research is considering a new objective function to maximize of "the collection of returned batteries covered by the opened facilities".

Summary of supply chain network design literature is shown in Table 1. In Table 1 F, A, c, b and x are objective function target, coefficients of constraints, coefficients of the objectives, right hand side of constraints and decision variables, respectively. According to the presented literature and Table 1 some drawbacks are found in closed loop supply chain network design area, such as ambiguous of decision variables, considering of more echelons, making tactical decisions, regarding transportation mode selection and so on. As it was mentioned in end of "Introduction", in this research it is tried to overcome these drawbacks and the suggestions of researchers. In this study, both forward and reverse flows are considered and moreover, strategic level decisions in a long-term horizon and tactical level decision in a medium term horizon are examined. In addition, it is more real because all of the parameters and decision variables are considered as triangular fuzzy numbers (it is a fully fuzzy model). We believe that by considering of tactical decision variables with fuzzy values the resulted strategic decisions will be more accurate according to the real-world conditions.

The proposed closed loop supply chain network design

The concerned integrated closed loop supply chain network in this research is a multi-echelon, single product logistics network type including suppliers, plants, distribution centers and customer zones in forward flows and also collection centers and disposal center in revers flows.

As it is demonstrated in Fig. 1 the structure of the closed loop supply chain network contains both forward and reverse flows. The manufacturers purchase some of the components from suppliers and assemble the components and produce the products, and then products are transported to distribution centers (DCs) and to customers. In the

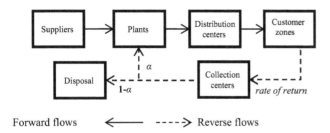

Forward flows   ⟵   ----> Reverse flows

Fig. 1 The structure of the proposed closed loop supply chain network

reverse flow, customers give back the percentage of used products to collection centers (CCs). The collected products are disassembled in collection centers and they are retuned into the components. The rate of the old components (e.g., α %) are reusable and are sold to plants by collection centers, and other $(1-\alpha)$ % are shipped to disposal, because they are not suitable for remanufacturing.

The supplier selection problem is considered here, it has been acknowledged an important and considerable challenge in supply chain and it has a growing effect on the success or failure of a business (Bevilacqua et al. 2006). Supplier selection problem receives attention of so many researchers, so that (Chai et al. 2012) selected and reviewed 123 journal articles and (Igarashi et al. 2013) examined 60 green supplier selection articles. There are also several models considering supplier selection decisions such as (Torabi and Hassini 2008). We regard to this and location of the other facilities in the mathematical model.

The following assumptions are considered in the proposed SCND problem.

- The long life of the components is more than the longevity of the products. So components of a used product can be reused in the other products.
- All demands of customers should be satisfied.
- There is a pull mechanism in the forward side of network while there is a push mechanism in the reverse side (Pishvaee and Torabi 2010).
- There are different transportation modes in each link. Once a selection is made about the transport modes, a decision should be made regarding the type and size of the transportation unit (Dekker et al. 2012).
- There is a single disposal.
- Buying cost of components from collection centers includes buying and transportation costs.
- Buying components from the collection centers is more economical than suppliers.
- Collection cost consists of collection and disassemble of the product.
- Some tactical decisions such as determination of transportation modes in each link, material flow

quantity, etc. in the presence of strategic decisions are made.

Formulation of the model

Indices

s: set of suppliers $s = 1,\ldots, S$; m: set of manufacturers/plants $m = 1,\ldots, M$; d: set of distribution centers $d = 1,\ldots, D$; c: set of customer zones $c = 1,\ldots, C$; k: set of collection centers $k = 1,\ldots, K$; t: transportation types $t = 1,\ldots, T$.

Parameters

The parameters are assumed as fuzzy triangular numbers for example $\tilde{f}_m$ is considered as $(f1_m, f2_m, f3_m)$.

$\tilde{f}_m$: fixed or setup cost for making products in plant m. $\tilde{g}_d$: fuzzy fixed cost for opening of distribution center d. $\tilde{l}_k$: fuzzy fixed cost for opening of collection center k. $\tilde{a}_{smt}$: fuzzy cost of buying and shipping components from supplier s to plant m with transportation mode t. $\tilde{b}_{kmt}$: fuzzy cost of buying and shipping components from collection center k to plant m with transportation mode t. $\tilde{\rho}_m$: fuzzy production cost of products by plant m $\tilde{\eta}_{dct}$: products fuzzy distribution cost from distribution center d to customer zone c with transportation mode t. $\tilde{\beta}_{mdt}$: fuzzy transportation cost of the manufactured products from plant m to distribution center d with transportation mode t. $\tilde{\pi}_k$: fuzzy contractual cost relevant to collection and disassembling of returned products in collection center k. $\tilde{\varepsilon}_k$: disposal related fuzzy cost for shipped unusable components from collection center k to disposal. $\tilde{\gamma}_m$: fuzzy capacity of plant m. $\tilde{\phi}_d$: fuzzy capacity of distribution center d. $\tilde{\sigma}_k$: fuzzy capacity of collection center k. $\tilde{\delta}_c$: fuzzy mean demand of customer zone c for products. $\tilde{r}_c$: fuzzy rate of return (%) from customer zone c. α: mean disposal fraction.

Decision variables

The decision variables are assumed as fuzzy triangular numbers for example $\tilde{x}_{smt}$ is considered as $(x1_{smt}, x2_{smt}, x3_{smt})$.

$\tilde{x}_{smt}$: fuzzy quantity of components shipped from supplier s to plant m with transportation mode t. $\tilde{y}_{smt}$: fuzzy quantity of components shipped from collection center k to plant m with transportation mode t. $\tilde{u}_{smt}$: fuzzy quantity of products shipped from plant m to distribution center d with transportation mode t. $\tilde{v}_{dct}$: fuzzy quantity of products shipped from distribution center d to customer zone c with transportation mode t. $\tilde{w}_{ck}$: fuzzy quantity of products shipped from customer zone c to collection center k. $\tilde{p}_k$: fuzzy quantity of components shipped from collection center k to the disposal. $\tilde{q}_m$: fuzzy quantity of product manufactured in plant m. n_m: 1, if the plant m is opened, 0 otherwise. e_d: 1, if the distribution center d is opened, 0 otherwise. o_k: 1, if the collection center k is opened, 0 otherwise.

Formulation

According to above notations, the formulation of the discussed closed loop supply chain network design is as following:

$$\min \quad TC = \sum_m \tilde{f}_m n_m + \sum_d \tilde{g}_d e_d + \sum_k \tilde{l}_k o_k$$

$$+ \sum_s \sum_m \sum_t \tilde{a}_{smt} \tilde{x}_{smt} + \sum_k \sum_m \sum_t \tilde{b}_{kmt} \tilde{y}_{kmt}$$

$$+ \sum_m \tilde{\rho}_m \tilde{q}_m + \sum_d \sum_c \sum_t \tilde{\eta}_{dct} \tilde{v}_{dct}$$

$$+ \sum_m \sum_d \sum_t \tilde{\beta}_{mdt} \tilde{u}_{mdt} + \sum_k \tilde{\varepsilon}_k \tilde{p}_k$$

$$+ \sum_k \tilde{\pi}_k \sum_c \tilde{w}_{ck} \tag{1}$$

s.t.

$$\sum_k \sum_t \tilde{y}_{kmt} + \sum_s \sum_t \tilde{x}_{smt} \le n_m \gamma_m \quad \forall m \tag{2}$$

$$\sum_m \sum_t \tilde{u}_{mdt} \le e_d \tilde{\phi}_d \quad \forall d \tag{3}$$

$$\sum_c \tilde{w}_{ck} \le o_k \tilde{\sigma}_k \quad \forall k \tag{4}$$

$$\sum_k \sum_t \tilde{y}_{kmt} + \sum_s \sum_t \tilde{x}_{smt} \ge \tilde{q}_m \quad \forall m \tag{5}$$

$$\tilde{q}_m \ge \sum_d \sum_t \tilde{u}_{mdt} \quad \forall m \tag{6}$$

$$\sum_m \sum_t \tilde{u}_{mdt} = \sum_c \sum_t \tilde{v}_{cdt} \quad \forall d \tag{7}$$

$$\sum_d \sum_t \tilde{v}_{cdt} \ge \tilde{\delta}_c \quad \forall c \tag{8}$$

$$\sum_k \tilde{w}_{ck} \ge \tilde{r}_c \tilde{\delta}_c \quad \forall c \tag{9}$$

$$\sum_m \sum_t \tilde{y}_{kmt} = \sum_c \alpha \times \tilde{w}_{ck} \quad \forall k \tag{10}$$

$$\tilde{p}_k = \sum_c (1 - \alpha) \times \tilde{w}_{ck} \quad \forall k \tag{11}$$

$$n_m, e_d, o_k \varepsilon \{0, 1\}$$
$$\tilde{x}_{smt}, \tilde{y}_{kmt}, \tilde{u}_{mdt}, \tilde{v}_{dct}, \tilde{w}_{ck}, \tilde{p}_k \ge 0 \quad \forall m, d, s, k, c \tag{12}$$

The objective function is minimization of the total cost (1), the total cost includes sum of the fixed, buying, manufacturing, distribution, transportation, collection & disassemble and disposal costs, respectively. Constraint (2) limits the quantity of manufactured product based on capacity of the plants. Constraints (3) and (4) are similar to the constraint (2) and they restrict the transported shipments to DCs and collection centers, respectively, based on their capacities. According to constraint (5), the quantities of input components to plants should be larger than the used components in products because some of them are useless. Constraint (6) assures that the production should be more than or equal to the final products, because they may be not accepted by the quality control unit. Constraint (7) is similar to constraint (6) and it balances the quantity transported to DCs with the distributed quantity to customer zones by DCs. Constraint (8) ensures that all of the demands are satisfied. Constraint (9) guarantees that the used products are returned to the collection centers (Pishvaee and Razmi 2012). Constraint (10) assures balancing of components volume after disassembling the products between plant and collection center. The same constraint for the disposal and collection center is guaranteed by constraint (11). Constraint (12) defines decision variables types.

The proposed fuzzy optimization approach

In this section a new fuzzy optimization approach is proposed for the aforementioned supply chain configuration in a fuzzy environment based on (Lai and Hwang 1992)'s approach, (Torabi and Hassini 2008) (TH) and (Kabak and Ülengin 2011). It is worth to mention that in this approach decision variables will remain fuzzy triangular variables during optimization stages.

(Kabak and Ülengin 2011) considered a mathematical model containing the resources allocation and outsourcing decisions which only two decision variables are considered to be fuzzy such as the input parameters. The advantage of this current research in comparison to (Kabak and Ülengin 2011) is that, the proposed model is fully fuzzy programming and it is applied to minimize the total cost in a supply chain network design problem. As authors' best knowledge, it is the first study which considerers SCND model containing fuzzy decision variables. The model presented by (Kabak and Ülengin 2011) is related to resources allocation and outsourcing decisions. Moreover, that model is extended to thirty objective function here, so the better solution can be find using this proposed approach.

(Torabi and Hassini 2008) presented a comprehensive supply chain master planning model which integrates the procurement, production and distribution plans. All of the parameters are handled as fuzzy number; however, the decision variables are assumed as crisp. (Torabi and Hassini 2008) proposed a novel solution approach in order to find an efficient compromise solution for a fuzzy multi-objective mixed-integer program with crisp decision variables.

None of the considered base papers regard to a supply chain network design problem.

In (Lai and Hwang 1992)'s approach, it is assumed that total objective function is a triangular fuzzy number $(\tilde{Z}^l, \tilde{Z}^m, \tilde{Z}^r)$, and $\tilde{Z}^m$ should be minimized, $(\tilde{Z}^m - \tilde{Z}^l)$ should be maximized and $(\tilde{Z}^r - \tilde{Z}^m)$ should be minimized simultaneously to obtain better solution in the minimization problems (Torabi and Hassini 2008) used the (Lai and Hwang 1992)'s approach, and proposed interactive fuzzy programming with multiple objectives.

(Kabak and Ülengin 2011) presented a new approach to solve possibilistic linear programming with fuzzy decision variables. The authors introduced the entropy for profit objectives and tried to minimize this entropy while the profit objective should be maximized. The entropy is the difference between the upper bound and the lower bound of the profit objective. In the mentioned approach, (Kabak and Ülengin 2011) proposed two linear programming, named LP-1 and LP-2. In LP-1 it is assumed that fuzzy variables are as crisp variables and the problem is optimized twice. In the first, the fuzzy parameters are set at their right value of triangular fuzzy number. After optimization of a crisp model upper bound of objective function is obtained which is called $\widehat{Z}$. In the second stage, the fuzzy parameters are set at their left value of triangular fuzzy number. Lower bound of objective function is obtained after optimization and is called $\widecheck{Z}$. The obtained values for objective function will be used to make two normalized measures presented in Eqs. (13) and (14).

$$\lambda_1 = \frac{z^m - \widecheck{z}}{\widehat{z} - \widecheck{z}} \tag{13}$$

$$\lambda_2 = 1 - \frac{z^r - z^l}{\widehat{z} - \widecheck{z}} \tag{14}$$

where Z^m is the median value of the fuzzy objective function, and λ_1 and λ_2 are the normalized measures which will be used for the LP-2 model. It is formulated as follows:

$LP - 2$

$$\max z = \lambda + \delta(\lambda_1 + \lambda_2)$$

$$s.t. \quad \lambda \le \lambda_1 = (z^m - \widecheck{z})/(\widehat{z} - \widecheck{z}) \tag{15}$$

$$\lambda \le \lambda_2 = ((\widehat{z} - \widecheck{z}) - (z^r - z^l))/(\widehat{z} - \widecheck{z})$$

where δ is a small number to give priority to the first part.

The proposed method in this research is more complete than (Kabak and Ülengin 2011)'s approach (hereafter KÜ approach), because in this study all parameters and decision variables are assumed fuzzy.

In this paper, based on the introduced three objective functions in (Lai and Hwang 1992)'s approach, thirty objective functions are proposed because of the fuzziness of decision variables expressed as following.

$$\min \quad H_{i,k} = c_i \cdot x_k \quad i = 1,2,3 \quad k = 1,2,3 \tag{16}$$

$$\min \quad H_{i-j,k} = (c_i - c_j) \cdot x_k \quad i = 3 \quad j = 1,2 \quad k = 1,2,3 \tag{17}$$

$$\min \quad H_{i,k-l} = c_i \cdot (x_k - x_l) \quad i = 1,2,3 \quad k = 3 \quad l = 1,2 \tag{18}$$

$$\min \quad H_{i-j,k-l} = (c_i - c_j) \cdot (x_k - x_l) \quad i = k = 3 \quad j = l = 1,2 \tag{19}$$

$$\max \quad H_{i-j,k} = (c_i - c_j) \cdot x_k \quad i = 2 \quad j = 1 \quad k = 1,2,3 \tag{20}$$

$$\max \quad H_{i,k-l} = c_i \cdot (x_k - x_l) \quad i = 1,2,3 \quad k = 2 \quad l = 1 \tag{21}$$

$$\max \quad H_{i-j,k-l} = (c_i - c_j) \cdot (x_k - x_l) \quad i = k = 2 \quad j = l = 1 \tag{22}$$

where $H = C \cdot X$ and also $C = (c_1, c_2, c_3)$ and $X = (x_1, x_2, x_3)$ are triangular fuzzy numbers. For example, $H_{2-1,1}$ will be constructed as Eq. (23):

$$\begin{aligned}
\min \quad H_{2-1,1} = &\sum_m (f2_m - f1_m) \times n_m \\
&+ \sum_d (g2_d - g1_d) \times e_d \\
&+ \sum_k (l2_k - l1_k) \times o_k \\
&+ \sum_s \sum_m \sum_t (a2_{smt} - a1_{smt}) \times x1_{smt} \\
&+ \sum_k \sum_m \sum_t (b2_{kmt} - b1_{kmt}) \times y1_{kmt} \\
&+ \sum_m (\rho2_m - \rho1_m) \times q1_m \\
&+ \sum_d \sum_c \sum_t (\eta2_{dct} - \eta1_{dct}) \times v1_{dct} \\
&+ \sum_m \sum_d \sum_t (\beta2_{mdt} - \beta1_{mdt}) \times u1_{mdt} \\
&+ \sum_k (\varepsilon2_k - \varepsilon1_k) \times p1_k \\
&+ \sum_k (\pi2_k - \pi1_k) \sum_c w1_{ck}. \tag{23}
\end{aligned}$$

Some of aforementioned objective functions can be deleted because some of them have the same direction.

Then, a linear membership function is defined for each objective and the problem continues as follows based on TH approach:For minimization objectives:

$$\mu_g = \begin{cases} 1 & \text{if} \quad H_{i,k} < H_{i,k}^p \\ \dfrac{H_{i,k}^n - H_{i,k}}{H_{i,k}^n - H_{i,k}^p} & \text{if} \quad H_{i,k}^p \leq H_{i,k} \leq H_{i,k}^n \quad g = 1,\ldots,9 \\ 0 & \text{if} \quad H_{i,k} > H_{i,k}^n \end{cases} \tag{24}$$

$$\mu_g = \begin{cases} 1 & \text{if} \quad H_{i-j,k} < H_{i-j,k}^p \\ \dfrac{H_{i-j,k}^n - H_{i-j,k}}{H_{i-j,k}^n - H_{i-j,k}^p} & \text{if} \quad H_{i-j,k}^p \leq H_{i-j,k} \leq H_{i-j,k}^n \quad g = 10,\ldots,15 \\ 0 & \text{if} \quad H_{i-j,k} > H_{i-j,k}^n \end{cases} \tag{25}$$

$$\mu_g = \begin{cases} 1 & \text{if} \quad H_{i,k-l} < H_{i,k-l}^p \\ \dfrac{H_{i,k-l}^n - H_{i,k-l}}{H_{i,k-l}^n - H_{i,k-l}^p} & \text{if} \quad H_{i,k-l}^p \leq H_{i,k-l} \leq H_{i,k-l}^n \quad g = 16,\ldots,21 \\ 0 & \text{if} \quad H_{i,k-l} > H_{i,k-l}^n \end{cases} \tag{26}$$

$$\mu_g = \begin{cases} 1 & \text{if} \quad H_{i-j,k-l} < H_{i-j,k-l}^p \\ \dfrac{H_{i-j,k-l}^n - H_{i-j,k-l}}{H_{i-j,k-l}^n - H_{i-j,k-l}^p} & \text{if} \quad H_{i-j,k-l}^p \leq H_{i-j,k-l} \leq H_{i-j,k-l}^n \quad g = 22,23 \\ 0 & \text{if} \quad H_{i-j,k-l} > H_{i-j,k-l}^n \end{cases} \tag{27}$$

where $H_{i,k}^n$, $H_{i-j,k}^n$, $H_{i,k-l}^n$ and $H_{i-j,k-l}^n$ are maximum of the Eqs. (16, 17, 18, 19) and $H_{i,k}^p$, $H_{i-j,k}^p$, $H_{i,k-l}^p$ and $H_{i-j,k-l}^p$ are minimum of them, respectively.

For maximization objectives:

$$\mu_g = \begin{cases} 1 & \text{if} \quad H_{i-j,k} > H_{i-j,k}^p \\ \dfrac{H_{i-j,k} - H_{i-j,k}^n}{H_{i-j,k}^p - H_{i-j,k}^n} & \text{if} \quad H_{i-j,k}^n \leq H_{i-j,k} \leq H_{i-j,k}^p \quad g = 24,\ldots,26 \\ 0 & \text{if} \quad H_{i-j,k} < H_{i-j,k}^n \end{cases} \tag{28}$$

$$\mu_g = \begin{cases} 1 & \text{if} \quad H_{i,k-l} > H_{i,k-l}^p \\ \dfrac{H_{i,k-l} - H_{i,k-l}^n}{H_{i,k-l}^p - H_{i,k-l}^n} & \text{if} \quad H_{i,k-l}^n \leq H_{i,k-l} \leq H_{i,k-l}^p \quad g = 27,\ldots,29 \\ 0 & \text{if} \quad H_{i,k-l} < H_{i,k-l}^n \end{cases} \tag{29}$$

$$\mu_g = \begin{cases} 1 & \text{if} \quad H_{i-j,k-l} > H_{i-j,k-l}^p \\ \dfrac{H_{i-j,k-l} - H_{i-j,k-l}^n}{H_{i-j,k-l}^p - H_{i-j,k-l}^n} & \text{if} \quad H_{i-j,k-l}^n \leq H_{i-j,k-l} \leq H_{i-j,k-l}^p \quad g = 30 \\ 0 & \text{if} \quad H_{i-j,k-l} < H_{i-j,k-l}^n \end{cases} \tag{30}$$

where $H_{i-j,k}^n, H_{i,k-l}^n$ and $H_{i-j,k-l}^n$ are minimum of the (20)-(22) and $H_{i-j,k}^p, H_{i,k-l}^p$ and $H_{i-j,k-l}^p$ are maximum of them, respectively.

Hence, the problem is performed as follows:

$$\max \quad \lambda = \gamma\lambda_0 + (1-\gamma)\sum_g \theta_g \cdot \mu_g \tag{31}$$
$$s.t. \quad \lambda_0 \leq \mu_g \quad g = 1,\ldots,30$$

$$\sum_k \sum_t y1_{kmt} + \sum_s \sum_t x1_{smt} \leq n_m\gamma1_m \quad \forall m \tag{32}$$

$$\sum_k \sum_t y2_{kmt} + \sum_s \sum_t x2_{smt} \leq n_m\gamma2_m \quad \forall m \tag{33}$$

$$\sum_k \sum_t y3_{kmt} + \sum_s \sum_t x3_{smt} \leq n_m\gamma3_m \quad \forall m \tag{34}$$

$$\sum_m \sum_t u1_{mdt} \leq e_d\phi1_d \quad \forall d \tag{35}$$

$$\sum_m \sum_t u2_{mdt} \leq e_d\phi2_d \quad \forall d \tag{36}$$

$$\sum_m \sum_t u3_{mdt} \leq e_d\phi3_d \quad \forall d \tag{37}$$

$$\sum_c w1_{ck} \leq o_k\sigma1_k \quad \forall k \tag{38}$$

$$\sum_c w2_{ck} \leq o_k\sigma2_k \quad \forall k \tag{39}$$

$$\sum_c w3_{ck} \leq o_k\sigma3_k \quad \forall k \tag{40}$$

$$\sum_k \sum_t y1_{kmt} + \sum_s \sum_t x1_{smt} \geq q1_m \quad \forall m \tag{41}$$

$$\sum_k \sum_t y2_{kmt} + \sum_s \sum_t x2_{smt} \geq q2_m \quad \forall m \tag{42}$$

$$\sum_k \sum_t y3_{kmt} + \sum_s \sum_t x3_{smt} \geq q3_m \quad \forall m \tag{43}$$

$$q1_m \geq \sum_d \sum_t u1_{mdt} \quad \forall m \tag{44}$$

$$q2_m \geq \sum_d \sum_t u2_{mdt} \quad \forall m \tag{45}$$

$$q3_m \geq \sum_d \sum_t u3_{mdt} \quad \forall m \tag{46}$$

$$\sum_m \sum_t u1_{mdt} = \sum_c \sum_t v1_{cdt} \quad \forall d \tag{47}$$

$$\sum_m \sum_t u2_{mdt} = \sum_c \sum_t v2_{cdt} \quad \forall d \tag{48}$$

$$\sum_m \sum_t u3_{mdt} = \sum_c \sum_t v3_{cdt} \quad \forall d \tag{49}$$

$$\sum_d \sum_t v1_{cdt} \geq \delta1_c \quad \forall c \tag{50}$$

$$\sum_d \sum_t v2_{cdt} \geq \delta2_c \quad \forall c \tag{51}$$

$$\sum_d \sum_t v3_{cdt} \geq \delta3_c \quad \forall c \tag{52}$$

$$\sum_d \sum_t v1_{cdt} \geq \delta1_c \quad \forall c \tag{53}$$

$$\sum_d \sum_t v2_{cdt} \geq \delta2_c \quad \forall c \tag{54}$$

$$\sum_d \sum_t v3_{cdt} \geq \delta3_c \quad \forall c \tag{55}$$

$$\sum_m \sum_t y1_{kmt} = \sum_c \alpha \times w1_{ck} \quad \forall k \tag{56}$$

$$\sum_m \sum_t y2_{kmt} = \sum_c \alpha \times w2_{ck} \quad \forall k \tag{57}$$

$$\sum_m \sum_t y3_{kmt} = \sum_c \alpha \times w3_{ck} \quad \forall k \tag{58}$$

$$p1_k = \sum_c (1 - \alpha) \times w1_{ck} \quad \forall k \tag{59}$$

$$p2_k = \sum_c (1 - \alpha) \times w2_{ck} \quad \forall k \tag{60}$$

$$p3_k = \sum_c (1 - \alpha) \times w3_{ck} \quad \forall k \tag{61}$$

$$x1_{smt} \leq x2_{smt} \leq x3_{smt} \quad \forall s, m, t \tag{62}$$

$$y1_{kmt} \leq y2_{kmt} \leq y3_{kmt} \quad \forall k, m, t \tag{63}$$

$$u1_{mdt} \leq u2_{mdt} \leq u3_{mdt} \quad \forall m, d, t \tag{64}$$

$$v1_{smt} \leq v2_{smt} \leq v3_{smt} \quad \forall d, c, t \tag{65}$$

$$w1_{ck} \leq w2_{ck} \leq w3_{ck} \quad \forall c, k \tag{66}$$

$$p1_k \leq p2_k \leq p3_k \quad \forall k \tag{67}$$

$$n_m, e_d, o_k \varepsilon \{0, 1\} \quad \forall m, d, k \tag{68}$$

$$x1_{smt}, x2_{smt}, x3_{smt}, y1_{kmt}, y2_{kmt}, y3_{kmt},$$
$$u1_{mdt}, u2_{mdt}, u3_{mdt}, v1_{dct}, v2_{dct}, v3_{dct},$$
$$w1_{ck}, w2_{ck}, w3_{ck}, p1_{ke}, p2_{ke}, p3_{ke} \geq 0 \quad \forall m, d, k, s, c \tag{69}$$

where λ is total satisfaction degree and indicates the final objective function of the proposed model, μ_g is the satisfaction degree of the mentioned objective functions and λ_0 is the minimum satisfaction degree of objectives. γ and θ_g denote coefficient of compensation and importance of corresponding objective, respectively, and they will be determined according to the decision maker preference $\left(\sum_g \theta_g = 1 \right)$.

Based on the introduced approach, in the current model, there is an objective function while $30 + 3 \times (3 m + 2d + 3 k + 2c) +$ (number of fuzzy decision variables) are taken into account. The problem involves ($s \times m \times t + k \times m \times t + m \times d \times t + d \times c \times t + c \times k + k + m$) fuzzy decision variables and ($m + d + k$) binary variables.

Numerical examples

In this section, several hypothetical numerical examples are analyzed to verify the model and the application of the proposed fuzzy optimization method. In the first example, it is assumed that there are two potential locations for plants, two potential locations for DCs, two potential locations for collection centers, two suppliers, five customer zones and a disposal. The shipments are transported with three transportation modes. The parameters are simulated by random numbers which some of them have been reported in Tables 2, 3, 4. The first example includes 80 fuzzy decision variables, 6 binary variables and 188 constraints.

The first numerical example is modeled and solved using proposed fuzzy mathematical programming. Firstly the membership functions are obtained and they are set in the problem. For example, for $H_{2,2}$ the following values are computed.

$$H_{2,2}^n = 437417.1 \tag{69}$$

$$H_{2,2}^p = 91672.82 \tag{70}$$

$$\mu_5 = \begin{cases} 1 & if \quad H_{2,2} < 91672.82 \\ \dfrac{437417.1 - H_{2,2}}{437417.1 - 91672.82} & if \quad 91672.82 \leq H_{2,2} \leq 437417.1 \\ 0 & if \quad H_{2,2} > 437417.1 \end{cases} \tag{71}$$

Finally the optimal configuration has been depicted in Fig. 2.

Table 2 Demands of customers in the first numerical example

Customer	Demands
c1	(130, 140, 150)
c2	(125, 135, 145)
c3	(135, 145, 155)
c4	(125, 135, 145)
c5	(130, 140, 150)

Table 3 Manufacturing cost in the first numerical example

Plant	Manufacturing cost
m1	(18, 20, 22)
m2	(20, 25, 28)

Table 4 Opening fixed cost of distribution centers in the first numerical example

Distribution center	Fixed cost
d1	(2000, 2200, 2300)
d2	(2200, 2400, 2500)

The first numerical example is optimized using KÜ and TH approaches and Table 5 illustrates the obtained results by the mentioned approaches. For this example related parameters for the problem have been assumed as following; $\gamma = 0.1$, also $\theta_1 = \theta_2 = \theta_3 = 0.3$ and the others are 0.0037. As it is shown in Table 5, the minimization objectives in the proposed approach are less than KÜ method and the maximization objectives are more.

Also Table 5 confirms that the proposed method has obtained more appropriate results especially in total satisfaction degree in comparison to both KÜ and TH methods.

Moreover the comparison of results of the proposed approach and TH approach shows that the transportation (and distribution) cost of TH method is further than the proposed approach. The reason of this increase is that the activated links and also facilities are greater in exact approach (as it is illustrated in Fig. 3). It should be noted that in fuzzy environment the flexibility is further and solution space is wider, consequently the superior solution can be find. This can be recognized by comparison of Figs. 2 and 3. In addition, as it was mentioned if decisions are assumed fuzzy in tactical level, more flexibility and proper decisions can be made for the strategic decisions. To do more sensitivity anlysis of the proposed research, it is tried to consider fuzzy parameters with least fuzziness, so each parameter were set as $(a - 0.01, a, a + 0.01)$ and the problem was optimized using our proposed method. Then the problem was solved by classic approach (with crisp values of parameters and variables).

Table 6 shows the results and the comparisons confirm that the proposed approach in the scenario in which parameters are handled with least fuzziness can obtain similar results of classic approach with crisp parameters and variables. Also the results of Table 5 show that

considering of ambiguous for parameters will lead to have a network with least network links and transportation modes. Moreover comparisons of Table 6 confirm that results of the proposed approach contain results of the problem with crisp parameters.

More examples are considered to assess the usefulness and performance of the proposed model and the solution method. Four numerical examples with different dimensions are generated and then optimized using proposed and KÜ approaches. The numerical examples and obtained total cost are illustrated in Table 7. As it is shown in Table 1 in all of the numerical examples, the total costs obtained by proposed approach have more appropriate values than the KÜ approach significantly. Moreover, the differences between upper and lower bounds (the entropies) in proposed approach are less than the corresponding values in KÜ approach.

Case study and managerial implications

If the supply chain network is designed properly and accurately and the entities are integrated and work cooperatively, managerial capabilities are promoted and some of the critical and major problems can be solved. It means the supply chain network design plays a vital and crucial role in overall performance of the supply chain. So in this study it is tried to develop a novel solution methodology in closed loop supply chain network configuration models in order to optimize the network configuration and help to the managers. This research can help to the stakeholders and improve the whole of the supply chain. Therefore, to verify this claim, in this section, the proposed supply chain network design model and the proposed solution algorithm are implemented for a real case study adapted from (Soleimani

Fig. 2 The optimal supply chain network of the example by the proposed approach

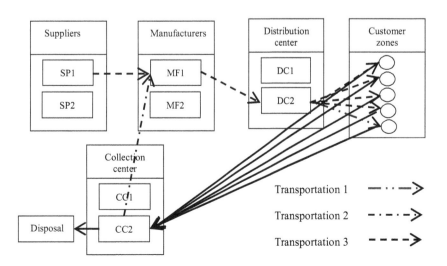

Fig. 3 The optimal supply
chain network by crisp decision
variables (TH method)

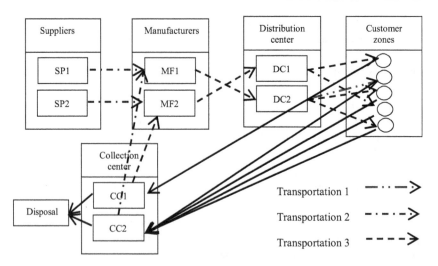

Table 5 Some optimal values
of decision variables using
proposed, KÜ and TH
approaches and crisp parameters

	Proposed approach	KÜ approach	TH approach
x_{smt}	$x_{113} = (371, 391, 412)$	$x_{113} = (412, 412, 412)$	$x_{112} = 227$
	Others equal to zero	Others equal to zero	$x_{222} = 243$
			Others equal to zero
y_{kmt}	$y_{212} = (274, 304, 333)$	$y_{112} = (274, 274, 274)$	$y_{123} = 52$
	Others equal to zero	$y_{212} = (0, 30, 59)$	$y_{212} = 223$
		Others equal to zero	Others equal to zero
q_m	$q_1 = (645, 695, 745)$	$q_1 = (686, 695, 745)$	$q_1 = 450$
	Others equal to zero	Others equal to zero	$q_2 = 295$
			Others equal to zero
$H_{1,1}$	68,216	77,487	110,810
$H_{2,1}$	85,633	94,729	136,590
$H_{3,2-1}$	15,838	12,503	0
$H_{2-1,2-1}$	1765	1085	0
Transportation cost	(52,345; 56,073; 59,820)	(57,907; 59,033; 61,333)	83,740
Number of facilities	5	6	9
Number of links	14	21	16
Total satisfaction degree	0.952	0.323	0.43

Table 6 Comparison of
obtained results using the
proposed approach with least
fuzziness and classic method

Obtained results with crisp parameters	Obtained results using proposed approach with least fuzziness
$x_{113} = 391$	$x_{113} = (395, 395, 395)$
$y_{212} = 304$	$y_{212} = (299.853, 304.05, 307.547)$
$q_1 = 695$	$q_1 = (694.95, 695.05, 695.05)$
$u_{123} = 695$	$u_{123} = (694.95, 695, 695.058)$
$p_2 = 304$	$p_2 = (299.853, 304.05, 307.547)$
Transportation cost = 56,073	Transportation cost = (56,235; 56,364; 56,469)

and Kannan 2015). It is related to Mehran Hospital Industries (Mehran Teb Med Co., http://www.Mehranmed. com), the first manufacturer of hospital industrial in Iran with 50 years of experience. Some hospital furniture such as beds, stretchers, trolleys, and so on are manufactured and sold to the hospitals, and after a while these used furniture are collected from the hospitals, so the network should be considered as closed loop.

Table 7 The numerical examples with the dimension and the total cost using proposed approach and KÜ approach

Test Problems	Number of suppliers	Number of manufacturers	Number of DC	Number of CC	Number of customers	Obtained total cost using proposed approach	Obtained total cost using KÜ approach
No. 1	2	2	2	2	5	(57,907; 59,180; 61,627)	(65,073; 69,071; 69,863)
No. 2	3	4	3	2	6	(71,727; 75,036; 79,602)	(75,456; 139,752; 142,476)
No. 3	4	5	4	3	7	(110,261; 113,887; 116,874)	(138,131; 200,743; 203,768)
No. 4	5	5	4	4	8	(123,354; 127,463; 130,843)	(219,121; 287,069; 290,446)

Table 8 Total cost in the case study obtained using proposed, KÜ and TH approaches (in millions of Rials)

Proposed approach	KÜ approach	Differences to proposed approach	TH approach	Differences to proposed approach
23,433.37	24,842.03	5.67 %	31,498.15	25.61 %

Since the proposed model is single product problem, so just hospital bed is considered here as a product.

After formulation and optimization of the studied case, the optimal solutions obtained by the proposed, TH and KÜ approaches can be seen in Table 8. In the proposed and KÜ approaches, since the solutions are obtained as a triangular fuzzy number (a, b, c), so they are converted to a single number by centroid method ($\frac{a+b+c}{3}$) (Lam et al. 2010), to compare of the results obtained by various approaches.

As it is seen in the obtained solutions, the proposed approach can present the better solutions and proposed supply chain network design model and also the proposed solution algorithm are confirmed in a real case study.

The main suggestion to the managers is extension of the capacity of the reverse supply chain to improve profits. Moreover, another recommendation is encouraging the customers to sell/replace their used products to the manufacturer. This improves the current rate of return of products and reduces the network total cost.

Finally, it can be concluded that the proposed hybrid approach provides an acceptable and satisfactory performance in real cases of a hospital furniture leader company that can prove and confirm its applicability in real-world situations. Moreover, comparison results confirm that by consideration of fuzzy tactical decision variables, total transportation cost and other related results will be improved because of flexibility of decisions for strategic stage.

Conclusion

This study proposes a novel fuzzy programming model to formulate a closed loop supply chain network design problem integrating the strategic and tactical decisions in a multi-echelon supply chain network. Moreover fuzzy programming procedure has been developed to optimize of the proposed closed loop supply chain network design model. In the proposed fuzzy approach, all parameters and decision variables (except binary variables) have ambiguous and it is fully fuzzy programming. As some researches mentioned, since we consider strategic level decisions, hence it is more satisfactory that the decision variables are not handled as crisp and certain and they are assumed as fuzzy triangular numbers. The environment is uncertain, fuzziness of decision variables in the supply chain models causes that the presented solutions have more flexibility and the managers can make more satisfactory decisions especially in tactical and strategic plans. Some numerical example are applied to verify the proposed fuzzy approach and the comparisons of results in the examples indicate that the proposed method is very promising fuzzy optimization approach because it presents more appropriate results than the previous approaches. Moreover comparison results confirm that by consideration of fuzzy tactical decision variables, total transportation cost and other related results will be improved because of flexibility of decisions for strategic stage.

This research can help to the stakeholders and improve the whole of the supply chain. The proposed model and also methodology are expected to provide an important guide to supply chain managers in making their decisions, taking into account the fuzziness of decision variables as it is shown in a real studied case. In the case study such as the numerical instances, the developed hybrid algorithm can achieve most appropriate solutions in comparison with the TH and KÜ.

As future research extension of the proposed model and designing of a dynamic closed loop supply chain network and also a sustainable supply chain network considering environmental and social aspects are suggested. Moreover, the researchers can apply the efficient proposed approach to solve the fully fuzzy mathematical programming models due to its computational advantages.

References

Aliev RA, Fazlollahi B, Guirimov B, Aliev RR (2007) Fuzzy-genetic approach to aggregate production–distribution planning in supply chain management. Inf Sci 177(20):4241–4255

Badri H, Bashiri M, Hejazi TH (2013) Integrated strategic and tactical planning in a supply chain network design with a heuristic solution method. Comput Oper Res 40(4):1143–1154

Bashiri M, Sherafati M (2012) A three echelons supply chain network design in a fuzzy environment considering inequality constraints. Paper presented at the International Constraints in Industrial Engineering and Engineering Management (IEEM), Hong Kong, p. 10–13

Baykasoğlu A, Göçken T (2008) A review and classification of fuzzy mathematical programs. J Intell Fuzzy Sys 19(3):205–229

Bevilacqua M, Ciarapica F, Giacchetta G (2006) A fuzzy-QFD approach to supplier selection. J Purch Supply Manag 12(1):14–27

Biehl M, Prater E, Realff MJ (2007) Assessing performance and uncertainty in developing carpet reverse logistics systems. Comput Oper Res 34(2):443–463

Chai J, Liu JN, Ngai EW (2012) Application of decision-making techniques in supplier selection: a systematic review of literature. Expert Syst Appl 40:3872–3885

Chen CL, Lee WC (2004) Multi-objective optimization of multi-echelon supply chain networks with uncertain product demands and prices. Comput Chem Eng 28(6):1131–1144

Chen CL, Yuan TW, Lee WC (2007) Multi-criteria fuzzy optimization for locating warehouses and distribution centers in a supply chain network. J Chin Inst Chem Eng, 38(5):393–407

Dekker R, Bloemhof J, Mallidis I (2012) Operations Research for green logistics–An overview of aspects, issues, contributions and challenges. Eur J Oper Res 219(3):671–679

Dhouib D (2014) An extension of MACBETH method for a fuzzy environment to analyze alternatives in reverse logistics for automobile tire wastes. Omega 42(1):25–32

Fazlollahtabar H, Mahdavi I, Mohajeri A (2012) Applying fuzzy mathematical programming approach to optimize a multiple supply network in uncertain condition with comparative analysis. Appl Soft Comput

French ML, LaForge RL (2006) Closed-loop supply chains in process industries: an empirical study of producer re-use issues. J Oper Manag 24(3):271–286

Giri PK, Maiti MK, Maiti M (2015) Fully fuzzy fixed charge multi-item solid transportation problem. Appl Soft Comput 27:77–91

Goetschalckx M, Vidal CJ, Dogan K (2002) Modeling and design of global logistics systems: a review of integrated strategic and tactical models and design algorithms. Eur J Oper Res 143(1):1–18

Govindan K, Soleimani H, Kannan D (2015) Reverse logistics and closed-loop supply chain: a comprehensive review to explore the future. Eur J Oper Res 240(3):603–626

Gumus AT, Guneri AF, Keles S (2009) Supply chain network design using an integrated neuro-fuzzy and MILP approach: a comparative design study. Expert Syst Appl 36(10):12570–12577

Hasani A, Zegordi SH, Nikbakhsh E (2012) Robust closed-loop supply chain network design for perishable goods in agile manufacturing under uncertainty. Int J Prod Res 50(16):4649–4669

Igarashi M, de Boer L, Fet AM (2013) What is required for greener supplier selection? A literature review and conceptual model development. J Purch Supply Manag 19(4):247–263

Jiménez M, Arenas M, Bilbao A (2007) Linear programming with fuzzy parameters: an interactive method resolution. Eur J Oper Res 177(3):1599–1609

Jouzdani J, Sadjadi SJ, Fathian M (2013) Dynamic dairy facility location and supply chain planning under traffic congestion and demand uncertainty: a case study of Tehran. Appl Math Model 37(18):8467–8483

Kabak Ö, Ülengin F (2011) Possibilistic linear-programming approach for supply chain networking decisions. Eur J Oper Res 209(3):253–264

Krikke H (2011) Impact of closed-loop network configurations on carbon footprints: a case study in copiers. Resour Conserv Recycl 55(12):1196–1205

Kulak O, Kahraman C (2005) Fuzzy multi-attribute selection among transportation companies using axiomatic design and analytic hierarchy process. Inf Sci 170(2):191–210

Kumar A, Kaur J, Singh P (2010) Fuzzy optimal solution of fully fuzzy linear programming problems with inequality constraints. Int J Math Comput Sci 6:37–41

Kusumastuti RD, Piplani R, Hian Lim G (2008) Redesigning closed-loop service network at a computer manufacturer: a case study. Int J Prod Econ 111(2):244–260

Lai YJ, Hwang CL (1992) A new approach to some possibilistic linear programming problems. Fuzzy Sets Syst 49(2):121–133

Lam KC, Tao R, Lam MCK (2010) A material supplier selection model for property developers using Fuzzy principal component analysis. Autom Constr 19(5):608–618

Liang TF (2011) Application of fuzzy sets to manufacturing/distribution planning decisions in supply chains. Inf Sci 181(4):842–854

Mahata GC, Goswami A (2013) Fuzzy inventory models for items with imperfect quality and shortage backordering under crisp and fuzzy decision variables. Comput Ind Eng 64(1):190–199

Moghaddam KS (2015) Fuzzy multi-objective model for supplier selection and order allocation in reverse logistics systems under supply and demand uncertainty. Expert Syst Appl 42(15):6237–6254

Nagurney A, Toyasaki F (2005) Reverse supply chain management and electronic waste recycling: a multitiered network equilibrium framework for e-cycling. Transp Res Part E: Logist Transp Rev 41(1):1–28

Nenes G, Nikolaidis Y (2012) A multi-period model for managing used product returns. Int J Prod Res 50(5):1360–1376

Olugu EU, Wong KY (2012) An expert fuzzy rule-based system for closed-loop supply chain performance assessment in the automotive industry. Expert Syst Appl 39(1):375–384

Özceylan E, Paksoy T (2013a) Fuzzy multi-objective linear programming approach for optimising a closed-loop supply chain network. Int J Prod Res 51(8):2443–2461

Özceylan E, Paksoy T (2013b) A mixed integer programming model for a closed-loop supply-chain network. Int J Prod Res 51(3):718–734

Ozgen D, Gulsun B (2014) Combining possibilistic linear programming and fuzzy AHP for solving the multi-objective capacitated multi-facility location problem. Inf Sci 268:185–201

Özkır V, Başlıgil H (2012) Multi-objective optimization of closed-loop supply chains in uncertain environment. J Clean Product

Paksoy T, Yapici Pehlivan N (2012) A fuzzy linear programming model for the optimization of multi-stage supply chain networks with triangular and trapezoidal membership functions. J Frankl Inst 349(1):93–109

Paksoy T, Pehlivan NY, Özceylan E (2012) Application of fuzzy optimization to a supply chain network design: a case study of an edible vegetable oils manufacturer. Appl Math Model 36(6):2762–2776

Pinto-Varela T, Barbosa-Póvoa APF, Novais AQ (2011) Bi-objective optimization approach to the design and planning of supply chains: economic versus environmental performances. Comput Chem Eng 35(8):1454–1468

Pishvaee MS, Razmi J (2012) Environmental supply chain network design using multi-objective fuzzy mathematical programming. Appl Math Model 36(8):3433–3446

Pishvaee M, Torabi S (2010) A possibilistic programming approach for closed-loop supply chain network design under uncertainty. Fuzzy Sets Syst 161(20):2668–2683

Pishvaee MS, Jolai F, Razmi J (2009) A stochastic optimization model for integrated forward/reverse logistics network design. J Manuf Sys 28(4):107–114

Pishvaee M, Razmi J, Torabi S (2012a) Robust possibilistic programming for socially responsible supply chain network design: a new approach. Fuzzy Sets Syst 206:1–20

Pishvaee M, Torabi S, Razmi J (2012b) Credibility-based fuzzy mathematical programming model for green logistics design under uncertainty. Comput Ind Eng 62(2):624–632

Pohlen TL, Farris MT (1992) Reverse logistics in plastics recycling. Int J Phy Distrib Logist Manag 22(7):35–47

Qin Z, Ji X (2010) Logistics network design for product recovery in fuzzy environment. Eur J Oper Res 202(2):479–490

Ramezani M, Bashiri M, Tavakkoli-Moghaddam R (2013) A new multi-objective stochastic model for a forward/reverse logistic network design with responsiveness and quality level. Appl Math Model 37(1):328–344

Sadeghi J, Sadeghi S, Niaki STA (2014) Optimizing a hybrid vendor-managed inventory and transportation problem with fuzzy demand: an improved particle swarm optimization algorithm. Inf Sci 272:126–144

Salema MIG, Barbosa-Povoa AP, Novais AQ (2010) Simultaneous design and planning of supply chains with reverse flows: a generic modelling framework. Eur J Oper Res 203(2):336–349

Santoso T, Ahmed S, Goetschalckx M, Shapiro A (2005) A stochastic programming approach for supply chain network design under uncertainty. Eur J Oper Res 167(1):96–115

Sarkar A, Mohapatra PK (2006) Evaluation of supplier capability and performance: a method for supply base reduction. J Purch Supply Manag 12(3):148–163

Selim H, Ozkarahan I (2008) A supply chain distribution network design model: an interactive fuzzy goal programming-based solution approach. Int J Adv Manuf Technol 36(3–4):401–418

Singh A (2014) Supplier evaluation and demand allocation among suppliers in a supply chain. J Purch Supply Manag 20(3):167–176

Snyder LV (2006) Facility location under uncertainty: a review. IIE Trans 38(7):547–564

Soleimani H, Kannan G (2015) A hybrid particle swarm optimization and genetic algorithm for closed-loop supply chain network design in large-scale networks. Appl Math Model 39(14):3990–4012

Subulan K, Baykasoğlu A, Özsoydan FB, Taşan AS, Selim H (2014) A case-oriented approach to a lead/acid battery closed-loop supply chain network design under risk and uncertainty. J Manuf Sys 37:340–361

Tabrizi BH, Razmi J (2013) Introducing a mixed-integer non-linear fuzzy model for risk management in designing supply chain networks. J Manuf Sys 32(2):295–307

Torabi S, Hassini E (2008) An interactive possibilistic programming approach for multiple objective supply chain master planning. Fuzzy Sets Syst 159(2):193–214

Tsai WH, Hung SJ (2009) Treatment and recycling system optimisation with activity-based costing in WEEE reverse logistics management: an environmental supply chain perspective. Int J Prod Res 47(19):5391–5420

Tsakiris G, Spiliotis M (2004) Fuzzy linear programming for problems of water allocation under uncertainty. Eur Water 7(8):25–37

Vahdani B, Tavakkoli-Moghaddam R, Modarres M, Baboli A (2012) Reliable design of a forward/reverse logistics network under uncertainty: a robust-M/M/c queuing model. Transp Res Part E Logist Transp Rev 48(6):1152–1168

Vahdani B, Tavakkoli-Moghaddam R, Jolai F, Baboli A (2013) Reliable design of a closed loop supply chain network under uncertainty: an interval fuzzy possibilistic chance-constrained model. Eng Optim 45(6):745–765

Xu R, Zhai X (2008) Optimal models for single-period supply chain problems with fuzzy demand. Inf Sci 178(17):3374–3381

Xu J, Liu Q, Wang R (2008) A class of multi-objective supply chain networks optimal model under random fuzzy environment and its application to the industry of Chinese liquor. Inf Sci 178(8):2022–2043

Xu J, He Y, Gen M (2009) A class of random fuzzy programming and its application to supply chain design. Comput Ind Eng 56(3):937–950

Yang P, Wee H, Chung S, Ho P (2010) Sequential and global optimization for a closed-loop deteriorating inventory supply chain. Math Comput Model 52(1):161–176

Zarandi MHF, Sisakht AH, Davari S (2011) Design of a closed-loop supply chain (CLSC) model using an interactive fuzzy goal programming. Int J Adv Manuf Technol 56(5–8):809–821

9

On the use of back propagation and radial basis function neural networks in surface roughness prediction

Angelos P. Markopoulos[1] · Sotirios Georgiopoulos[1] · Dimitrios E. Manolakos[1]

Abstract Various artificial neural networks types are examined and compared for the prediction of surface roughness in manufacturing technology. The aim of the study is to evaluate different kinds of neural networks and observe their performance and applicability on the same problem. More specifically, feed-forward artificial neural networks are trained with three different back propagation algorithms, namely the adaptive back propagation algorithm of the steepest descent with the use of momentum term, the back propagation Levenberg–Marquardt algorithm and the back propagation Bayesian algorithm. Moreover, radial basis function neural networks are examined. All the aforementioned algorithms are used for the prediction of surface roughness in milling, trained with the same input parameters and output data so that they can be compared. The advantages and disadvantages, in terms of the quality of the results, computational cost and time are identified. An algorithm for the selection of the spread constant is applied and tests are performed for the determination of the neural network with the best performance. The finally selected neural networks can satisfactorily predict the quality of the manufacturing process performed, through simulation and input–output surfaces for combinations of the input data, which correspond to milling cutting conditions.

Keywords Artificial neural networks · Training algorithms · Radial basis function · Surface roughness · Milling

Introduction

Computational methods for modeling and simulation are significant in contemporary manufacturing sector, where quality plays a key role. The industrial and academic interest in the role of computers in manufacturing technology is increasing since modeling and simulation are able to lead to optimization of processes and at the same time reduce expensive and time consuming experimental work. This is especially applicable for material removal processes that are involved in numerous final products that demand high quality, which usually is quantified through surface roughness of the final product; surface roughness influences several attributes of a part such as fatigue behaviour, wear, corrosion, lubrication and surface friction.

Modelling and simulation techniques that are more popular for the analysis of manufacturing processes are the Finite Element Method (FEM), soft computing techniques including Artificial Neural Networks (ANNs) (Galanis and Manolakos 2014; Szabó and Kundrák 2014; Niesłony et al. 2015; Markopoulos et al. 2006) and statistical methods (Raissi et al. 2004; Farsani et al. 2007; Shokuhfar et al. 2008; Saeidi et al. 2013; Ghasemi et al. 2013). In the past years ANNs have emerged as a highly flexible modeling tool applicable in numerous areas of manufacturing discipline (Ezugwu et al. 2005; Al-Hazza et al. 2013). An artificial neural network is defined as "a data processing system consisting a large number of simple, highly interconnected processing elements (artificial neurons) in an architecture inspired by the structure of the cerebral cortex

✉ Angelos P. Markopoulos
amark@mail.ntua.gr

[1] Section of Manufacturing Technology, School of Mechanical Engineering, National Technical University of Athens, Heroon Polytechniou 9, 15780 Athens, Greece

of the brain" (Tsoukalas and Uhrig 1997). Actually, ANNs are models intended to imitate functions of the human brain using its certain basic structures. ANNs have shown to be effective as computational processors for various associative recall, classification, data compression, combinational problem solving, adaptive control, modeling and forecasting, multisensor data fusion and noise filtering. ANNs have been used in connection to milling in various papers in an effort to predict cutting forces, tool wear and cutting temperatures (Zuperl et al. 2006; Ghosh et al. 2007; Adesta et al. 2012). More specifically, many researchers have made several efforts to predict the surface roughness in milling. Statistical and empirical models have been proposed (Zhang et al. 2007). Furthermore, soft computing techniques are quite common, including ANNs (Kohli and Dixit 2005; Özel and Karpat 2005), genetic algorithms (Oktem et al. 2006) and fuzzy logic (Chen and Savage, 2001).

Back Propagation (BP) ANNs are more common than any other kind of network (Al Hazza and Adesta 2013) to be found in the relevant literature. Although this kind of models may be approached in many ways pertaining to the characteristics of the training procedure used, the problem of selecting the most suitable scheme is not addressed satisfactorily by the researchers. In this paper, an attempt is made to test several training BP algorithms to test their characteristics and suitability for the at hand problem. Feed-forward perceptrons are trained with three different back propagation algorithms, namely the adaptive back propagation algorithm of the steepest descent, the back propagation Levenberg–Marquardt algorithm and the back propagation Bayesian algorithm. This way, several models with different characteristics are built and tested and the optimum is selected. The analysis results indicate that the proposed model can be used to predict surface roughness in end milling with a less that 10 % error, even for tests with cutting conditions that were not used in the training of the system. Furthermore, Radial Basis Function (RBF) neural networks are tested with the same problem and the results are compared to the previous ones. A comparative study indicates the advantages and disadvantages of each approach. Furthermore, the models are used for the prediction of surface roughness in milling. The best models in terms of their performance are stored and can be used for the prediction of surface roughness of random input data, confined of course in the extremes of the input data used in the training of the models, but also 3D surfaces are produced for the a priori determination and selection of optimum cutting conditions, based on experimental results and simulation output. It can be concluded that the proposed novel models prove to be successful, resulting in reliable predictions, therefore providing a possible way to avoid time- and money-consuming experiments.

Artificial neural networks

Artificial neural networks are parallel systems which consist of many special, non-linear processors, known as neurons. Like human cerebrum, they are able to learn from examples, they possess generalization capabilities and fault tolerance and they can respond intelligently to new triggers. Each neuron is a primary processing unit, which receives one or more external inputs and uses them to produce an output. It consists of three basic elements: a number of synapses, an adding node and an activation or transfer function. Each synapse is characterized by a specific weight w_i with which the respective input signal x_i is multiplied. The node adds the resulting numbers and finally the activation function "squashes" the neuron output in the normalized $(0,1)$ or $(-1,1)$ range. Neuron model also includes a threshold θ that practically demotes input to the activation function. The activation function input can be increased if a bias term b is used, which is equal to the negative of the threshold value, i.e. $b = -\theta$.

The whole system is perceived as parallel because many neurons can implement calculations simultaneously. The most important feature of a neural network is the structure of the connected neurons because it determines the way the calculations are performed. Apart from the source layer which receives the inputs and the output layer on which the input layer is mapped, a neural network can have one or more intermediate hidden layers. Furthermore, the types of inter-neuron connections determine the characteristics of a network and consequently the tasks for which it is designed to be used. According to this, in feed forward networks, data run exclusively from the input units to the output units while in recurrent networks, feedback connections act beneficially for the training process and the behaviour of the network. A typical feed forward ANN with one hidden layer consisting of four units, six source units and two output units is shown in Fig. 1.

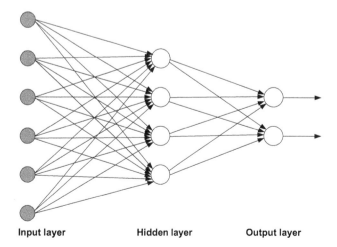

Input layer   **Hidden layer**   **Output layer**

Fig. 1 Single hidden layer feed forward ANN 6-4-2

Finally, the algorithm used for the network training affects its performance and effectiveness. The training of neural networks refers to the procedure adopted to achieve a desired behaviour by modifying the synaptic weights, which allows them to learn from their environment and improve their performance through time. The various methods of adjusting the connections weights constitute different training algorithms. Each algorithm comprises a well-defined set of rules for solving the training problem and has specific advantages and disadvantages. Depending on the environment of the neural network, three different training methods may be distinguished, namely supervised, unsupervised and reinforcement learning. In supervised learning the training is based on examples of desired behaviour. The parameters are adjusted according to the training vector and the error signal between the desired $y_d(t)$ and the calculated values $y(t)$ of the network outputs. The adjustment is implemented with an error correction algorithm like Least Mean Square (LMS). It must be pointed out that typical supervised learning neural networks are the feed forward and the radial basis function networks. On the other hand, in unsupervised learning the weights are modified in response to network inputs only. There are no target outputs available. Most of these algorithms perform clustering operations. They categorize the input patterns into a finite number of classes. This is especially useful in applications such as vector quantization. Finally, reinforcement learning trains the network with the use of a feedback signal called reinforcement signal, which "awards" the right behaviour or "punishes" the wrong behaviour of the network. It can be easily understood that the network receives only a training data set without the respective desired outputs and during the training it tries to find a set of weights which tend to avoid negative reinforcement signals.

In this study, feed forward networks with one or two hidden layers and RBF networks were employed. The output layer had constantly a single neuron corresponding to the predicted value of the surface roughness. The most known neural networks with one or more hidden layers are the multilayer perceptrons (MLP). These networks, unlike simple perceptron, are capable of classifying linearly inseparable patterns and can solve complicated problems. They can handle satisfactorily problems like pattern classification, generalization and functions approximation. They are trained with back propagation algorithms and the transfer function employed is a differentiable sigmoid function like hyperbolic tangent.

Back propagation algorithm

Back propagation algorithm is a supervised learning algorithm which adapts the synaptic weights, aiming to minimize the Mean Squared Error (MSE) between the desired and the actual network outputs after each input vector presentation. Standard back propagation is a gradient descent algorithm in which the network weights are moved along the negative of the gradient of the performance function. The main idea of the algorithm is that the errors of the hidden layers are specified by the back propagation of the output neurons errors. The algorithm includes a forward and a backward phase. During the forward phase the input signals "travel" through the network from input to output, layer by layer, generating eventually a certain response and during the backward phase the error signals are back propagated, from the output layer to the input layer resulting in the adjustment of the network parameters by minimizing the MSE. Therefore, the synaptic weights are adjusted to minimize the criterion:

$$E_p(t) = \frac{1}{2} \sum_k e_k^2(t), e_k(t) = y_{d_k}(t) - y_k(t) \qquad (1)$$

and the summation includes all k neurons of the output layer after the presentation of each training pattern p at the input, consequently the minimization of $E_p(t)$ must be done pattern to pattern.

According to the aforementioned, the steps of the back propagation algorithm are the following:

1. Selection of the initial weights using small, positive values.
2. Presentation of the training vector to the network and forward calculation of the input weighted sums and all the neurons output values, neuron by neuron, up to the output layer where the output vector is produced.
3. Calculation of the difference between the actual output vector and the desired output vector as well as of the necessary weights modification.
4. Calculation of the hidden neurons error and the respective change of weights.
5. Adjustment of all the weights, beginning from the output neurons and continuing backwards to the input layer, by adding the corresponding values calculated before and using the equation $w_{ji}(t+1) = w_{ji}(t) + \gamma \delta_j(t) y_i(t)$, where γ is the learning rate parameter. In previous equations, to achieve faster convergence, a momentum term $a[w_{ji}(t) - w_{ji}(t-1)]$, $0 < \alpha < 1$, is added. The term momentum is used to determine the effect of the previous weight modification to the next one.

It worth noticing that there are two different ways in which the gradient descent algorithm can be implemented: incremental or pattern mode and batch mode. In the incremental mode, the gradient is computed and the weights are updated after each input is applied to the network. In the batch mode all inputs are applied to the network before the weights are updated; in other words this mode is based on the whole set of examples known as epoch. In the incremental mode, BP algorithm does not always converge to a local minimum because the gradient used for the weights adjustment is computed for a single training example. BP algorithm needs many repetitions to converge; however there is always the possibility to get stuck in a local minimum, often due to false weight dimension choice. So, the weights selection is very important for the successful training and usually randomly small, negative values, uniformly distributed in a narrow range are appointed.

Gradient descent and gradient descent with momentum are two methods often too slow for practical problems. In these cases high performance algorithms that can converge from 10 to 100 times faster than the algorithms discussed previously, like Levenberg–Marquardt algorithm, are used. All these algorithms operate in batch mode.

Back propagation Levenberg–Marquardt algorithm

Levenberg–Marquardt algorithm (LMA) provides arithmetical solution to a sum of squares of linear functions minimization problem. The functions depend on a common set of parameters and the algorithm is an alternative solution of the Gauss–Newton algorithm (GNA) and the gradient descent method. LMA is more robust than GNA; it uses the second derivatives of the cost function resulting in better and faster convergence and in many cases it is able to give solution even when it starts far away from the final minimum. However, when the functions are "well behaved" and the initial parameters vary between logical values Levenberg–Marquardt is generally slower than Gauss–Newton algorithm. In gradient descent method only the first derivatives are calculated and the information change parameter includes only the direction in which the cost function is minimized. Therefore many solutions leading to convergence may arise, increasing the possibility the time needed for a solution to be found to become prohibitive.

The algorithm can be described by the following equations:

$$e = d - F(\phi.u) \tag{2}$$

$$J = \frac{1}{2}e^2 \tag{3}$$

$$\Delta\phi = -\left(\nabla^2 J(\phi)\right)^{-1}\nabla J(\phi) \tag{4}$$

where e is the observed output error, d is the target output, u is the system output, F is the fuzzy system function, ϕ is a general parameter of the fuzzy system, J is the cost function, $\nabla^2 J(\phi)$ is the Hessian matrix and $\nabla J(\phi)$ is the gradient relative to the cost function (3).

The observed output error is used for the minimization of the cost function by calculating the value of Eq. (4). The target is the minimization of the instant cost of Eq. (3). If the expansion of the Taylor series to the error $e(\phi)$ is applied around the function point then the first derivatives produce the Jacobian matrix:

$$J_s = \begin{bmatrix} \dfrac{\partial e_1}{\partial \phi_1} & \cdots & \dfrac{\partial e_1}{\partial \phi_B} \\ \cdots & \cdots & \cdots \\ \cdots & \cdots & \cdots \\ \cdots & \cdots & \cdots \\ \dfrac{\partial e_L}{\partial \phi_1} & \cdots & \dfrac{\partial e_L}{\partial \phi_B} \end{bmatrix} \tag{5}$$

where B is the number of the adaptive parameters and L is the outputs number. Finally, the parameters adjustment algorithm is given by:

$$\Delta\phi = N_\phi = -\left(J_s^T J_s + qI\right)^{-1} J_s^T e \tag{6}$$

when q is large, the adjustment method shown above becomes the gradient descent method with step $1/q$ and when q is small it actually becomes Newton's method. Thus, by importing a term of this kind a smooth transition between the methods of gradient descent and Newton–Gauss is achieved and the invertibility problem is eliminated.

The parameters which can be modified, aiming to optimize the training procedure with Levenberg–Marquardt method of each network are the adaptive Marquardt value mu, the mu decrease and mu increase factors, the final target mean squared error, the minimum gradient value and the maximum mu value. The parameter mu is the initial value for q. This value is multiplied by mu decrease value whenever the performance function is reduced by a step. It is multiplied by mu increase whenever a step would increase the performance function. If mu becomes larger than maximum mu, the algorithm is stopped. Generally, training stops when one of the following conditions is fulfilled: the maximum epoch number is reached, the maximum training time is reached, the desired mean squared error is achieved, the gradient value becomes smaller than its minimum value and as mentioned above when the adaptive Marquardt overruns its maximum value.

LMA appears to be the fastest method for training moderate-sized feed forward neural networks. It also has a very efficient Matlab implementation, since the solution of the matrix equation is a built-in function; thus its attributes become even more pronounced in a Matlab setting.

Back propagation Bayesian algorithm

Back propagation Bayesian algorithm updates the weight and bias values according to Levenberg–Marquardt optimization. Bayesian regularization minimizes initially a linear combination of squared errors and weights and then determines the correct combination so as to produce a network that generalizes well. The parameters which can be modified to optimize each network's training procedure with the Bayesian method are the adaptive Marquardt mu, the mu decrease and mu increase factors, the final target mean squared error, the minimum gradient value and the maximum mu value. In Bayesian regularization back propagation is used to calculate the Jacobian jX of the network performance with respect to the weight and bias values X. Each variable is adjusted according to LMA as shown below:

$$jj = jX \cdot jX \tag{7}$$

$$je = jX \cdot E \tag{8}$$

$$dX = \frac{-(jj + I \cdot mu)}{je} \tag{9}$$

where E is the errors sum and I is the identity matrix. The adaptive value mu is increased by mu increase value until the change of X results in a reduced performance value. The change is then made to the network and mu is decreased by mu decrease value. Training stops whenever one of the conditions mentioned in Levenberg–Marquardt method is fulfilled.

Radial basis function neural networks

Radial basis function neural networks are trained with supervised learning algorithms and can be perceived as improvements of the multilayer feed forward back propagation networks. Many of their characteristic features are similar to those of feed forward neural networks because they perform linear representations and weights summations. However, the transformations performed are local, resulting in their much faster training. Radial basis networks may require more neurons than standard feed forward networks, but often they can be designed to take a fraction of time it takes to train standard feed forward networks. It is proven that RBFs with one hidden layer can approximate any function; as a result they are called universal approximators. Perceptrons also have the capability of universal approximation but only RBFs possess the ability of optimum approximation. Even if structurally they are less complicated than feed forward back propagation networks they can achieve better arbitrary functions approximations with only one hidden layer. Also, it has been observed that they work best when many training vectors are available. When for every input which must be classified there is a basis function $\Phi(\|x - y\|)$ the network will give a function, adaptive to every pattern.

The basic structure of an RBF neural network includes an n dimension input layer, a fairly larger dimension m hidden layer ($m > n$) and the output layer. Typical radial basis functions are Gaussian and logistic function, which are shown below:

$$R_t(\mathbf{x}) = \Phi(\|\mathbf{x} - \mathbf{c_i}\|) = \exp\left[\frac{-\|\mathbf{x} - \mathbf{c_i}\|^2}{2\sigma_i^2}\right] \tag{10}$$

$$R_t(\mathbf{x}) = \Phi(\|\mathbf{x} - \mathbf{c_i}\|) = \frac{1}{1 + \exp\left[\frac{-\|\mathbf{x}-\mathbf{c_i}\|^2}{\sigma_i^2}\right]} \tag{11}$$

where c_i is the i centre of $R_t(\mathbf{x})$ function. The right placement of the centres c_i is very important for the achievement of great learning rates. The σ_i parameter is called amplitude and determines the value of the maximum distance between two inputs that the node has a prominent impact. The radial basis functions selection is not decisive for the network efficiency.

The typical structure of an RBF network is shown in Fig. 2. Input units distribute the values to the hidden layer units uniformly, without multiplying them with weights. Hidden units are known as RBF units because their transfer function is a monotonous radial basis function. Hidden layer outputs are led to the output units and summed with appropriate weights.

For training, the number of RBF units is primarily selected, which is very important for the success of the procedure, usually by a "trial and error" method. Afterwards, four steps are followed:

1. Input training data are grouped and the centres c_i, $i = 1, 2, ..., M$ of these groups are defined as centres of the M hidden units. Afterwards, k-centres algorithm is implemented and as a result a certain number of clustering centres is considered and a separation of the whole set of patterns into subsets is performed.

2. The amplitudes σ_i are defined, usually with the p-closest neighbour method which alters the values

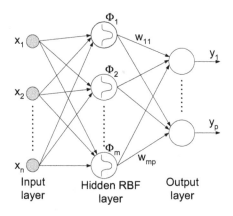

Fig. 2 Basic structure of an RBF neural network

to achieve overlapping of the response of every hidden and its neighbouring unit. The function used is:

$$\sigma_i = \left\{ \frac{1}{p} \sum_{j=1}^{p} \left\| x_i - x_j \right\|^2 \right\}^{1/2} \tag{12}$$

where x_j is the p-closest neighbour of x_i.

3. Transfer functions using Eqs. (10) and (11) are computed for the cases of Gaussian and Logistic function respectively.

4. Weights vector $w = [w_1, w_2,..., w_m]^T$ of the output units is computed, utilizing the minimum squares method:

$$J = \frac{1}{2} \sum_{k=1}^{p} \left\| y_k - a_{ki}^T w \right\|^2 = \frac{1}{2} (y - \mathbf{A}w)^T (y - \mathbf{A}w) \tag{13}$$

where $y = [y_1, y_2,..., y_p]^T$ is the output training vector, $\alpha_{\kappa l}$ is the transfer function vector and $\mathbf{A}$ is the hidden units' activation matrix.

Table 1 Training set data

No of Training data	Speed (min^{-1})	Feed (mm/min)	Depth of cut (mm)	Vibrations (μV)	Surface roughness (μm)
1	1500	152.4	1.27	0.10168	1.4224
2	1500	457.2	0.254	0.13581	3.048
3	1500	609.6	0.762	0.19091	2.6162
4	1500	304.8	0.254	0.11231	2.2352
5	1250	304.8	0.254	0.1448	2.54
6	1250	609.6	1.27	0.18291	3.0734
7	1250	152.4	1.27	0.096899	1.8034
8	1000	609.6	1.27	0.18417	3.6068
9	1000	152.4	0.762	0.10976	1.9812
10	1000	304.8	1.27	0.18001	2.3368
11	1000	457.2	0.762	0.16149	3.1496
12	750	457.2	0.762	0.14068	3.7338
13	750	304.8	0.762	0.12654	2.5908
14	750	152.4	1.27	0.089752	1.8288
15	750	609.6	0.762	0.17928	4.3434
16	1500	228.6	0.254	0.08833	1.3462
17	1250	228.6	0.762	0.13814	2.0828
18	1000	533.4	0.254	0.10338	3.7846
19	750	228.6	0.254	0.093096	2.7686
20	750	533.4	0.254	0.11352	4.5212
21	750	533.4	1.27	0.16586	3.81

Table 2 Test set data

No of Test data	Speed (min^{-1})	Feed (mm/min)	Depth of cut (mm)	Vibrations (μV)	Surface roughness (μm)
1	1500	609.6	1.27	0.17874	2.794
2	1250	457.2	0.254	0.14558	2.921
3	1250	381	0.254	0.13378	2.7178
4	1000	533.4	0.762	0.16794	3.683
5	1500	381	0.254	0.14637	2.794
6	1250	533.4	0.254	0.13001	3.2766
7	1000	228.6	0.254	0.091113	2.3368
8	1000	381	0.762	0.14862	2.7432
9	750	533.4	0.762	0.16241	4.1402
10	750	381	1.27	0.15298	2.6416

Networks' modelling

The aim of the study is to test various training algorithms in both BP and RBF networks for the purpose of predicting the surface roughness in milling. Then the results are compared and useful conclusions are drawn. In all the proposed models the same set of input and output data is used. For the application of the method, experimental results from the relevant literature were exploited (Lou 1997). The experiments pertain to the CNC end milling 6061 aluminium blocks. The tool used was a four-flute 3/4 inch diameter milling cutter of HSS. During the machining an accelerometer sensor was used to measure tool vibrations. Spindle speed, feed rate, depth of cut and vibrations were selected as independent variables in this study. Vibrations depend partly on the other three independent

variables and thus they could be treated as a dependent variable. However, due to the complex structural system consisting of workpiece, fixture, cutting tool and machine tool the vibrations and consequently the surface roughness cannot be described quite accurately by the limited set of independent variables. Therefore, vibrations are treated as an independent variable, as well.

Two sets of experimental data were obtained: training data set and testing data set. The training data set was obtained on the basis of four levels of spindle speed (750, 1000, 1250, 1500 rpm), six levels of feed rate (152.4, 228.6, 304.8, 457.2, 533.4, 609.6 mm/min) and three levels of depth of cut (0.254, 0.762, 1.27 mm). For each combination of spindle speed, feed rate and depth of cut, the corresponding vibration data (in μV) were recorded. The corresponding value of the roughness average R_α (in μm), the dependent output, was collected for each measurement. The training data used for the analysis are presented in Table 1.

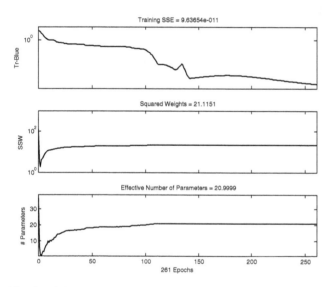

Fig. 3 MSE during training

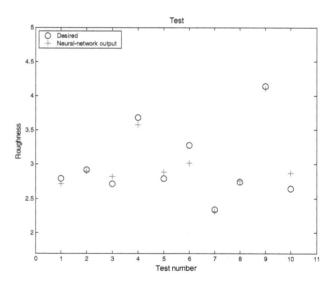

Fig. 5 Experimental values of surface roughness versus the NN predicted ones

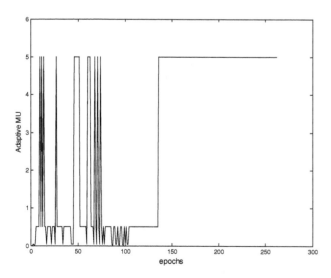

Fig. 4 Adaptive mu during training

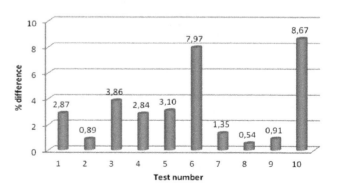

Fig. 6 The discrepancy of the predicted value from the experimental in percentage, for test data

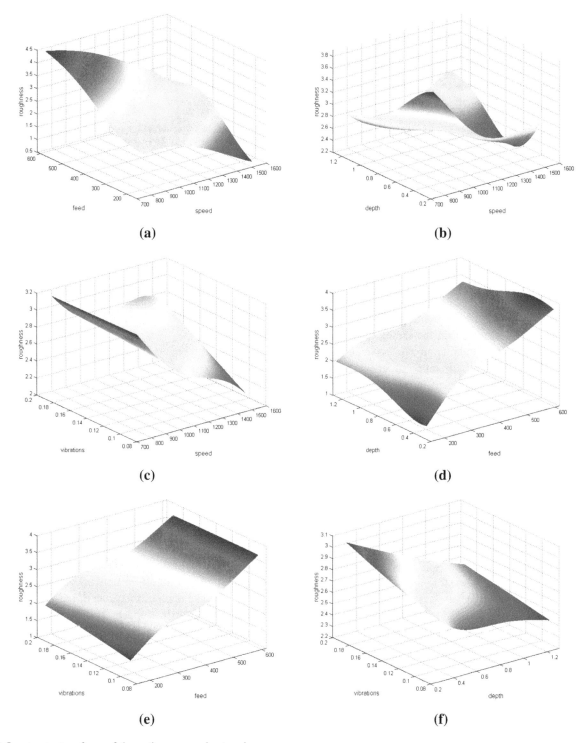

Fig. 7 Input-output surfaces of the optimum neural network

In this work, training data comprised 21 measurements selected randomly out of the 400 measurements originally presented by Lou (1997). The test data set was obtained on the basis of four levels of spindle speed (750, 1000, 1250, 1500 rpm), seven levels of feed rate (152.4, 228.6, 304.8, 381, 457.2, 533.4, 609.6 mm/min) and three levels of depth of cut (0.254, 0.762, 1.27 mm). Also for the test data set the data on vibrations and surface were recorded. The test data set comprised 10 measurements that are shown in Table 2. Note that in the test data set a value for the feed rate, namely 381 mm/min that has not been used in the training data set was also considered. This was chosen to check whether the constructed system could predict correctly the value of the surface roughness when it has as input, values that it has not been trained for. The aim of this work was to create a system that could predict the surface roughness quite accurately; it is quantified as a small value of Mean Squared Error of training and test data, respectively.

Results and discussion

BP networks

For BP networks three different network training methods are employed, namely the adaptive back propagation algorithm of the steepest descent with the use of momentum term, the back propagation Levenberg–Marquardt algorithm and the back propagation Bayesian algorithm. In the proposed models, regardless of the training algorithm used, some modelling parameters remain the same. More specifically, it is set that the final training error to be smaller than 1×10^{-10} and the maximum number of epochs during which the network could be trained to be equal to 15000. However, the training process stops whenever the desired MSE is achieved or the designated epoch number is reached or finally when one of the parameters of each training algorithm reaches the maximum or minimum value.

It must be noticed that at the input layer all the values need to be of the same size so as to achieve faster and better convergence. Thus, modified values of the experimental data were used; depth of cut and vibration data are multiplied by 5 and 10, respectively and the spindle speed and feed rate data are divided by 500 and 100, respectively. All neural networks used have a four neurons input layer, because there are four types of input data, i.e. spindle

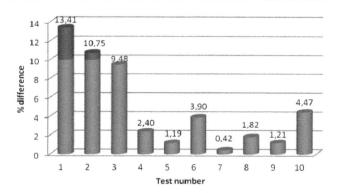

Fig. 9 The discrepancy of the predicted value from the experimental in percentage, for test data

speed, feed rate, depth of cut and vibrations, one or two hidden layers with variable number of neurons and a single neuron output layer for the surface roughness which is the system output. The hidden layers use the hyperbolic tangent sigmoid transfer function, which is expressed as:

$$f(u) = \tanh\left(\frac{u}{2}\right) = \frac{1 - e^{-u}}{1 + e^{-u}} \tag{14}$$

After training, the best neural network that managed to be fully trained and produced the smallest test mean squared error was the 4-6-1-1 network, meaning with 2 hidden layers with 6 and 1 neurons in the first and second hidden layer, respectively, trained with the back propagation Bayesian algorithm. For this specific network, it took 262 epochs to terminate the training procedure, achieving MSE of training data equal to 4.59×10^{-12}, while the respective MSE of test data was 0.01599 and all test data produced error smaller than 10 %. Obviously, the values of both the mean squared errors of training and test data are significantly small. Also, this network produced the smaller value of mean squared error of all the networks that were trained.

In Fig. 3 the alteration of the value of mean squared error of training data versus the epochs is depicted and in Fig. 4 the alteration of the value of adaptive training factor *mu* versus the epochs is plotted. In Fig. 5 the experimental values of surface roughness and the corresponding calculated values of surface roughness by the neural network for the test data are shown. Finally, Fig. 6 shows the difference of the computed to the experimental surface roughness values for the test data, indicating in most cases a very good prediction.

In Fig. 7a, b, c, d, f, the total surfaces which describe the input–output space, produced when only two of the input variables are altered each time, are shown. The input vector used was (speed = 1125 rpm, feed = 381 mm/min, depth of cut = 0.762 mm and vibrations = 0.1392 μV), therefore, in each figure the variables that are not mentioned take their pre-assigned values from the vector above. The two input variables that are altered each time take all the possible values between their widths of rate.

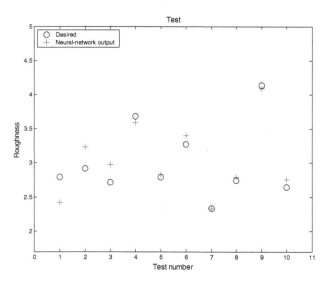

Fig. 8 Surface roughness for test data

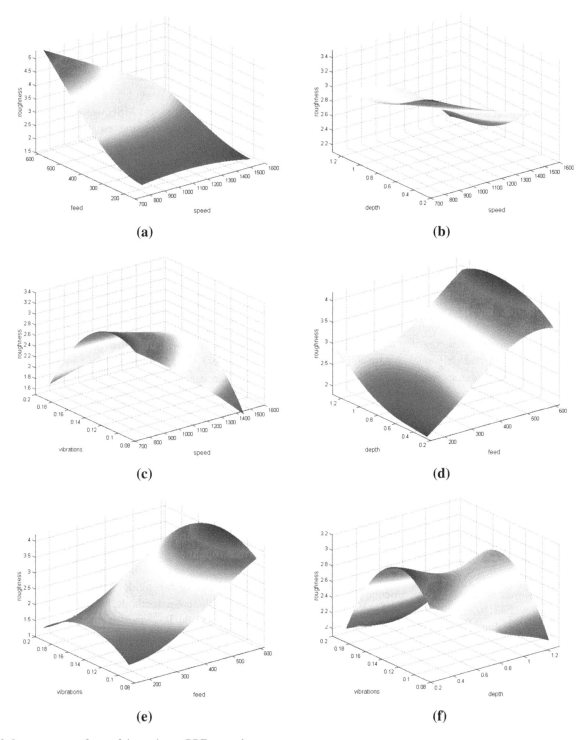

Fig. 10 Input-output surfaces of the optimum RBF network

It is worth noticing that generally neural networks are not stable; each time a network is trained, the initial weights as well as the initial bias values are chosen randomly from the programme. This random selection strongly affects the training procedure and the final error of the network. The same network can achieve complete training for a specific set of initial weights and biases and afterwards can fail to be trained for another set of weight and bias values. For this reason, for each network examined, the network ran 15 times and then the most possible value one could take with only one "running" of the network was chosen.

Table 3 Comparison of optimum BP and RBF networks

Network type	Structure	MSE training	MSE Test	Time (s)	Test data > 10 %
Feed forward	4-6-1-1	4.59×10^{-12}	0.01599	8.041	0
Radial	4-21-1	8.85×10^{-20}	0.034973	0.31	2

RBF networks

In the second part of this work radial basis function networks are examined, which could be trained and then predict surface roughness. For RBFs, the target training mean squared error was chosen equal to 10^{-30}. Once again, all values at the input layer need to be of the same size, trying to achieve faster and better convergence. Thus, the modified values of the experimental data, as used in BP networks, were employed. Furthermore, in the proposed RBFs, the value of spread constant needs to be determined; an inappropriate spread constant could cause RBF overfitting or underfitting. For this purpose an algorithm is developed that can test all the resultant networks for spread values from 0.1 to 100, with changing step equal to 0.1. All RBF networks produced had only one hidden layer. The number of hidden layer neurons was modified from the program with respect to the spread value, so as the target mean squared error to be achieved. Similarly to BP networks, the networks presented in this paragraph have four input neurons, one for each input variable and one output neuron, corresponding to surface roughness. Hidden layer neurons used the radial basis activation function and the output neuron used the linear transfer function.

For every network, the results obtained were the training MSE, the test MSE and the training time. By comparing all the results it was concluded that the optimum network was the one with spread value equal to 15. This network produces the smallest test data MSE and only two of its test data produce difference between experimental and predicted data greater than 10 %; the one was hardly greater than the target value, thus substantially only for one data point there was high percentage error.

Figure 8 depicts the experimental values of surface roughness and the corresponding calculated values of surface roughness by the RBF network for the test data used. Figure 9 shows the discrepancies of the predicted values from the experimental ones in percentage, for the test data.

In Fig. 10a, b, c, d, f, the total input–output surfaces which are produced when only two of the input variables are altered each time, are shown. The input vector used was (speed = 1125 rpm, feed = 381 mm/min, depth = 0.762 mm, vibrations = 0.139 μV) therefore, in each figure the variables that are not mentioned take their pre-assigned values from the vector above.

From all the results taken it is concluded that training times for all networks were very small. Furthermore, as

spread value was rising, test MSE was decreasing up to the moment spread value became equal to 15. For farther increase of spread value the error was increasing, too. Finally, RBF networks are absolutely stable; when the values of their parameters are kept constant they give the same results, no matter how many times they are trained.

In general, test data which produced most frequently error greater than 10 % were these placed in rows 2, 7 and 10 in test data table. Test data 10 contains a feed rate value (381 mm/min) for which it had not been trained. For test data 7, the corresponding feed rate (2228.6 mm/min) was used during training but not combined with the spindle speed value of 1000 rpm. Test data 2 consists of a combination of spindle speed (1250 rpm) and feed rate (457.2 mm/min) for which the network had not been trained. Table 3 tabulates some parameters connected to the performance of the finally selected BP and RBF networks, for comparison.

Conclusions

In this work, various training algorithms for BP networks and RBF networks were put to test for the prediction of surface roughness in end-milling. Four independent variables were used as inputs, namely spindle speed, feed rate, depth of cut and vibrations and the output of the networks was surface roughness. The first system was a feed forward network, which was examined for various numbers of neurons, with one or two hidden layers and for the training process three different methods, steepest descent, Levenberg–Marquardt and Bayesian, were used, with the latter giving the best results. The second system under examination was a radial basis network. Both systems can exhibit advantages and disadvantages when compared to one another and in both cases the results were quite satisfying.

Certain remarks concerning these two approaches are the following:

- RBFs can be trained much faster than perceptrons.
- Test data MSEs of perceptrons trained with the Bayesian method were generally smaller than the errors resulting from RBFs.
- The smallest training error was achieved with radial basis networks.
- Radial basis networks were very stable. Also, the networks trained with the Bayesian algorithm were generally stable contrary to the networks trained with

the Levenberg–Marquardt and steepest descent algorithms.

It can be finally concluded that artificial neural networks can satisfactorily predict surface roughness in milling.

References

Adesta EYT, Al Hazza MHF, Suprianto MY, Riza M (2012) Prediction of cutting temperatures by using back propagation neural network modeling when cutting hardened H-13 steel in CNC end milling. Adv Mater Res 576:91–94

Al Hazza MHF, Adesta EYT (2013). Investigation of the effect of cutting speed on the surface roughness parameters in CNC end milling using artificial neural network. IOP Conf. Series: Materials Science and Engineering 53. doi:10.1088/1757-899X/53/1/012089

Al Hazza MHF, Ndaliman MB, Hasan MH, Ali MY, Khan AA (2013) Modeling the electrical parameters in EDM process of Ti6Al4 V alloy using neural network method. Int Rev Mech Eng 7(7):1464–1470

Chen JC, Savage M (2001) A fuzzy net based multilevel in process surface roughness recognition system in milling operations. Int J Adv Manuf Technol 17:670–676

Ezugwu EO, Fadare DA, Bonney J, Da Silva RB, Sales WF (2005) Modelling the correlation between cutting and process parameters in high-speed machining of Inconel 718 alloy using an artificial neural network. Int J Mach Tools Manuf 45(12–13):1375–1385

Farsani R, Raissi S, Shokuhfar A, Sedghi A (2007) Optimization of carbon fibers made up of commercial polyacrylonitrile fibers using screening design method. Mater Sci Pol 25(1):113–120

Galanis NI, Manolakos DE (2014) Finite element analysis of the cutting forces in turning of femoral heads from AISI 316 l stainless steel. Lect Notes Eng Comp Sci 2:1232–1237

Ghasemi FA, Raissi S, Malekzadehfard K (2013) Analytical and mathematical modeling and optimization of fiber metal laminates (FMLs) subjected to low-velocity impact via combined response surface regression and Zero-One programming. Lat Am J Solids Struct 10(2):391–408

Ghosh N, Ravi YB, Patra A, Mukhopadhyay S, Paul S, Mohanty AR, Chattopadhyay AB (2007) Estimation of tool wear during CNC milling using neural network-based sensor fusion. Mech Syst Signal Process 21(1):466–479

Kohli A, Dixit US (2005) A neural network based methodology for the prediction of surface roughness in a turning process. Int J Adv Manuf Technol 25(1–2):118–129

Lou S (1997) Development of four in-process surface recognition systems to predict surface roughness in end milling. PhD Thesis, Iowa State University, Iowa, USA

Markopoulos A, Vaxevanidis NM, Petropoulos G, Manolakos DE (2006) Artificial neural networks modeling of surface finish in electro-discharge machining of tool steels. In: Proceedings of ESDA 2006, 8th Biennial ASME Conference on Engineering Systems Design and Analysis, 847–854

Niesłony P, Grzesik W, Chudy R, Habrat W (2015) Meshing strategies in FEM simulation of the machining process. Arch Civil Mech Eng 15(1):62–70

Oktem H, Erzurumlu T, Erzincanli F (2006) Prediction of minimum surface roughness in end milling mold parts using neural network and genetic algorithm. Mater Des 27(9):735–744

Özel T, Karpat Y (2005) Predictive modeling of surface roughness and tool wear in hand turning using regression and neural networks. Int J Mach Tools Manuf 45(4–5):467–479

Raissi S, Eslami FR, Shokuhfar A, Sedghi A (2004) Improving carbon fibers quality characteristic by using statistical modeling. Int J Appl Math Stat 2(4):60–72

Saeidi RG, Amin GR, Raissi S, Gattoufi S (2013) An efficient DEA method for ranking woven fabric defects in textile manufacturing. Int J Adv Manuf Technol 68(1–4):349–354

Shokuhfar A, Khalili SMR, Ashenai Ghasemi F, Malekzadeh K, Raissi S (2008) Analysis and optimization of smart hybrid composite plates subjected to low-velocity impact using the response surface methodology (RSM). Thin Walled Struct 46:1204–1212

Szabó G, Kundrák J (2014) Investigation of residual stresses in case of hard turning of case hardened 16MnCr5 Steel. Key Eng Mater 581:501–504

Tsoukalas LH, Uhrig RE (1997) Fuzzy and neural approaches in engineering. Wiley Interscience, New York

Zhang JZ, Chen JC, Kirby ED (2007) Surface roughness optimization in an end-milling operation using the Taguchi design method. J Mater Process Technol 184(1–3):233–239

Zuperl U, Cus F, Mursec B, Ploj T (2006) A generalized neural network model of ball-end milling force system. J Mater Process Technol 175(1–3):98–108

10

Combination of real options and game-theoretic approach in investment analysis

Abdollah Arasteh[1]

Abstract Investments in technology create a large amount of capital investments by major companies. Assessing such investment projects is identified as critical to the efficient assignment of resources. Viewing investment projects as real options, this paper expands a method for assessing technology investment decisions in the linkage existence of uncertainty and competition. It combines the game-theoretic models of strategic market interactions with a real options approach. Several key characteristics underlie the model. First, our study shows how investment strategies rely on competitive interactions. Under the force of competition, firms hurry to exercise their options early. The resulting "hurry equilibrium" destroys the option value of waiting and involves violent investment behavior. Second, we get best investment policies and critical investment entrances. This suggests that integrating will be unavoidable in some information product markets. The model creates some new intuitions into the forces that shape market behavior as noticed in the information technology industry. It can be used to specify best investment policies for technology innovations and adoptions, multi-stage R&D, and investment projects in information technology.

Keywords Investment analysis · Real options · Game theory · Information technology

✉ Abdollah Arasteh
arasteh@nit.ac.ir

[1] Industrial Engineering Department, Babol Noshirvani University of Technology, Shariati Av., P.O. Box: 484, Babol, Mazandaran, Iran

Introduction

Investments in information technologies

Historically, creating infrastructure needed huge investment. In the change from an industrial economy to an information-based one, companies today invest huge quantities of resources in new information technologies (IT) and connected infrastructures. In the information era, the necessary assets for business success are no longer factories, but knowledge assets and the allowing technological infrastructures (Albuquerque and Miao 2014; Berghman et al. 2012).

From a single firm's viewpoint, an early investment in IT infrastructure may result in getting a "power" that would let the firm take better advantage of future growth opportunities. This is mainly important for information thorough firms, where a firm's information infrastructure gets progressively essential to its ability to apply new business strategies. An ordinary benefit of IT investments is the ability to engage in the product markets at lower incremental cost or better customer attraction. Particularly, a firm that has already made such IT-increasing infrastructure investments might launch new business strategies that create or support competitive benefits at lower cost compared with other firms that have not made similar investments (Alexandrov and Deb 2012; Amram and Kulatilaka 1999; Gao et al. 2013).

The difficulty in evaluating IT investments

Proof has shown that businesses have problems in assessing investment decisions in IT field. Part of these problems were presented as the "productivity paradox" (Brynjolfsson 1993; Dewan and Kraemer 1998). So far, the assessment problem is basic to the continuous innovation and application of IT in business. To assess IT investment is hard because available assessment methods have not developed at the same speed as the needs of present practice. IT investments provide firms with growth chances to change to new business events, or get business growth through the exercise of IT-based strategies. Besides, strategic IT investments often affect the behavior of participants. From these causes, assessment of an investment project relied on potential competitive effect is different from assessment relied on cash flows (Dimitrios et al. 2013; Merali et al. 2012).

Assessing IT investment projects creates a few problems that investing in the traditional assets does not introduce. The emphasis shifts from calculating the cash flows to assessing strategic effects that IT investments give: the value of real-time information, managerial flexibility, the ability to answer to unpredicted moves by opponents, and an improved information infrastructure that may have a long-term suggestion for the competitiveness of the firm. To express the value of the decision flexibility set in technology investments, we need searching for new methods to assess technology investment projects (Berghout and Tan 2013; Khallaf 2012).

Technology investments as real options

Investment projects in IT can be analyzed as sets of real options: a firm with an opportunity to invest in a technology is taking an option similar to a financial call option—it has the right, but not the need, to get the asset at some future time. So, making an investment is similar to exercising a call option with an exercise price proper for the investment expenses, and the underlying asset is the new technology. From a real options viewpoint, also, IT investment is about real options. Real options can be either "simple" options or combined options. A "simple" option is almost like a call option where the exercise of the option guides to the gain of the underlying asset. In combined options, the exercise of one option guides to the gain of another option. Most of the sequential investments can be analyzed as combined options in the feeling the investment in one period gives the firm the option to continue to the next period. Today's investments may have characteristics that will allow a firm to exercise a particular strategy in the future (Fernandes et al. 2013; McIntyre and Chintakananda 2013; Rohlfs and Madlener 2013).

As discussed earlier, investing in growth options rather than cash flows is one of the key characteristics of technology investment. Many multiperiod strategic investments have a negative NPV when analyzed without relation to

others, even though they may have significant growth option value. The NPV and options valuation methods may give different results.

The likeness between financial and real options gives the potential the options-pricing theory could be expanded to assessing investment decisions on technological assets? Nonetheless studies of real-options based method for IT investments are still rare; the literature appears in this area, specifying the increasing attention paid to real options.

Its benefit over other capital budgeting methods like DCF analysis has been broadly identified in considering the strategic investment decision under uncertainties (Amram and Kulatilaka 1999; Luehrman 1998a, b). Smith and McCardle (1998, 1999) moreover show that option pricing can be combined with a standard decision analysis framework to get the best of the both worlds. Some previous IS researches have identified many IT investment projects hold some option—like characteristics (Clemons; Dos Santos 1991; Kumar 1996). Benaroch and Kauffman (1999) and Taudes et al. (2000) have applied the real options theory to real-world business cases and assessed this approach's benefits as a tool for IT investment planning. Kim and Sanders (2002) expand a framework of strategic actions relied on real option theory. Some researchers use real options combined with game theory to analyze strategic technology adoption. For example Huisman and Kort (2004) find out a dynamic duopoly in which firms take part in the adoption of new technologies. Smit and Trigeorgis (2006) illustrate the use of real options valuation and game theory concepts to consider original investment opportunities including important strategic decisions under uncertainty. It uses innovation cases, unions and gains to discuss strategic and competitive features, applicable in industries like consumer electronics and telecom. Wu and Ong (2008) in an interesting paper used real options analysis in association with classical financial theory, specifically, the Mean–Variance (MV) model to give new viewpoints on project selection. Pendharkar (2010) used the market asset disclaimer supposition and expand a binomial lattice based real options model to involve cash flow interrelations between multiperiod IT investments. Wu et al. (2012) use a combination of real options and game theory to consider the investment strategies of a case company in the TFT-LCD industry. Martzoukos and Zacharias (2013) demonstrate to decision makers how to optimally make costly strategic pre-investment R&D decisions in the existence of full results in an option pricing structure with logical tractability. van Zee and Spinler (2014) illustrates a real options method for valuing public-sector research and development projects, using a down-and-out barrier option.

Competition in technology investment

Viewing an IT investment project as a real option puts greater importance on the possibilities and benefits of postponing investment to wait for more information to resolve uncertainty. Although, investment opportunities for new technologies are scarcely dedicated, as thorough competition and low barriers of entrance are distinguishing features of the IT industry. Competition over limited investment opportunities may decrease a firm's option value, or still force the option to run out too early. Therefore, the timing of the investment decision could have notable results on the recognized value of the project (Bos et al. 2013; Chaton and Guillerminet 2013).

When to exercise a real option is a strategic decision. If the option is non-dedicated, the firm and its competitors hold an option on the identical asset, and whoever exercises first may get the fundamental asset. The problem is that provided uncertainty stays on the market or technology, no one can be certain that they want the asset. The problem is naturally that no firm knows what condition the game will be in at future times. Also, in a market of incomplete competition, one firm's decision could change the market price and structure. This may have extra strategic impacts on competitor's behavior.

Another typical supposition made in the most of the literature is that information is symmetric that means each firm has complete information about the other's profit structures and that they split similar opinions about future market demands. That is, market demand may continue stochastically, but this is supposed to be public information. So firms are critically supposed to be consistently informed, and no personal and incomplete information is included. Although, in the real world, competition often happens in a situation of information asymmetry. That is, companies have incomplete and asymmetrically assigned information on boundaries like development costs or market demands (Lestage et al. 2013; Wang et al. 2012; Wrzaczek and Kort 2012).

In this paper analyzing the common results of both uncertainty and competition, this paper expands a method to assess investment decisions in an oligopolistic structure. It combines the real options structure with strategic limits of game theory, and prepares a greater comprehension of the results of uncertainty and competition on the strategic exercise of real options inserted in technology investments.

Major contributions of the paper

The majority of real options researches, has concentrated on business sector situations without strategic collaborations. The extensive greater part of the models of capital speculations, which utilize a real options methodology, has regularly been founded on two particular suspicions: (a) the firm has an imposing business model control over a venture opportunity; and (b) the item market is consummately aggressive. Accordingly, speculation not influence either costs or business sector structure. Strategic issues have rather fallen in the space of modern association. The strategic methodology in modern association writing endogenizes business sector structure; be that as it may, it regularly disregards instability and along these lines the option value of adaptability.

Considering the joint impacts of both uncertainty and competition, this paper adds to a system to assess speculation choices in an oligopolistic business sector structure. The technique separates itself by demonstrating options exercise under endogenous, multi-period competition in the setting of innovation speculation. Through building up a balance model of a dynamic venture diversion, the paper makes a few particular commitments:

First, our study amplifies the ordinary single-specialists improvement models to a game-theoretic setting that consolidates numerous, contending firms. Under the weight of aggressive acquisition, firms race to practice their alternatives early. This significantly dissolves the option value of holding up and changes the key conduct of capital speculation. The model's consequences help clarify some forceful venture examples saw in the IT industry. The second contribution is identified with the assessment of real options when the suspicions for monetary option valuing hypothesis no more hold. On the off chance that the future settlements and the dangers of an innovation speculation undertaking can be reproduced by exchanged resources, the valuation of real option is the clear utilization of the money related option valuation models. Be that as it may, difficulty arises in most of the real asset investment projects. We propose a technique taking into account diversion hypothesis and dynamic programming to assess speculation projects when suspicions for financial option pricing hypothesis don't hold. Third, we derive optimal venture approaches and basic speculation limits. The model is further stretched out to the multi-period setting. Dynamic programming is utilized to handle between worldly speculation choices. Forth, our work amplifies the full-data models in the researches to a more practical uneven data connection.

In summarize, the differences between our present model and the real options models in the literature are:

- Our model includes strategic interactions and competitive risks;
- These strategic elements influence the equilibrium and investment behavior (such as early exercise and violent investment);
- The option value is lower but more realistic than that examined in the real options literature;
- Our model analyzes asymmetric information.
- Our model identifies the option value of waiting to better resolve uncertainty;
- This option value is not only conceptualized but also quantified in the present model;
- We continue the consideration to multi-period setting with the linkage existence of continuous uncertainty and competition through a method called "dynamic programming with externalities";
- We endogenize the timing and the leader–follower sequence of investment.

The impact of competition on investment: a simple model

In this section, we use a simple model to show the basic ideas of investment under competition. First, let us explain the particular problem. Assume firm X encounters a decision whether to take on an investment project. For a cost of C, firm X can commercialize a new technology and begin a product into the market. The market demand is uncertain. To remain it simple, suppose the market demand could be "low" with probability p or "high" with probability $1 - p$. The net edge of purchasing one unit of the product is s dollars. Suppose the firm uses a discount rate of i. The project life is n years; technology grows outdated next. For clarity, suppose that if the firm waits a year it will be able to analyze uncertainty. If we follow the traditional DCF analysis, we may calculate NPV as follows:

$$
\text{NPV (invest now)} = -C + p \left[\sum_{t=0}^{n-1} \frac{1}{(1+i)^t} D_L s \right]
$$
$$
+ (1-p) \left[\sum_{t=0}^{n-1} \frac{1}{(1+i)^t} D_H s \right] \quad (1)
$$

The NPV rule would tell us to drop the project if NPV < 0, invest if NPV > 0, and be unconcerned if NPV = 0.

The option to defer investment

A key defect of the above static analysis is that it ignores the option to "wait and see". The investment opportunity is

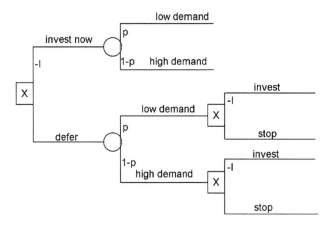

Fig. 1 The option to defer investment

not a "now-or-never" selection. The firm could wait to get more information about the market uncertainty. Figure 1 shows this "wait and see" option.

The NPV of this "wait and see" option is

$$
\text{NPV (wait)} = \frac{p}{1+i} \left\{ \max \left[0, -C + \left(\sum_{t=0}^{n-2} \frac{1}{(1+i)^t} \right) D_L s \right] \right\}
$$
$$
+ \frac{1-p}{1+i} \left\{ \max \left[0, -C + \left(\sum_{t=0}^{n-2} \frac{1}{(1+i)^t} \right) D_H s \right] \right\}
$$
$$
(2)
$$

The difference between these values in (1) and (2) show the value of the "wait and see" option. Therefore,

$$
O = |\text{NPV}_{\text{wait}} - \text{NPV}_{\text{invest now}}| \quad (3)
$$

The options-based analysis expresses the value of waiting, which is a development over the static NPV method. Although, it is right only if the investment opportunity stays obtainable for the firm during the period of waiting (Meyer and Rees 2012). This is to suppose the investment opportunity is dedicated, i.e., only one firm has the ability to go into the market. So, the above analysis may have overestimated the option value of waiting as it ignores the risk of competitive entrance.

Investment opportunity under competition

To cure the above consideration, our model would have to include the risk of competitive entrance. Assume market research leads us to think that with probability q one of the participating firms will enter the market during the first period. Figure 2 shows a changed model that involves competition.

From the real options' viewpoint, the option may be pressure to expire rashly because of competitive entrance. This alters the NPV of the "wait and see" alternative to:

Fig. 2 Investment option under competition

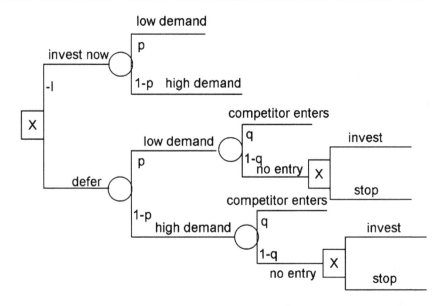

NPV (wait, w/competition)

$$= \frac{p(1-q)}{1+i} \max\left[0, -C + \left(\sum_{t=0}^{n-2} \frac{1}{(1+i)^t}\right) D_L s\right]$$

$$+ \frac{(1-p)(1-q)}{1+i} \max\left[0, -C + \left(\sum_{t=0}^{n-2} \frac{1}{(1+i)^t}\right) D_H s\right]$$

$$= (1-q)\text{NPV (wait, no competition)}$$

$$(4)$$

Similar to the description in (3), the option value is the distinction between the two NPV values in (1) and (4), i.e.,

$$O_{\text{with competition}} = \text{NPV}_{\text{wait. with competition}} - \text{NPV}_{\text{invest now}}$$

$$(5)$$

It is easy to prove that (4) is smaller than (2), as long as $q > 0$. That is to say, the option value of waiting with competition is smaller than that without competition, as long as there is a positive probability of competitive entrance. By analyzing the result of competition on the real options value, we could say that competition abrade the option value of waiting.

An example

To put the above examination in the connection of a particular illustration, we accept the numbers are: $C = \$7500$, $p = 0.75$, $i = 15\%$, $s = \$12$, $D_L = D$, $D_H = 2D$. Suppose D is known and $D = 97.5$. Subsequent to connecting to these numbers, Eqs. (1) and (2) yield NPV (invest now) = 408 and NPV (wait) = 1083, respectively. Hence, conceding the choice has an advantage. In the wake of watching the business sector and getting more data, it may choose to contribute if the business sector interest ends up being "high", and not to contribute if the interest is "low".

In this illustration, the alternative to hold up gives the firm an extra estimation of 675.

If D is unknown, the value of immediate investment becomes

$$\text{NPV (invest now)} = -7500 + 0.5\left[\sum_{t=0}^{9} \frac{1}{1.1^t} 12D\right]$$

$$+ 0.5\left[\sum_{t=0}^{9} \frac{1}{1.1^t} 24D\right]$$

$$= -7500 + 121.68D \qquad (6)$$

For $D > 92.475$, this NPV will be positive, and the customary NPV rule would recommend contribute promptly. Note, then again, that project will wind up being a cash washout if interest ends up being D_L as opposed to D_H. For this situation, the "invest now" option will be productive with no danger of misfortune just if $D > 138.705$.

Under symmetric information, we may expect that it is normal learning that the venture will be productive without a doubt if $D > 138.705$. Firm A knows this through the estimation we simply did. Firm B can obviously realize this by comparative examination. Firm A ought to expect that firm B will enter the business sector amid first period if firm A not so that the "wait" option will have an estimation of 0 for $D > 138.705$. Thus, the hold up's estimation option gets to be

$$\text{NPV (wait)} = \begin{cases} 0; & D < 74.055 \\ -2.3865 + 48.345D; & 74.055 < D < 138.705 \\ 0; & D > 138.705 \end{cases}$$

$$(7)$$

For the interest region $D \in (74.055, 138.705)$, there is a likelihood that the contender will enter the business sector in the first place, and firm A will lose its option.

Insights

When uncertainties exist about the values of key limits, companies often postpone their investment decisions until the key uncertainties have been (fairly) resolved. Although, as we have noticed in real technology investments, companies sometimes do perpetrate investment at an early period despite their ability to postpone their decision. Companies that do so must believe the cost of postponing the investment is greater than the value abandoned from initial exercise (Bacchiega et al. 2012; Briglauer et al. 2013; Koetter and Noth 2013).

In summary, under the NPV, companies do not see the value of waiting, while under the options theory; the value of waiting is engaged but overestimated because of the lack of competition. By correctly including competition in our model, we have an option value that is lower but more practical. The existence of competition abrades the option value of waiting, because chip investments by the competition can abrade or even prevent benefits. Although how precisely competition abrades benefits or prevents investment options will rely on the market framework and each firm's strategic calculation. To engage this, we have to model competition internally by using a game-theoretic method, a topic we are now turning to.

Strategic exercise of growth options under imperfect competition: a game-theoretic model

In a real world, the investment decisions are affected by a private firm with individual favorites and inconsistent motivations. Besides, each competitor's investment decision is dependent on and sensitive to the other's moves. Game theory provides the method to determine how the players will act when each requests to maximize his own benefit. In such a game-theoretic situation, the value of a real option can be engaged only if it is exercised in a best way, which is fairly dependent on the right expectation of competitors' motions (Flaig et al. 2013; Li et al. 2013).

Model assumptions

To analyzing competition, we assume (1) competitors make logical tradeoffs in specifying when to exercise their options, so showing optimizing behavior; (2) each player decides by seeing a continuous uncertain state variable and expecting competitor's motions; and (3) the payoffs rely on the resulting equilibrium.

Subgame equilibrium

The option exercise game

To highlight the applicability of the method, we concentrate on a special real-world problem: the investment decisions of two firms that are analyzing investing in a new technology. At any time t, a firm can spend $C_{i,t}$ to get the technology, for which expected future cash flows dependent on tackling the project have a present value $V_{i,t}$. This is a 2-stage decision.

Normally, $C_{i,t}$ and $I_{i,t}$ are stochastic. We stress the value of $V_{i,t}$ could be notably influenced by the competitor's decisions. So, this two-period model is representative of a wide category of technology investment problems: one first invests in capacities, then gets some extra information, and eventually uses capacities dependent on the displayed information. More accurately, we explain the dynamic option-exercise game below:

- Players: Firm X and firm Y.
- Strategies: At the investment period, both firms determine either to invest or postpone in an indivisible technology that requires a rough investment cost, $C_{i,t}$. If a firm resolves invest, it also needs to determine, at the commercialization period, how much to produce, i.e., a quantity q_i $(i = X, Y)$ that maximizes its expected payoff. Therefore each firm has a strategic space $\sigma_i = (C, D; q_i | C), (i = X, Y)$.
- Payoffs: The payoff to firm i is a function of the strategies selected by it and its competitor. If both firms X and Y invest without watching each other's decision, they will divide the market as stated by Nash–Cournot equilibrium. If one firm invests first and the other does later, their payoffs will be mentioned through Stackelberg leader–follower equilibrium. If one firm invests first, but the other never does, then the earlier will enjoy a monopoly position. We suppose that a firm's payoff is directly the present value of its profit stream, $s_i(q_i, q_j)$.

Subgame equilibrium results

To solve the game, we first get the equilibrium quantities and payoffs for the commercialization period by cost and demand boundaries from an optimization process. These will serve as creating blocks in our following analysis of Nash–Cournot equilibrium under internal competition. Assume the reverse demand function is given by

$$P(\alpha_t, Q) = \alpha_t - (b_X q_X + b_Y q_Y) \tag{8}$$

where α_t is the stochastic demand-shift parameter, depicting the uncertainty in market demand, with expected value $E_0[\alpha_t] = \alpha_0 > 0$. In this model, α_t is supposed to develop as stated by a binomial process. $Q = q_X + q_Y$ is the total quantity on the market, where q_X and q_Y are the quantities provided by firms X and Y respectively. Without loss of generality, assume $b_X = b_Y = b$; then (6) becomes $P(\alpha_t, Q) = \alpha_t - b(q_X + q_Y)$. Show Γ_i as firm i's cost function, i.e.,

$$\Gamma_i(q_i) = \gamma_i q_i + F \qquad (9)$$

where F is the fixed cost, and γ_i is the marginal cost. Without loss of generality, suppose $F = 0$.

Concurrent decisions If firms X and Y make their decisions without noticing each other, each would have incomplete information about the other's real motions. This is equal to the situation in which they decide at the same time. Then each firm specifies its optimal quantity to maximize its profit:

$$\max_{q_i} s_i(q_i, q_j) = \max_{q_i} \left[P(\alpha_t, (q_i + q_j)) q_i - \gamma_i q_i \right] \qquad (10)$$

where $s_i (i = X, Y)$ is firm i's profit, and q_i, q_j are quantities of firms i and j separately. Solving this problem gives the equilibrium quantity:

$$q_i^* = \frac{1}{3b}(\alpha_i - 2\gamma_i + \gamma_j) \qquad (11)$$

The related equilibrium profit for each firm is

$$s_i = \frac{1}{9b}(\alpha_t - 2\gamma_i + \gamma_j)^2 \qquad (12)$$

It is easy to show $\frac{\partial s_i^2(q_i, q_j)}{\partial q_i^2} < 0$, so these quantities selects maximize profit. If the two firms have similar cost structures, i.e., $\gamma_i = \gamma_j = \gamma$ then the equilibrium quantity and profit will be symmetric:

$$q_i^* = q_j^* = \frac{1}{3b}(\alpha_t - \gamma) \qquad (13\text{-}1)$$

$$s_i = s_j = \frac{1}{9b}(\alpha_t - \gamma)^2 \qquad (13\text{-}2)$$

Sequential decisions If two firms move sequentially, the game would begin in an information structure that one firm can notice the other's move. Assume firm X invests first and firm Y, on seeing X's move, follows up. We use the backward method to solve the game. Supposing the leader is already in the market, the follower's decision is

$$\max_{q_Y} s_Y(q_X^*, q_Y) = \max_{q_Y} \left[P(\alpha_t, (q_X^* + q_Y)) - \gamma_Y \right] q_Y \qquad (14)$$

Expecting the follower's move, the leader's decision is

$$\max_{q_X} s_X(q_X, q_Y^*(q_X)) = \max_{q_X} \left[P(\alpha_t, (q_X + q_Y^*(q_X))) - \gamma_X \right] q_X \qquad (15)$$

Solving the optimization problems in (13) and (14) results the optimal quantities:

$$q_i^* = \frac{1}{2b}(\alpha_t - 2\gamma_t + \gamma_j)$$
$$q_j^* = \frac{1}{4b}(\alpha_t - 3\gamma_j + 2\gamma_i) \qquad (16)$$

Their related equilibrium profits will then be

$$s_i = \frac{1}{8b}(\alpha_t - 2\gamma_i + \gamma_j)^2$$
$$s_j = \frac{1}{16b}(\alpha_t - 3\gamma_j + 2\gamma_i)^2 \qquad (17)$$

where the subscript i depicts the leader, j the follower.

If the technology is good for multiple periods, we require to reduce the future cash flows. Assuming the operating cash flows s_i last n periods, the NPV_i of the profit values will be

$$\text{NPV}_i = V_i - C_i = \sum_{i=1}^{n} \frac{s_t}{(1+i)^t} - C_i \qquad (18)$$

where s_i is the operating profit in each period and i is the discount rate. If the technology can produce incomes infinitely, the NPV_i of the constant cash flows would be

$$\text{NPV}_i = V_i - C_i = \frac{s_i}{i} - C_i \qquad (19)$$

The exercise of growth options under competition

We now analyze the decision whether to make the strategic investment in the first stage. Without the beginning investment, the two firms would take their existing technologies (and related costs) as given.

Model assumptions

We analyze the investment decision in two conditions: (1) one developing firm has a devoted option to make the beginning investment, but two firms contend directly in the second period; and (2) the option is shared by the two firms, i.e., both firms can invest in the new technology even lessen future costs.

Devoted investment by the developing firm

Consider first the case where firm X makes no beginning investment, so ex post it has no strategic benefit over its competitor. If both firms select to sell on the market, they encounter the same marginal cost γ. The equilibrium quantity and profit are precisely the same as in (13-1) and (13-2), with γ being returned by Γ, i.e.,

$$q_i^* = q_j^* = \frac{1}{3b}(\alpha_t - \bar{\gamma}) \qquad (20)$$

$$s_i = s_j = \frac{1}{9b}(\alpha_t - \bar{\gamma})^2 \qquad (21)$$

On the other hand, if firm X makes the strategic investment therefore reducing its marginal cost to $\gamma_X = \gamma < \bar{\gamma} = \gamma_Y$ the market interaction is influenced by its technological benefit, which is admitted by firm Y when making its output decision. Firms X and Y will select the following quantities, respectively:

$$q_X^* = \frac{1}{3b}(\alpha_t - 2\gamma + \bar{\gamma}) \qquad (22\text{-}1)$$

$$q_Y^* = \frac{1}{3b}(\alpha_t - 2\bar{\gamma} + \gamma) \qquad (22\text{-}2)$$

Similarly, the related benefits for firms X and Y are, respectively:

$$s_X^* = \frac{1}{9b}(\alpha_t - 2\gamma + \bar{\gamma})^2 \qquad (23\text{-}1)$$

$$s_Y^* = \frac{1}{9b}(\alpha_t - 2\bar{\gamma} + \gamma)^2 \qquad (23\text{-}2)$$

since $\gamma < \bar{\gamma}$, then

$$q_X^* > q_Y^* \qquad (24\text{-}1)$$

$$s_X^* > s_Y^* \qquad (24\text{-}2)$$

It is now optimal for firm Y to select a lower quantity, resulting a lower benefit and smaller market share, because of the strategic result of firm X's investment.

As we can see, the cost benefit got from the strategic investment grows firm X's benefits and market share. So, the strategic investment creates a competitive advantage. It may be valuable to decay the growth option got by strategic investment in two pieces. First, it results in a lower "unit exercise price" ($\gamma < \bar{\gamma}$) for future expansion. Second, the optimal output q_X^*, "the number of unit production options that are optimally exercised", also grows, as other competitors select to limit their own output to make room for the stronger firm. The optimal investment policy is summarized in the following proposition:

Proposition 1 *There exists a supposed demand entrance, such that*

$$\alpha_X^{C^*} = \inf\{\alpha_t : \text{NPV}_X^C \geq \text{NPV}_X^D\} \qquad (25)$$

Strategic investment is optimal when demand is more than this entrance.

[Here, we accept the proposition and other presented propositions without proof. For considering the complete proof and further reading of these propositions, please refer to Azevedo and Paxson (2010, 2014) and Nishihara (2011)].

Concurrent investments by both firms

We now expand the basic model in the last section of the case when neither firm likes to devote protection (license) on the strategic investment, meaning the investment opportunity is open to all competitors.
Model assumptions
Both firms X and Y can invest in the new technology to lessen their future costs to γ. The final market results could be a monopoly, symmetric or asymmetric Cournot equilibrium, or no investment. Figure 3 shows the four possible combinations: (I, I), (I, D), (D, I), (D, D) where I means "invest" and D "defer".

Particularly, the first-stage investment game may result in a second-period commercialization period with the following possible results: symmetric but lower costs for both firms (both invested), asymmetric production costs (one firm invested) and the same the existing state of affairs costs of the existing technology (neither invested in the new technology; Huang and Behara 2013).

We have the following result:

Proposition 2 *The equilibria to exercise the investment option are*

$$\begin{array}{ll}(I, I), & \text{if } \alpha > \gamma + 3\sqrt{biC} \\ (D, D), & \text{if } \alpha \leq \gamma + 2\sqrt{biC}\end{array} \qquad (26)$$

Mixed strategy (I, D) or (D, I), if $\gamma + 2\sqrt{biC} < \alpha \leq \gamma + 3\sqrt{biC}$.

That is, optimal investment strategy is concurrent investment by both firms if $\alpha > \gamma + 3\sqrt{biC}$, mixed strategy by either firm if $\gamma + 2\sqrt{biC} < \alpha \leq \gamma + 3\sqrt{biC}$, and no investment by both if $\alpha \leq \gamma + 2\sqrt{biC}$. The demand entrance for concurrent investment is

$$\alpha^{II^*} = \gamma + 3\sqrt{biC} \qquad (27)$$

Multi-period model

We use a multi-period game tree structure as a thorough-form representation of the option exercise game. Firms X and Y determine either to invest (I) or defer (D) in each period. Then Nature (N), which shows the external uncertainty, determines the market demand will be either moved up to $u\alpha$ or down to $d\alpha$ similar to a binomial process, where u and d are the binomial parameters. On noticing the decisions made in the previous period and developing the market demand, each firm determines once more to invest or defer in the next period. The game can continue as many periods as required. In a multi-period setting, dynamic programming and backward induction allow us with the

Fig. 3 Concurrent investment in growth options (the *dotted line* shows the information structure that firm Y cannot see the firm's action)

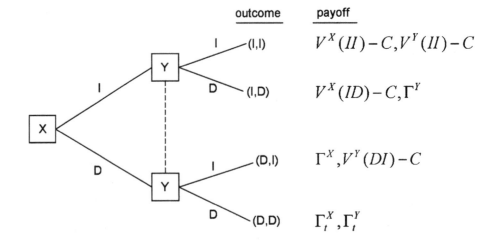

outcome	payoff
(I,I)	$V^X(II) - C, V^Y(II) - C$
(I,D)	$V^X(ID) - C, \Gamma^Y$
(D,I)	$\Gamma^X, V^Y(DI) - C$
(D,D)	Γ_t^X, Γ_t^Y

mathematical tools to solve multiple period problems (Huang and Qiao 2012). Specially, the value function of the investment project can be shown by the "Bellman" equation:

$$V(x) = \max\left\{ E[s_i] - C, \frac{1}{1+i} E[V'(x)|x] \right\} \quad (28)$$

where $V(x)$ is the value of the investment project, x the state variable, C the investment cost, s the expected cash flows dependent on the investment has been made, and i the discount rate. $V'(x)$ is the future extension value dependent on the present state variable.

The first term in (26) depicts the value of exercising the option, while the second term chooses the value of extension (i.e., holding the option). In every period, each firm would distinguish these two terms, taking analysis what the other firm would do.

Two-period equilibrium

We now appeal the dynamic structure in the two-period case, where the option to make the strategic investment stays obtainable for two periods. The demand entrance and the investment strategy are summarized in the following proposition:

Proposition 3 *Both firms exercise their options concurrently when the expected demand is more than the entrance:*

$$\alpha_0^{2-p^*} = e^{\sigma\sqrt{\Delta t}}(\gamma + 3\sqrt{biC}) \quad (29)$$

Investment entrance (27) specifies that $\alpha_0^{2-p^*}$ is an increasing function of the volatility of the market demand, interest rate, marginal cost, and the investment cost. So firms would incline to wait longer if the market is more unpredictable, if the technology costs more to install, or if

the cost demotion is small. Reasons like lower uncertainty, shorter option life, and more thorough competition would incline to lower the investment entrance. A lower investment entrance suggests the lower option value of waiting and more violent investment strategy. Note that

$$e^{\sigma\sqrt{\Delta t}} > 1 \quad (30)$$

Consequently

$$\alpha_0^{II^*}(1 \text{ period}) = \gamma + 3\sqrt{biC} < \alpha_0^{2-p^*}(2 \text{ period})$$
$$= e^{\sigma\sqrt{\Delta t}}(\gamma + 3\sqrt{biC}) \quad (31)$$

where σ measures the volatility of the market uncertainty. Higher volatility σ suggests a higher investment entrance, α^*. So, if the uncertainty is high, firms incline to wait longer. In other words, higher volatility suggests greater option value to defer investment. From (29), we have the following corollary:

Corollary *The investment entrance of the two periods is higher than that of one period.*

More periods

When the game lasts 3, 4, 5,… and n periods, the same method appeals often. One can always work backwards all the way to the beginning period. One just roll the equilibrium payoffs of the period t back to the previous period $(t - 1)$ until one achieves the present period of the game. Relied on the NPV_i's of different chances, each firm resolves its best strategies. Attention the game revives itself if it gets a (D, D) division.

Effects of competition on investment

The option value of waiting in a two-period game is

$$O = \begin{cases} \dfrac{q}{1+i}\left[\dfrac{(u\alpha_0 - \gamma)^2}{9ib} - C_0(1+i)\right], & \alpha_0 \le \gamma + 3\sqrt{ibC_o} \\ \dfrac{q}{1+i}\dfrac{(u\alpha_0 - \gamma)^2}{9ib} - \dfrac{(\alpha_0 - \gamma)^2}{9ib} + (C - q)C_0, & \alpha_0 > \gamma + 3\sqrt{ibC_0} \end{cases}$$

(32)

where q shows the probability in a risk-neutral world, i the discount rate, and u the binomial factor "up". Equation (30) illustrates the option value is a growing function of u and q, meaning the firms would be more possible to wait if the extent and chance of demand rising motion are larger than if they are small.

It can be proved the option value under competition in (30) is lower than that without competition (i.e., the option value of waiting for a monopoly as in Dixit and Pindyck 1994). That is, competition abrades the option value. The real options literature has explored the option value of waiting for a firm when payoffs are stochastic and investment irreparable. It has been illustrated in these studies that firms will typically delay investing until well after the point at which supposed discounted benefits identical beginning costs. In so doing, they use the option value of waiting. Although, the option value of waiting may have been overestimated when the risk of competitive abrasion or prevention is excluded (Smit and Trigeorgis 2006; Aleksandrov et al. 2013; Levaggi et al. 2012; Podoynitsyna et al. 2013). Certainly, the option value without competition is only an upper limit of the option value with competition. So, including competition is important to get a practical option valuation.

Discussions and extensions

New and existing technologies

The vital benefit of the investment is a more logical technology with lower marginal cost. That is, the marginal cost of selling one unit to the market will be lower with the investment than that without the investment. As a result, the first-period investment may result in a second-period competitive benefit about its competitor, with the following possible results: symmetric but lower costs for both firms (if both invested), asymmetric production costs (if one firm invested), and the same base-case costs related to the existing technology (if neither invested in the new technology).

In the above consideration, we have supposed there was an existing technology. Firms invest to improve or return the old technology. Another case is there is no existing technology so far, firms invest to enter this new product market. Without the new technology, firms would have to stay out the market and make zero income.

These two conditions can be summarized more accurately. The option value of continuation, Γ, is either zero or positive. That is,

$$\Gamma = \begin{cases} 0 & \text{no prior tech exist} \\ \dfrac{1}{9ib}(\alpha_t - \bar{\gamma})^2 & \text{there is an existing tech} \end{cases}$$

(33)

Asymmetric competition

We supposed earlier that, for both firms, the investment would result the same benefit (cost decline), i.e., the technologies got through beginning investment have the identical marginal costs. What if the investments may result in technologies with different costs?

Consider the cost first. Assume $\gamma_X > \gamma_Y$, from (21) and Proposition 3, we have

$$s_X^* = \frac{1}{9b}(\alpha_t - 2\gamma_X + \gamma_Y)^2 < s_Y^* = \frac{1}{9b}(\alpha_t - 2\gamma_Y + \gamma_X)^2$$
$$\alpha_X^* = e^{\sigma\sqrt{\Delta t}}(\Gamma_X + 3\sqrt{biC}) > \alpha_Y^* = e^{\sigma\sqrt{\Delta t}}(\gamma_Y + 3\sqrt{biC})$$

(34)

That is, if to implement the new technology results in a lower cost for firm Y than for firm X, then firm Y would have greater motivation to invest in the technology first. Also, if this cost distinction grows big enough to the extent that

$$\alpha_Y^* = e^{\sigma\sqrt{\Delta t}}(\gamma_Y + 3\sqrt{biC}) < e^{\sigma\sqrt{\Delta t}}(\gamma_X + 2\sqrt{biC})$$

(35)

then firm Y will invest first still in the middle area. Consequently the diversified investment strategy, (I, D) and (D, I), is substituted by (D, I), a sequential investment strategy where the firm with lower marginal cost (firm Y in this case) will invest first.

Second, the investment may result in different quality of product or service, as evaluated by the parameter b, in (7). Assuming $b_X > b_Y$ meaning firm Y's product is better recognized by the customers, the investment would result in a higher payoff function, and lower investment entrance, for firm Y. Eventually, all these factors are normally integrated. It is hard to see that a company with higher marginal cost and lower quality can continue in the market. Generally a firm with higher cost may like a better product.

Asymmetric information

Until now we have limited our consideration to option exercise under an information structure in which firms have symmetric information. Although in real world, investment and competition often happen in an environment of information asymmetry. That is, companies may have incomplete and asymmetrically scattered information on parameters like development costs or market demands. In this section, we emphasize on information asymmetry and its effect on option exercise. In an option-exercise game with full information, the firms' payoff functions are general knowledge. In a game with incomplete information, in contrast, at the minimum one firm is unsure about another firm's payoff or cost functions. An equilibrium pricing method must be used in a world of asymmetric information. The equilibrium method moderates the tradability suppositions required for arbitrage pricing (Batabyal 2012; Schwienbacher 2013). To avoid further problem, we assume risk neutrality, so prices are specified by discounting expected values, where the supposition is dependent on the obtainable information. Besides, we suppose the investment project under consideration is a small part of the firm's total assets.

Given the role of asymmetric information, changing the full-information supposition will add notable realism to the models of option exercise and technology investments. A few recent studies involve asymmetric information in their models. For example, Grenadier (1999) shows how information externalities may be created through noted option exercise decisions. Nadiminti et al. (2002) analyze intrafirm resource assignment under asymmetric information and negative externalities . These studies represent the fast evolution of the field.

Model assumptions of asymmetric information

We define the asymmetric information as follows:

(1)   Information is incomplete and asymmetric. Firm X knows its own cost function,

$$\Gamma_X(q_X) = \gamma_X q_X \qquad (36)$$

but has only incomplete information about firm Y's cost function. The following probability distribution shows firm X's opinion about firm Y's cost function:

$$\Gamma_Y(q_Y) = \begin{cases} \gamma_H q_Y & \text{with probability } \theta \\ \gamma_L q_Y & \text{with probability } 1 - \theta \end{cases} \qquad (37)$$

where $\gamma_L < \gamma_X < \gamma_H$ (to avoid unimportant cost benefit).

(2)   Firm Y realizes both firms' cost functions, therefore has better information. Firm Y could have just created a new technology, and its cost has not got public information so far. On the other hand, firm X continues to use the traditional technology of which the cost is generally known.

(3)   All of this is usual knowledge: firm X realizes that firm Y has better information, firm Y knows that firm X realizes this, and so on.

(4)   The inverse demand function and the stochastic demand-shift parameter are described.

In such a game with incomplete information, we say that firm Y has two possible types, γ_L and γ_H or its type space is $T_Y = \{\gamma_L, \gamma_H\}$. Firm X's type space is simply $T_X = \{\gamma_X\}$. Firm Y realizes its own type besides firm X's type, while firm X is uncertain about *the Y's type*. Formally,

$$\begin{aligned} P_X(t_Y = c_H | t_X = \gamma_X) &= \theta, \\ P_X(t_Y = c_L | t_X = \gamma_X) &= 1 - \theta \end{aligned} \qquad (38)$$

Sequential exercises

The sequencing of exercises is critical for an option-exercise game under asymmetric information, because decisions about exercise (and nonexercise) may release private information, as one firm can notice the other. Firms can gather information by moving later than others. The order of moves suggests each firm's calculated tradeoff between the strategic result of exercising early and the informational benefit of waiting to learn competitors' individual information through their disclosed actions. In the literature, the sequencing of actions has been normally supposed to be pre-determined. In contrast, we permit the sequencing of exercise to be internally mentioned through agents' optimizing decisions. Two sequences are possible:

Sequence 1   The less informed firm (X) moves first and the more informed firm (Y) follows.

In the spirit of backward induction, we first solve the follower's decision. Supposing the leader has already decided q_X^*, the follower will choose q_Y to maximize its profit dependent on its cost structures, i.e.,

$$\begin{aligned} &\max_{q_Y(\gamma_H)} s_Y(q_X^*, q_Y, \gamma_H) \\ &= \max_{q_Y(\gamma_H)} \left[P(\alpha_t, (q_X^*(q_Y) + q_Y(\gamma_H))) - \gamma_H \right] q_Y(\gamma_H) \end{aligned} \qquad (39)$$

$$\begin{aligned} &\max_{q_Y(\gamma_L)} s_Y(q_X^*, q_Y, \gamma_L) \\ &= \max_{q_Y(\gamma_L)} \left[P(\alpha_t, (q_X^*(q_Y) + q_Y(\gamma_L))) - \gamma_L \right] q_Y(\gamma_L) \end{aligned} \qquad (40)$$

The leader's decision, expecting firm Y's above move, is to select q_X to maximize its payoff:

$$\begin{aligned} &\max_{q_X} s_X(q_X, q_Y) \\ &= \max_{q_X} \big\{ \theta \left[P(\alpha_t, (q_X + q_Y^*(\Gamma_H))) - c_X \right] q_X \\ &\quad + (1 - \theta) \left[P(\alpha_t, (q_X + q_Y^*(\gamma_L))) - c_X \right] q_X \big\} \end{aligned} \qquad (41)$$

The solutions to (39)–(41) are, respectively

$$q_X^* = \frac{1}{2b} \left[\alpha_t - 2\gamma_X + \theta \gamma_H + (1 - \theta)\gamma_L \right] \qquad (42)$$

$$q_Y^*(\gamma_H) = \frac{1}{4b}(\alpha_t - 3\gamma_H + 2\gamma_X) + \frac{1-\theta}{4b}(\gamma_H - \gamma_L) \qquad (43)$$

$$q_Y^*(\gamma_L) = \frac{1}{4b}(\alpha_t - 3\gamma_L + 2\gamma_X) - \frac{\theta}{4b}(\gamma_H - \gamma_L) \qquad (44)$$

Then the correlating equilibrium benefits are

$$s_X^* = \frac{1}{8b}[\alpha_t - 2\gamma_X + \theta\gamma_H + (1-\theta)\gamma_L]^2 \qquad (45)$$

$$s_Y^*(\gamma_H) = \frac{1}{16b}[(\alpha_t - 3\gamma_H + 2\gamma_X) + (1-\theta)(\gamma_H - \gamma_L)]^2 \qquad (46)$$

$$s_Y^*(\gamma_L) = \frac{1}{16b}[(\alpha_t - 3\gamma_L + 2\gamma_X) - \theta(\gamma_H - \gamma_L)]^2 \qquad (47)$$

Sequence 2 The more informed firm (Y) moves first and the less informed firm (X) follows.

If the firm with individual information moves first, the follower would have an opportunity to conclude the leader's individual information through disclosed effects. More precisely, firm X would notice Y's quantity selects $q_Y^*(\gamma_H)$, $q_Y^*(\gamma_L)$, and conclude firm Y's cost functions, γ_H or γ_L properly. The more informed firm would disclose its individual information through its exercise decisions. Because of this information disclosure, the information asymmetry may be reduced.

On learning firm Y's individual information about its cost function, firm X selects its quantity to maximize its benefit. The learning could disclose two possible results: firm Y's cost could be high ($\gamma_Y = \gamma_H$) and low ($\gamma_Y = \gamma_L$). Depending on that firm X learned $\gamma_Y = \gamma_H$, firm X's decision would be

$$\max_{q_X} s_X(q_X, q_Y^*(\gamma_H)) = \max_{q_X}[P(\alpha_t, (q_X + q_Y^*(\gamma_H))) - \gamma_X]q_X \qquad (48)$$

As well, dependent on that firm X learned $\gamma_Y = \gamma_L$, firm X's decision would be

$$\max_{q_X} s_X(q_X, q_Y^*(\gamma_L)) = \max_{q_X}[P(\alpha_t, (q_X + q_Y^*(\gamma_L))) - \gamma_X]q_X \qquad (49)$$

Expecting firm X's above reaction, firm Y solves for $q_Y(\gamma_H)$ when its true cost is γ_H, i.e.,

$$\max_{q_Y(\gamma_H)} s_Y(q_X^*, q_Y(\gamma_H)) = \max_{q_Y(\gamma_H)}[P(\alpha_t, (q_X^*(q_Y) + q_Y(\gamma_H))) - \gamma_H]q_Y(\gamma_H) \qquad (50)$$

By the similar analysis, firm Y solves for $q_Y(\gamma_L)$ when its true cost is γ_L, i.e.,

$$\max_{q_Y(\gamma_L)} s_Y(q_X^*, q_Y(\gamma_L)) = \max_{q_Y(\gamma_L)}[P(\alpha_t, (q_X^*(q_Y) + q_Y(\gamma_L))) - \gamma_L]q_Y(\gamma_L) \qquad (51)$$

Solving the optimization problems in (46)–(49) results the below equilibrium quantities:

Conditional on $\gamma_Y = \gamma_H$,

$$q_X^*(\gamma_H) = \frac{1}{4b}(\alpha_t - 3\gamma_X + 2\gamma_H) \qquad (52)$$

$$q_Y^*(\gamma_H) = \frac{1}{2b}(\alpha_t - 2\gamma_H + \gamma_X) \qquad (53)$$

and dependent on $\gamma_Y = \gamma_L$,

$$q_X^*(\gamma_L) = \frac{1}{4b}(\alpha_t - 3\gamma_X + 2\gamma_L) \qquad (54)$$

$$q_Y^*(\gamma_L) = \frac{1}{2b}(\alpha_t - 2\gamma_L + \gamma_X) \qquad (55)$$

The similar equilibrium benefits are, respectively

$$s_X^*(\gamma_H) = \frac{1}{16b}(\alpha_t - 3\gamma_X + 2\gamma_H)^2 \qquad (56)$$

$$s_Y^*(\gamma_H) = \frac{1}{8b}(\alpha_t - 2\gamma_H + \gamma_X)^2 \qquad (57)$$

$$s_X^*(\gamma_L) = \frac{1}{16b}(\alpha_t - 3\gamma_X + 2\gamma_L)^2 \qquad (58)$$

$$s_Y^*(\gamma_L) = \frac{1}{8b}(\alpha_t - 2\gamma_L + \gamma_X)^2 \qquad (59)$$

where (54) and (55) are conditional on $\gamma_Y = \gamma_H$ while (56) and (57) are dependent on $\gamma_Y = \gamma_L$.

Equilibrium analysis

When information is asymmetric, equilibrium exercise may be sequential, with the more informed firm exercising first and permitting the less informed to free sit on the information expressed by the exercise (or failure to exercise). Although, the information asymmetry is furthermore difficult by the presence of cost asymmetry. The firm with lower cost may have lower investment entrance and higher inducements to move early (Genc and Zaccour 2013). To make easier resemblance, we achieve the equilibrium analysis for two situations: (1) we first suppose that firm Y realizes that its true cost is $\gamma_Y = \gamma_L$ (firm X does not realize this) in "Equilibrium under asymmetric information ($\gamma_Y = \gamma_L$)" section; (2) we then turn to situating $\gamma_Y = \gamma_H$ in "Equilibrium under asymmetric information ($\gamma_Y = \gamma_H$)" section.

Equilibrium under asymmetric information ($\gamma_Y = \gamma_L$)

If firm Y realizes that its true cost is low (again firm X does not realize this because of information asymmetry), firm Y would exercise its option first to engage the payoff benefit. The entire demand range is divided into three areas, thus area *I*, waiting region (D, D), area *II*, sequential investment region (D, I) and area *III*, simultaneous investment region (I, I). This result is formalized in the following proposition:

Proposition 4 (Equilibrium under asymmetric information when $\gamma_Y = \gamma_L$) *Under asymmetric information, the option-exercise game has three equilibria*

$$(D,D), \alpha < 2\gamma_L - \gamma_X + \sqrt{8biC}$$
$$(D,I), 2\gamma_L - \gamma_X + \sqrt{8biC} \leq \alpha < 2\gamma_X - \theta\gamma_H$$
$$\qquad\qquad - (1-\theta)\gamma_L + 3\sqrt{biC} \qquad (60)$$
$$(I,I), \alpha \geq 2\gamma_X - \theta\gamma_H - (1-\theta)\gamma_L + 3\sqrt{biC}$$

Proposition 4 illustrates that when asymmetry exists and demand is in area *II* equilibrium exercise will be sequential and instructive. With asymmetric information, the option exercise is (D, I); the more informed firm moves first and engages higher benefits from being a leader. The less informed firm selects to wait and free sit on the information expressed by the leader's exercise. This permits the follower to conclude the leader's individual information through noticed exercise of options. So, in this equilibrium, the leader gets payoff compensation and the follower gets *informational benefits*.

Equilibrium under asymmetric information $(\gamma_Y = \gamma_H)$

If firm *Y* knows that its true cost is high (remember firm *X* does not know this because of information asymmetry), firm *Y* would become unwilling to exercise its option first because doing so may disclose to its competitor that it is really a high cost (therefore weak) player. This normally guides us to feel the equilibrium (D, I) in area *II* may no longer exist.

It appears, although, the equilibrium (D, I) still exists, but with more restrictive situations than in the $\gamma_Y = \gamma_L$ case. This result is formalized in the following proposition.

Proposition 5 (Equilibrium under asymmetric information when $\gamma_Y = \gamma_L$) *Under asymmetric information the option-exercise game may have the following the arealibria: (D, D) in area I and (I, I) in area III. In area II, the equilibrium is sequential (D, I) if the following situations are encountered:*

$$\gamma_X > \max\left\{\frac{1}{3}(2\gamma_H + \gamma_L + \theta(\gamma_H - \gamma_L)), \left(\gamma_H - \left(1 - \frac{\sqrt{2}}{2}\right)\sqrt{biC}\right)\right\}$$
$$(1-\theta)(\gamma_H - \gamma_L) < 2(3 - \sqrt{8})\sqrt{biC}$$
$$(61)$$

The equilibrium may be (I, D), in other respects. The three areas are explained as

$$C, \alpha < \min\left\{(2\gamma_H - \gamma_X + \sqrt{8biC}), (2\gamma_X - \theta\gamma_H - (1-\theta)\gamma_L + \sqrt{8biC})\right\}$$

$$III, \alpha \geq \max\left\{\left(2\gamma_X - \theta\gamma_H - (1-\theta)\gamma_L + 3\sqrt{biC}\right), \left(2\gamma_H - \gamma_X - \frac{1-\theta}{2}(\gamma_H - \gamma_L) + 3\sqrt{biC}\right)\right\}$$

II, the area between regions *I* and *III* $\qquad (62)$

Comparative statics

We have considered the existence and the economic rationality of equilibria for the two conditions above. It is interesting to distinguish the equilibrium area under incomplete information to those under full information. This will also permit us to measure the results of asymmetric information.

For firm *X*, the entrance demand levels with full information would be

$$\alpha_{NS}^*(X|\gamma_L, FI) = 2\gamma_X - \gamma_L + 3\sqrt{biC}$$
$$\alpha_{NS}^*(X|\gamma_H, FI) = 2\gamma_X - \gamma_H + 3\sqrt{biC} \qquad (63)$$

With incomplete information, the relating entrance (from Proposition 4) becomes

$$\alpha_{NS}^*(X|\theta, II) = 2\gamma_X - \gamma_L - \theta(\gamma_H - \gamma_L) + 3\sqrt{biC} \qquad (64)$$

where *FI* denotes "full information" and *II* "incomplete information". $\alpha_{NS}^*(X|\gamma_H, FI)$ stands for the entrance level that firm *X* will invest under Nash equilibrium, dependent on firm *X*'s having full information and assuming firm *Y*'s cost is γ_H with a probability $\theta = 1$. As well, $\alpha_{NS}^*(X|\theta, II)$ depicts the entrance level that firm *X* will invest under Nash equilibrium, depending on firm *X*'s having incomplete information and assuming firm *Y*'s cost is γ_H with probability θ. From (61), we have

$$\frac{\partial \alpha_{NS}^*(X|\theta, II)}{\partial \theta} = -(\gamma_H - \gamma_L) < 0 \qquad (65)$$

Thus $\alpha_{NS}^*(X|\theta, II)$ is a reducing function of θ, suggesting that firm *X* would invest at a lower entrance (so more violently) if it has a stronger opinion that its competitor is a high cost player. This is regular with what we have learned in previous sections. It can be confirmed that

$$\alpha_{NS}^*(X|\gamma_H, FI) < \alpha_{NS}^*(X|\gamma_L, FI) \qquad (66)$$

As a consequence, the full-information entrance levels are directly specific cases of the asymmetric-information entrance. More generally, the full-information equilibrium is a special case of the asymmetric-information equilibrium. From firm Y's view, the entrance levels are

$$\alpha^*_{NS}(Y|\gamma_L, II) = 2\gamma_L - \gamma_X + \frac{\theta}{2}(\gamma_H - \gamma_L)$$
$$+ 3\sqrt{biC} > \alpha^*_{NS}(Y|\gamma_L, FI)$$
$$= 2\gamma_L - \gamma_X + 3\sqrt{biC} \qquad (67)$$

$$\alpha^*_{NS}(Y|\gamma_H, II) = 2\gamma_H - \gamma_X - \frac{1-\theta}{2}(\gamma_H - \gamma_L)$$
$$+ 3\sqrt{biC} < \alpha^*_{NS}(Y|\gamma_H, FI)$$
$$= 2\gamma_H - \gamma_X + 3\sqrt{biC} \qquad (68)$$

In a world of asymmetric information, $\alpha^*_{NS}(Y|\gamma_L, II)$ is greater than $\alpha^*_{NS}(Y|\gamma_L, FI)$ and $\alpha^*_{NS}(Y|\gamma_H, II)$ is less than $\alpha^*_{NS}(Y|\gamma_H, FI)$. The difference is larger if the information asymmetry is notable (as measured by θ and $\Delta\gamma = \gamma_H - \gamma_L$). This happens because firm Y not only adjusts its entrance to its own cost but also replies to the fact that firm X has incomplete information and thus cannot do the same. If firm Y's costs are high, for example it waits longer and invest at an entrance that is higher than firm X would do if it knew with full information firm Y's costs to be high.

Conclusions

Most of the real options literature has concentrated on market environments without strategic interactions. On the other hand, the industrial organization literature endogenizes market structure; so far it usually neglects uncertainty and so the option value of flexibility. Considering the linkage effects of both uncertainty and competition, this paper extends a method for assessing technology investment decisions in an oligopolistic market structure. It combines the game-theoretic models of strategic market interactions with a real options approach to investment under uncertainty, and gives an improved comprehension of the results of uncertainty and competition on the strategic exercise of real options inserted in technology investments.

Through expanding an equilibrium model of a dynamic investment game, the paper makes several contributions. First, showing that investment strategies critically rely on competitive interactions, the study improves our comprehension of the linkage effects of competition and uncertainty on investment decisions. We have best investment policies and vital investment entrances. Besides, we have also taken analysis of different information structures.

One of the restrictions of the paper is the work is mostly methodological and theoretical—appealing economic models to technology investment under both uncertainty and competition. Although we have tried to link the theory

to fact, the work could be improved by adding some practical elements. However, the results got in the paper could be used to form theories to carry out practical testing.

References

Albuquerque R, Miao J (2014) Advance information and asset prices. J Econ Theory. doi:10.1016/j.jet.2013.06.001

Aleksandrov N, Espinoza R, Gyurkó L (2013) Optimal oil production and the world supply of oil. J Econ Dyn Control 37(7):1248–1263. doi:10.1016/j.jedc.2013.01.015

Alexandrov A, Deb J (2012) Price discrimination and investment incentives. Int J Ind Organ 30(6):615–623. doi:10.1016/j.ijindorg.2012.07.001

Amram M, Kulatilaka N (1999) Real options: managing strategic investment in an uncertain world. Harvard Business School, Cambridge, MA

Azevedo AF, Paxson DA (2010) Real options game models: a review. Real Options. http://realoptions.org/papers2010/109.pdf

Azevedo A, Paxson D (2014) Developing real option game models. Eur J Oper Res 237(3):909–920

Bacchiega E, Randon E, Zirulia L (2012) Strategic accessibility competition. Res Econ 66(2):195–212. doi:10.1016/j.rie.2011.12.001

Batabyal AA (2012) Project financing, entrepreneurial activity, and investment in the presence of asymmetric information. N Am J Econ Finance 23(1):115–122. doi:10.1016/j.najef.2011.11.006

Benaroch M, Kauffman RJ (1999) A case for using real options pricing analysis to evaluate information technology project investments. Inf Syst Res 10(1):70–86

Berghman L, Matthyssens P, Vandenbempt K (2012) Value innovation, deliberate learning mechanisms and information from supply chain partners. Ind Mark Manag 41(1):27–39. doi:10.1016/j.indmarman.2011.11.014

Berghout E, Tan C-W (2013) Understanding the impact of business cases on IT Investment decisions: an analysis of municipal e-government projects. Inf Manag. doi:10.1016/j.im.2013.07.010

Bos JWB, Kolari JW, van Lamoen RCR (2013) Competition and innovation: evidence from financial services. J Bank Finance 37(5):1590–1601. doi:10.1016/j.jbankfin.2012.12.015

Briglauer W, Ecker G, Gugler K (2013) The impact of infrastructure and service-based competition on the deployment of next generation access networks: recent evidence from the European member states. Inf Econ Policy 25(3):142–153. doi:10.1016/j.infoecopol.2012.11.003

Brynjolfsson E (1993) The productivity paradox of information technology. Commun ACM 36(12):66–77

Chaton C, Guillerminet M-L (2013) Competition and environmental policies in an electricity sector. Energy Econ 36:215–228. doi:10.1016/j.eneco.2012.08.014

Clemons EK Evaluation of strategic investments in information technology. Paper presented at the Communications of the ACM

Dewan S, Kraemer K (1998) Information technology and productivity: evidence from country level data. Graduate School of Management, University of California

Dimitrios NK, Sakas DP, Vlachos DS (2013) Analysis of strategic leadership models in information technology. Proc Soc Behav Sci 73:268–275. doi:10.1016/j.sbspro.2013.02.052

Dixit AK, Pindyck RS (1994) Investment under uncertainty. Princeton University Press, Princeton

Dos Santos B (1991) Justifying investments in new information technologies. Krannert Graduate School of Management, Purdue University

Fernandes R, Gouveia B, Pinho C (2013) A real options approach to labour shifts planning under different service level targets. Eur J Oper Res 231(1):182–189. doi:10.1016/j.ejor.2013.05.008

Flaig D, Rubin O, Siddig K (2013) Imperfect competition, border protection and consumer boycott: the future of the dairy industry in Israel. J Policy Model 35(5):838–851. doi:10.1016/j.jpolmod.2013.01.001

Gao X, Zhong W, Mei S (2013) A differential game approach to information security investment under hackers' knowledge dissemination. Oper Res Lett 41(5):421–425. doi:10.1016/j.orl.2013.05.002

Genc TS, Zaccour G (2013) Capacity investments in a stochastic dynamic game: equilibrium characterization. Oper Res Lett 41(5):482–485. doi:10.1016/j.orl.2013.05.012

Grenadier SR (1999) Information revelation through option exercise. Rev Financ Stud 12(1):95–129

Huang CD, Behara RS (2013) Economics of information security investment in the case of concurrent heterogeneous attacks with budget constraints. Int J Prod Econ 141(1):255–268. doi:10.1016/j.ijpe.2012.06.022

Huang X, Qiao L (2012) A risk index model for multi-period uncertain portfolio selection. Inf Sci 217:108–116. doi:10.1016/j.ins.2012.06.017

Huisman KJM, Kort PM (2004) Strategic technology adoption taking into account future technological improvements: a real options approach. Eur J Oper Res 159(3):705–728. doi:10.1016/S0377-2217(03)00421-1

Khallaf A (2012) Information technology investments and nonfinancial measures: a research framework. Account Forum 36(2):109–121. doi:10.1016/j.accfor.2011.07.001

Kim YJ, Sanders GL (2002) Strategic actions in information technology investment based on real option theory. Decis Support Syst 33(1):1–11. doi:10.1016/S0167-9236(01)00134-8

Koetter M, Noth F (2013) IT use, productivity, and market power in banking. J Financ Stab. doi:10.1016/j.jfs.2012.06.001

Kumar N (1996) The power of trust in manufacturer-retailer relationships. Harv Bus Rev 74(6):92

Lestage R, Flacher D, Kim Y, Kim J, Kim Y (2013) Competition and investment in telecommunications: Does competition have the same impact on investment by private and state-owned firms? Inf Econ Policy 25(1):41–50. doi:10.1016/j.infoecopol.2013.02.001

Levaggi R, Moretto M, Pertile P (2012) Static and dynamic efficiency of irreversible health care investments under alternative payment rules. J Health Econ 31(1):169–179. doi:10.1016/j.jhealeco.2011.09.005

Li S, Blake A, Thomas R (2013) Modelling the economic impact of sports events: the case of the Beijing Olympics. Econ Model 30:235–244. doi:10.1016/j.econmod.2012.09.013

Luehrman TA (1998a) Investment opportunities as real options: getting started on the numbers. Harv Bus Rev 76:51–66

Luehrman TA (1998b) Strategy as a portfolio of real options. Harv Bus Rev 76:89–101

Martzoukos SH, Zacharias E (2013) Real option games with R&D and learning spillovers. Omega 41(2):236–249. doi:10.1016/j.omega.2012.05.005

McIntyre DP, Chintakananda A (2013) A real options approach to releasing "network" products. J High Technol Manag Res 24(1):42–52. doi:10.1016/j.hitech.2013.02.007

Merali Y, Papadopoulos T, Nadkarni T (2012) Information systems strategy: Past, present, future? J Strateg Inf Syst 21(2):125–153. doi:10.1016/j.jsis.2012.04.002

Meyer E, Rees R (2012) Watchfully waiting: medical intervention as an optimal investment decision. J Health Econ 31(2):349–358. doi:10.1016/j.jhealeco.2012.02.002

Nadiminti R, Mukhopadhyay T, Kriebel CH (2002) Research report: intrafirm resource allocation with asymmetric information and negative externalities. Inf Syst Res 13(4):428–434. Retrieved from http://www.jstor.org/stable/23015723

Nishihara M (2011) A Real Options Game Involving Multiple Projects. In: Proceedings of the international multiconference of engineers and computer scientists

Pendharkar PC (2010) Valuing interdependent multi-stage IT investments: a real options approach. Eur J Oper Res 201(3):847–859

Podoynitsyna K, Song M, van der Bij H, Weggeman M (2013) Improving new technology venture performance under direct and indirect network externality conditions. J Bus Ventur 28(2):195–210. doi:10.1016/j.jbusvent.2012.04.004

Rohlfs W, Madlener R (2013) Investment decisions under uncertainty: CCS competing with green energy technologies. Energy Proc 37:7029–7038. doi:10.1016/j.egypro.2013.06.638

Schwienbacher A (2013) The entrepreneur's investor choice: the impact on later-stage firm development. J Bus Ventur 28(4):528–545. doi:10.1016/j.jbusvent.2012.09.002

Smit HT, Trigeorgis L (2006) Real options and games: competition, alliances and other applications of valuation and strategy. Rev Financ Econ 15(2):95–112

Smith JE, McCardle KF (1998) Valuing oil properties: integrating option pricing and decision analysis approaches. Oper Res 46(2):198–217

Smith JE, McCardle KF (1999) Options in the real world: lessons learned in evaluating oil and gas investments. Oper Res 47(1):1–15

Taudes A, Feurstein M, Mild A (2000) Options analysis of software platform decisions: a case study. MIS Q 24(2):227–243

van Zee RD, Spinler S (2014) Real option valuation of public sector R&D investments with a down-and-out barrier option. Technovation. doi:10.1016/j.technovation.2013.06.005

Wang B, Wang X, Wang J (2012) Construction and empirical analysis of agricultural science and technology enterprises investment risk evaluation index system. IERI Proc 2:485–491. doi:10.1016/j.ieri.2012.06.121

Wrzaczek S, Kort PM (2012) Anticipation in innovative investment under oligopolistic competition. Automatica 48(11):2812–2823. doi:10.1016/j.automatica.2012.08.007

Wu L-C, Ong C-S (2008) Management of information technology investment: a framework based on a Real Options and Mean–Variance theory perspective. Technovation 28(3):122–134. doi:10.1016/j.technovation.2007.05.011

Wu L-C, Li S-H, Ong C-S, Pan C (2012) Options in technology investment games: the real world TFT-LCD industry case. Technol Forecast Soc Change 79(7):1241–1253. doi:10.1016/j.techfore.2012.03.008

11

Biofuel supply chain considering depreciation cost of installed plants

Masoud Rabbani[1] · Farshad Ramezankhani[1] · Ramin Giahi[1] ·
Amir Farshbaf-Geranmayeh[1]

Abstract Due to the depletion of the fossil fuels and major concerns about the security of energy in the future to produce fuels, the importance of utilizing the renewable energies is distinguished. Nowadays there has been a growing interest for biofuels. Thus, this paper reveals a general optimization model which enables the selection of preprocessing centers for the biomass, biofuel plants, and warehouses to store the biofuels. The objective of this model is to maximize the total benefits. Costs of the model consist of setup cost of preprocessing centers, plants and warehouses, transportation costs, production costs, emission cost and the depreciation cost. At first, the deprecation cost of the centers is calculated by means of three methods. The model chooses the best depreciation method in each period by switching between them. A numerical example is presented and solved by CPLEX solver in GAMS software and finally, sensitivity analyses are accomplished.

Keywords Biomass · Biofuel supply chain ·
Multi-echelon · Depreciation costs

Introduction

By considering depletion of fossil fuel in the future, the importance of using renewable energy increases production (Petroleum 2015). One of the disadvantages of fossil fuels is air pollution. Greenhouse gases spread out in

environment via burning of these fuels and cause global warming. On the other hand, renewable energy has less global warming effects and increases the energy security. Renewable energy divides into solar, wind power, biomass, geothermal, and tidal energy. The types of biomass feedstock which are utilized for energy purposes are categorized as: agricultural, dedicated energy crops, forestry, industry, gardens residues (Tumuluru et al. 2011). In this study, the supply chain of the biomass is proposed as:

1. Procuring of the feedstock (i.e., purchasing biomass, importing, and cultivating them).
2. Transporting to preprocessing centers.
3. Preprocessing biomass.
4. Transporting the preprocessed biomass to plants.
5. Producing biofuel in the plant.
6. Transporting the biofuels to the warehouses.
7. Distributing the biofuels.

Literature review

Ayoub et al. (2007) proposed a general bioenergy decision system. They believe that planners have to consider social concerns, environmental and economic impacts related to establishing the biomass systems. Leduc et al. (2008) developed a model to determine the locations and sizes of methanol plants and gas stations in Austria. The objective function of the model consisted of plant and gas station setup cost, methanol production cost, and material transportation cost. Mele et al. (2009) proposed a model that simultaneously minimizes the total cost of the network and its environmental performance over the entire life cycle of the product. Zamboni et al. (2009) proposed the bioethanol supply chain optimization in which they presented a model

✉ Masoud Rabbani
 mrabani@ut.ac.ir

[1] School of Industrial and Systems Engineering, College of
 Engineering, University of Tehran,
 P.O. Box 11155-45632, Tehran, Iran

for the strategic design of biomass-based fuel supply networks. Finally, they applied the model for a case study in Italy. Jackson et al. (2009) found that firms using accelerated depreciation make significantly larger capital investments than firms that use straight line depreciation and found that there has been a migration away from accelerated depreciation to straight line depreciation over the past two decades. Finally, results suggest that a choice made for external financial reporting purposes influences managers' capital investment decisions. Ayoub et al. (2009) proposed an optimization model for designing and evaluating integrated system of bioenergy production supply chains. Their model was applied in a case study in Japan. Rentizelas and Tatsiopoulos (2010) utilized a hybrid optimization method to find the optimum location of a bioenergy generation facility considering the maximization of the net present value (NPV) of the investment for the project's lifetime. Velazquez-Marti and Fernandez-Gonzalez (2010) supposed two criteria for the location of established plants for producing the biofuel as: minimizing the transportation costs of biofuels and using all the energy produced by the plant. They applied the model to Spanish rural regions. Akgul et al. (2011) presented the model to optimize the locations and scales of the bioethanol production plants, biomass and bioethanol flows between regions. The purpose of this study is minimizing the total supply chain costs. Kim et al. (2011) formulated a model that enables the selection of fuel conversion technologies and capacities, biomass locations and the logistics of transportation from forestry resources to conversion, and from conversion to final markets. The objective function to be maximized was the overall profit. The revenue of the model includes selling various products in the final market and the credits for the utility energy produced at each plant location. The cost encompassed operating cost, annualized capital cost, transportation cost and biomass acquisition cost for each biomass type. Mobini et al. (2011) developed a simulation model to evaluate the cost of delivered forest biomass, the equilibrium moisture content, and carbon emissions from the logistics operations. Zhu and Yao (2011) proposed a multi-commodity network flow model to design the logistics system. They formulated a model to determine the locations of warehouses, the size of harvesting group, the types and amounts of biomass harvested or purchased, stored, and processed, and the transportation of biomass in the system. The objective function of Leão et al. (2011) consisted of investments for the production plants, transportation costs, agricultural production costs, processing costs and purchasing cost of any additional volumes of oil in the market to meet the demand of the plants. Chen and Fan (2012) developed a two-stage stochastic programming model to minimize the system cost. The system includes bioethanol production, feedstock procurement, fuel delivery, ethanol transportation and possible penalty on fuel shortage. The model was used to evaluate the economic possibility and system robustness in a case study of California. Finally, the model was solved by a Lagrange relaxation-based decomposition solution algorithm. Ayoub and Yuji (2012) utilized a demand-driven approach for optimizing biomass utilization networks cost by applying genetic algorithm to solve the network problem. Judd et al. (2012) proposed a mathematical programming to determine satellite storage locations and equipment routes to minimize the total cost of designing a feedstock logistics system. The feedstock logistics system includes transporting biomass from production fields to the bioenergy plant. Kostin et al. (2012) integrated bioethanol and sugar production supply chain under demand uncertainty. They considered several financial risk mitigation options in the supply chain model. They applied the model in the Argentinean sugarcane industry. Finally the problem was solved by applying the sample average approximation algorithm. Akgul et al. (2012) presented an optimization framework for the strategic design of a hybrid first/second-generation ethanol supply chain. The applicability of the model is demonstrated with a case study of ethanol production in the UK. The potential cost reductions of second-generation biofuel systems are likely to lead to the deployment of these technologies at a larger scale. Bio-based supply chain that was proposed by Pérez-Fortes et al. (2012) led to produce electricity or other bio-products. Their model took into account three main objectives: economic, environmental and social criteria. Biomass storage periods, location and capacity of plants, material transportation between echelons and biomass utilization to produce biofuel are determined in their model. They applied the model for a case study in Ghana. To produce a low-cost urban energy system, Keirstead et al. (2012) accomplished the trade-offs between the alternatives by considering the air pollution impacts. According to their exploration of the trade-offs, biomass energy system is the best choice. Supply chain that worked by Čuček et al. (2012) is included agricultural, preprocessing, processing, and distribution layers. Also they presented a multi-criteria optimization for the conversion of biomass to energy. Fazlollahi and Maréchal (2013) simultaneously minimized costs and CO_2 emission of integrated biomass resources using multi-objective evolutionary algorithms. Zhang et al. (2013) focused on switch grass as one of the best second-generation feedstock for bioethanol production. They proposed an integrated mathematical model to minimize the total switch grass-based bioethanol supply chain cost. The proposed model considered the impact of switch grass crop yield, switch grass densification, switch grass dry-matter loss during storage, and economies of scale in bio refinery capacities on the total SBSC cost.

Meier et al. (2005) evaluated the economic concept of the industrial solar production of lime. The three capital investment decision indicators used in economic analysis are: (1) the payback time (PBT), defined as time required for an investment project to recover its initial cost; (2) NPV, defined as the present value of the flow of net incomes subtracted by the present value of the flow of investments; and (3) the internal rate of return (IRR), defined as the discount rate at which NPV equals zero.

Mahmoudi et al. (2014) investigated the problem of source selection of competitive power plants under government intervention. Kumar et al. (2015) investigated the impact of various factors affecting coal-fired power plant economics for electricity generation.

There are plentiful papers in biofuel supply chain and a lot of mathematics models are presented in this field but there are not any papers which regard the depreciation cost as an important element of the model. The significant part of our model is considering the depreciation cost within supply chain model. In this case, depreciation cost is defined as a crucial element of any supply chain design. Our study is the extension of Akgul et al. (2012) and the contributions of our study are as follows:

- Considering penalty on fuel shortage.
- Considering environmental impact of biofuel plants such as CO_2 emission.
- Considering two manners for plants; purchasing or renting.
- Calculating NPV of the project.
- Considering total depreciable capital and salvage value of the network.
- Revenue of selling the fuels in the market.

The remainder of this paper is organized as follows. The description, assumptions and the mathematical model are introduced in "Problem description and assumption". "Computational results" embraces a numerical example, and also the results of the solved model are presented here. In "Sensitivity analysis", sensitivity analysis is applied to verify the accuracy of the model. Finally, "Conclusion" represents the conclusions of the paper.

Problem description and assumption

There are many influencing factors in biofuel supply chain which impact on each other. The whole system may change unpredictably by changing any of these factors. Given that the mathematical model can calculate these very detailed interactions, in this study, a mathematical model is applied for designing biofuel supply chain. Biofuel supply chain consists of the following echelons:

1. Biomass centers.
2. Biomass preprocessing center.
3. Plants for biofuel production.
4. Biofuel warehouses.
5. Demand points.

Three types of biomass exists generally; woody source, non-woody source, and animal fat and waste. In this paper, we consider woody source of biomass as an input to biofuel supply chain. At the first echelon, we have three ways to procure biomass from the biomass centers: cultivating the biomass, purchasing them from domestic supplier and importing them from abroad. When the biomass is procured, we need a place for storing and drying them; therefore, echelon 2 is assigned to these warehouses. Echelon 3 represents plants of biofuel production. Fourth echelon states warehouses for biofuel storage, and the last echelon is the demand center (customer), as shown in Fig. 1.

We considered three capable regions for warehouses of biomass and K capable regions for plants. The model chooses j, k and l regions to establish warehouses for biomass, plants and warehouses of biofuels. We can purchase or rent the warehouses needed for biomass, plants and biofuel warehouses. The plants can be established in three sizes (small, medium and large). The interest rate is monthly. In the process of plants, α percent of biomass has become biofuel, β percent of the biomass are dried. In addition, we have inventory costs in each warehouse.

Mathematical programming

There are so many papers which considered mathematical programming for modeling the problems in various areas (Mousavi et al. 2014; AriaNezhad et al. 2013; Alimardani et al. 2013; Seifbarghy et al. 2015). In this section, we develop mixed integer linear programming (MILP) model. For modeling the problem we need to present the indices, parameters, and variables which are introduced in Tables 1, 2 and 3 respectively.

To simplify the problem, two models are introduced as follows. First model selects the best depreciation method from sum-of-the-years-digits method (SOYD), straight line and double declining balance (DDB) to determine the best switch points to maximize the cash flow. The second model calculates all costs of biofuel supply chain by considering

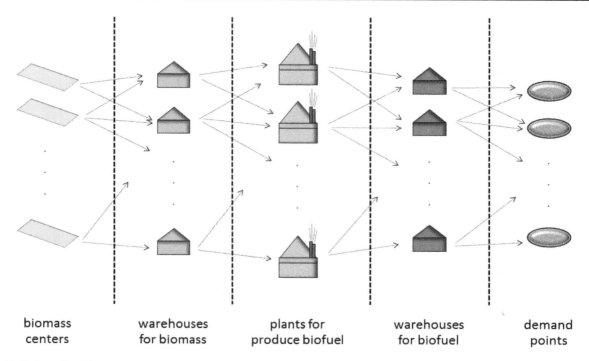

biomass   warehouses   plants for   warehouses   demand
centers   for biomass   produce biofuel   for biofuel   points

Fig. 1 Biofuel supply chain

Table 1 The indices of the model

Indices	Description	Set
$i \in I$	Biomass center	$I = 1,2, ..., I$
$j \in J$	Preprocessing center of biomass	$J = 1,2, ..., J$
$k \in K$	Biofuel production plants	$K = 1,2, ..., K$
$p \in P$	Plant size	$P = 1,2,3$
$l \in L$	Warehouse for biofuel	$L = 1,2, ..., L$
$t \in T$	Time period	$T = 1,2, ..., T$
$w \in W$	Demand point	$W = 1,2, ..., W$

the depreciation cost of installed plants which is obtained from the first model.

First model

According to the fact that each organization desires to select the best depreciation method to reduce their costs, this model facilitates selecting the depreciation cost by which they could be able to choose the best one with regards to the net present value of depreciation in every year.

$$\min z1 = \sum_{t=1}^{T} \sum_{p=1}^{P} \frac{D_{pt}T}{(1 + ir)^t} \qquad (1)$$

$$D_{pt} \geq \frac{BV_{t-1} - SV_p}{n - t + 1} \qquad \forall p, t \qquad (2)$$

$$D_{pt} \geq BV_0 \left(\frac{\alpha}{N}\right)\left(1 - \frac{\alpha}{N}\right)^{t-1} \qquad \forall p, t \qquad (3)$$

$$D_{pt} \geq \frac{2(BV_0 - SV_p)(N - t + 1)}{N(N + 1)} \qquad \forall p, t \qquad (4)$$

$$BV_{t+1} = BV_t - D_{pt} \qquad \forall p, t \qquad (5)$$

The objective function (1) calculates the total net present value of depreciations of the all plants for all periods of time. Constraints (2–5) represent the depreciation methods which could be utilized to calculating the depreciation. Almost always the owner of any factory would like to state that the depreciation of the equipment in the factory is a lot, to pay as little tax as they can. So the straight line, SOYD and DDB methods are introduced as the depreciation methods.

Second model

In the second model, at first, biomass is provided through three different ways, purchasing, importing and harvesting the provided inputs maintained in preprocessing centers. The plants producing biofuels could be established in three sizes: small, medium and large. Biofuels are sold to the customers from biofuel warehouses where biofuels are kept. The second model is presented as follows:

$$\max z2 = \sum_{l=1}^{L}\sum_{w=1}^{W}\sum_{t=1}^{T}(1-T)/(1+ir)^t \cdot s'_{lwt} \cdot P'_t - \left[\sum_{k=1}^{K}\sum_{p=1}^{P}(\text{BC}_{kp} \cdot z^1_{kp} + \text{RC}_{kp} \cdot z^{1'}_{kp}) + \sum_{j=1}^{J}(\text{BC}_j \cdot z^2_j + \text{RC}_j \cdot z^{2'}_j)\right.$$

$$+ \sum_{l=1}^{L}(\text{BCC}_l \cdot z^3_l + \text{RCC}_l \cdot z^{3'}_l) + \sum_{t=1}^{T}\sum_{j=1}^{J}(1-T)/(1+ir)^t \cdot (\text{CC}_t \cdot x_{1jt} + \text{BCB}_t \cdot x_{2jt} + \text{ICB}_t \cdot x_{3jt})$$

$$+ \sum_{i=1}^{I}\sum_{j=1}^{J}\sum_{t=1}^{T}(1-T)/(1+ir)^t \cdot C_{ijt} \cdot d_{ijt} \cdot x_{ijt} + \sum_{k=1}^{K}\sum_{p=1}^{P}\sum_{j=1}^{J}\sum_{t=1}^{T}(1-T)/(1+ir)^t \cdot C_{jkpt} \cdot d_{jkpt} \cdot y_{jkpt}$$

$$+ \sum_{k=1}^{K}\sum_{p=1}^{P}\sum_{l=1}^{L}\sum_{t=1}^{T}(1-T)/(1+ir)^t \cdot C_{kplt} \cdot d_{kplt} \cdot s_{kplt} + \sum_{w=1}^{W}\sum_{l=1}^{L}\sum_{t=1}^{T}(1-T)/(1+ir)^t \cdot C_{lwt} \cdot d_{lwt} \cdot s'_{lwt} \quad (6)$$

$$+ \sum_{j=1}^{J}\sum_{t=1}^{T}(1-T)/(1+ir)^t \cdot \text{SC}_{jt} \cdot I_{jt} + \sum_{l=1}^{L}\sum_{t=1}^{T}(1-T)/(1+ir)^t$$

$$\cdot \text{SC}'_{lt} \cdot \text{II}_{lt} + \sum_{k=1}^{K}\sum_{p=1}^{P}\sum_{j=1}^{J}\sum_{t=1}^{T}(1-T)/(1+ir)^t \cdot \alpha \cdot \text{PC}_t \cdot y_{jkpt}$$

$$+ \sum_{w=1}^{W}\sum_{t=1}^{T}\sum_{l=1}^{L}(1-T)/(1+ir)^t \cdot \rho \cdot B_{lwt} + \left.\sum_{k=1}^{K}\sum_{p=1}^{P}\sum_{j=1}^{J}\sum_{t=1}^{T}\gamma \cdot \text{EM}_{kpt} \cdot \alpha \cdot y_{jkpt}\right]$$

Subject to:

$$z^{1'}_{kp} + z^1_{kp} \leq 1 \quad \forall k, p \qquad (7)$$

$$z^{2'}_j + z^2_j \leq 1 \quad \forall j \qquad (8)$$

$$z^{3'}_l + z^3_l \leq 1 \quad \forall l \qquad (9)$$

$$\sum_{p=1}^{P}\sum_{k=1}^{M}(z^{1'}_{kp} + z^1_{kp}) = k \qquad (10)$$

$$\sum_{p=1}^{P}(z^{1'}_{kp} + z^1_{kp}) = 1 \quad \forall k \qquad (11)$$

$$\sum_{j=1}^{J}\text{EM}_{kpt} \cdot \alpha \cdot y_{jkpt} \leq \text{EMMAX} \quad \forall k, p, t \qquad (12)$$

$$\sum_{j=1}^{J}x_{ijt} \leq \text{capr}_{it} \quad \forall i, t \qquad (13)$$

$$\beta\sum_{t=1}^{T}\sum_{i=1}^{I}x_{ijt} \geq \sum_{k=1}^{K}\sum_{p=1}^{P}\sum_{t=1}^{T}y_{jkpt} \quad \forall j \qquad (14)$$

$$I_{jt} \leq \text{capb}_{jt} \quad \forall j, t \qquad (15)$$

$$\sum_{j=1}^{J}\sum_{t=1}^{T}y_{jkpt} \cdot \alpha \geq \sum_{l=1}^{L}\sum_{t=1}^{T}s_{kplt} \quad \forall k, p \qquad (16)$$

$$\sum_{j=1}^{J}y_{jkpt} \leq (z_{kp} + z'_{kp}) \cdot M \quad \forall k, p, t \qquad (17)$$

$$\text{II}_{lt} \leq \text{capw}_{lt} \quad \forall l, t \qquad (18)$$

$$I_{j(t-1)} + \sum_{i=1}^{I}x_{ijt} - \sum_{k=1}^{K}\sum_{p=1}^{P}y_{jkpt} - I_{jt} \geq 0 \quad \forall j, t \qquad (19)$$

$$\text{II}_{l(t-1)} - \text{II}_{lt} + \sum_{w=1}^{W}(B_{lw(t-1)} - B_{lwt} - s'_{lwt})$$
$$+ \sum_{k=1}^{K}\sum_{p=1}^{P}s_{kpt} \geq 0 \qquad (20)$$
$$\forall l, t$$

$$\sum_{l=1}^{L}(s'_{lwt} + B_{lwt} - B_{lw(t-1)}) \geq dd_{wt} \quad \forall w, t \qquad (21)$$

$$\sum_{i=1}^{I}x_{ijt} \leq \left(z^2_j + z^{2'}_j\right) \cdot M \quad \forall j, t \qquad (22)$$

$$\sum_{k=1}^{K}\sum_{p=1}^{P}s_{kplt} \leq \left(z^3_l + z^{3'}_l\right) \cdot M \quad \forall l, t \qquad (23)$$

$$\sum_{l=1}^{L}s_{kplt} \leq \text{capp}_{kpt} \quad \forall k, p, t \qquad (24)$$

$$z^{1'}_{kp}, z^1_{kp}, z^{2'}_j, z^2_j, z^{3'}_l, z^3_l \in \{0, 1\} \qquad (25)$$

Objective function of second model (6) consists of two terms. First one states the present value of revenues and second is the costs. The model also demonstrates the revenue of the supply chain gained by selling the biofuels. The total costs are calculated by the cost of purchasing or renting preprocessing centers, plants and warehouses in $t = 0$, the total cost of biomass (consists of buying, importing from

Table 2 The parameters of the model

Notations	Description	Notations	Description
P'_t	Sale price of biofuel in period t	C_{klt}	Transportation cost for each unit of biofuel between plant k and warehouse l in period t
BC_{kp}	Buy cost for plant k with size p	C_{lwt}	Transportation cost for each unit of biofuel between warehouse l and demand point w in period t
RC_{kp}	Rent cost for plant k with size p	$capr_{it}$	Capacity of resource of biomass i in period t
BC_j	Buy cost for warehouse j	$capb_{jt}$	Capacity of the biomass of preprocessing center j in period t
RC_j	Rental cost for warehouse j	$capw_{lt}$	Capacity of the warehouse j in period t
BC'_l	Buy cost for warehouse l	$capp_{kpt}$	Capacity of plant k with size p in period t
RC'_l	Rental cost for warehouse l	SC'_{lt}	Storage cost of warehouse l for each unit of biofuel in period t
BCB_t	Buy cost for each unit of biomass in period t	PC	Process cost of each unit of biomass
ICB_t	Import cost for each unit of biomass in period t	d'_{wt}	Demand of demand center w in period t
CC_t	Cultivation cost for each unit of biomass in period t	EM_{kpt}	Amount of CO_2 that emission by plant k with size p for each unit of produced biofuel in period t
d_{ijt}	Distance between biomass center i and warehouse j in period t	α	Fraction of biomass conversion to biofuel
d_{jkt}	Distance between warehouse j and plant k in period t	β	Percentage of biomass dry in warehouse
d_{klt}	Distance between plant k and warehouse l in period t	ρ	Penalty cost of shortage in meeting the demands
d_{lw}	Distance between warehouse l and demand point w in period t	EMMAX	Maximum permissible amount of generating the gases in the plant
C_{ijt}	Transportation cost for each unit of biomass between biomass center i and warehouse j in period t	BV_t	The book value of plant in period t
C_{jkt}	Transportation cost for each unit of biomass between warehouse j and plant k in period t	SV_p	The salvage value of plant p

Table 3 The variable of the model

Variables	Descriptions
x_{ijt}	Flow of input biomass i to warehouse j in period t
y_{jkpt}	Flow of biomass from warehouse j to plant k with size p in period t
s_{kplt}	Flow of biofuel from plant k with size p to warehouse l in period t
s'_{lwt}	Flow of biofuel from warehouse l to demand point w in period t
z^1_{kp}	=1 if we purchase the plant k with size p and 0 if we do not purchase the plant k with size p
$z^{1'}_{kp}$	=1 if we rent the plant k with size p and 0 if we do not rent the plant k with size p
z^2_j	=1 if we buy the warehouse j and 0 if we do not buy the warehouse j
$z^{2'}_j$	=1 if we rent the warehouse j and 0 if we do not rent warehouse j
z^3_l	=1 if we buy the warehouse l and 0 if we do not buy the warehouse l
$z^{3'}_l$	=1 if we rent the warehouse l and 0 if we do not rent warehouse l
D_{pt}	Depreciation of plant with size p in period t
I_{jt}	Inventory level of warehouse j at the end of period t
Π_{lt}	Inventory level of warehouse l in the period t
B_{lwt}	Backorder of warehouse l in the period t

abroad or harvested one), the total cost of transportation in the whole supply chain, the inventory cost in all periods, the production cost of biofuel, the penalty cost of shortage in meeting the demands and the emission cost of the plants in all periods. As said before, the depreciation costs of installed plants are obtained from the first model.

Table 4 Inputs' capacity in period t

Capacity of inputs (ton)	Period											
	1	2	3	4	5	6	7	8	9	10	11	12
Purchasing	200	120	220	340	200	120	220	340	200	120	220	340
Importing	200	120	220	340	200	120	220	340	200	120	220	340
Harvesting	310	200	200	700	200	120	220	340	200	120	220	340

Constraints (7–9), respectively, represent that the preprocessing center of biomass, plants and the warehouse of biofuels can be purchased or rented. Constraint 10 indicates that k lactation of K candidates is selected to establish the plants. Constraint 11 shows that only one size of plants can be establish in each selected location. Constraint 12 shows the CO_2 emission in every plant should be less than maximum limitation of CO_2 generation. Constraint 13 represents the maximum capacity of the input biomass. Constraint 14 shows that the β percent of biomass dry in warehouse of biomass then is sent to plants. Constraint 14 and 15 state that the inventory level of preprocessing centers of biomass and warehouses of biofuels, respectively, should been less than maximum capacity. Constraint 16 states α percent of the preprocessed biomass transferred to the next stage. Constraints 17, 22 and 23 are logic constraints stating that no biofuel can be produced unless there is a plant operating at this location, no biomass can be utilized unless there is a preprocessing center operating at that location and no biofuel can be sold unless there is a warehouse operating at that location. Constraints 19 and 20 are the inventory constraint in each warehouse. Constraint 21 shows that the demands in place w should be met by supply.

Computational results

In this section, a hypothetical numerical example is presented to state the applicability of the model. The indices and parameters of the example are as follows: the numbers of biomass center, preprocessing centers of biomass, plants, warehouse for products and demand point are three; the preprocessing centers, plants and warehouses have constant capacity which is declared in Table 4.

The outputs of the model indicate the amount of all variables which are used there. The results indicate that all three plants should be made in small sizes. The decision variables about purchasing or renting the warehouses and plants are shown in Table 5. The biomass preprocessing centers which needed to be established are purchased as well as product warehouses. But, for plants two of them are rented and other one is purchased.

Table 5 Either purchasing or renting the warehouses and plants

		Purchase	Rent
Preprocessing centers of biomass	1	1	
	2	1	
	3	1	
Plant	1	1	
	2		1
	3		1
Warehouse for products	1	1	
	2	1	
	3	1	

To solve the second model, it is needed to specify the parameters, such as investment cost and salvage value, which are stated in Table 6. To increase the reality of the example, data are gathered through experts' opinions.

The flows of the biomass and biofuels in the supply chain network are represented in Tables 7, 8, 9 and 10.

The amount of each types of the biomass to each preprocessing centers in each of the months of the year is presented in Table 7.

Amount of biomass sent from warehouses to plants which are calculated through model is stated in Table 8 and depicted in Fig. 2 for clarifying these flows.

Amount of biofuel sent from plants to warehouses which are calculated through model is presented in Table 9.

Amount of biofuel sent from warehouses to demand points which are calculated through model is stated in Table 10.

According to the results, in all periods, there are flows between the echelons in the supply chain network from the biomass centers to demand points. So, twelve diagrams can be depicted related to each period (e.g., the flow of the network in period $t = 1$ is depicted in Fig. 2). Figure 2 is depicted to clarify the amount of flows of Figs. 7, 8, 9 and 10.

Regularly the assets and equipment depreciation are calculated by only one or a combination of DDB, straight line and SOYD. Given the fact that the owner of the facilities tend to pay as little tax as they can in the early years, the combination of these methods is utilized to state depreciation value. It is presumed that in the beginning of each period these three methods would be utilized. The

Table 6 The investment cost and salvages value of each plant and warehouse

		Investment cost (USD)	Salvages value (USD)
Preprocessing centers of biomass	1	10,000	3000
	2	15,000	4500
	3	10,400	3500
Plant	1	100,000	30,000
	2	90,000	20,000
	3	110,000	40,000
Warehouse for products	1	10,300	3000
	2	10,100	3000
	3	13,400	4000

Table 7 Amount of biomass i to preprocessing center j in period t

Biomass	Period											
	1	2	3	4	5	6	7	8	9	10	11	12
Preprocessing center 1												
1	53.7	•	•	256	•	•	•	•	•	•	220	•
2	•	86.3	•	193.8	•	•	•	•	•	•	220	•
3	•	•	•	•	•	•	•	•	•	•	•	•
Preprocessing center 2												
1	146.3	120	220	•	200	•	•	•	200	120	•	106.7
2	•	33.7	220	•	•	•	•	•	•	•	•	340
3	310	200	•	•	•	•	•	340	•	•	•	•
Preprocessing center 3												
1	•	•	•	84	•	120	220	340	•	•	•	233.3
2	•	•	•	•	•	120	•	340	•	•	•	•
3	•	•	80	•	•	120	•	•	200	•	•	•

Table 8 Amount of biomass from warehouse j to plant k with size $p = 1$ in period t

Preprocessing center	Period											
	1	2	3	4	5	6	7	8	9	10	11	12
Plant 1												
1	•	129.5	•	•	•	•	•	•	•	•	116.8	•
2	161	•	•	•	172.6	•	•	137.9	•	175.8	•	208.4
3	•	•	120	65.3	•	141.1	160	•	80.5	•	•	•
Plant 2												
1	161	129.5	•	•	172.6	•	•	•	•	•	116.8	•
2	•	•	•	130.5	.	141.1	160	137.9	•	•	•	208.4
3	•	•	120	•	•	•	•	•	161.1	175.8	•	•
Plant 3												
1	•	•	•	•	172.6	•	•	•	•	•	•	•
2	143	129.5	120	•	•	•	•	•	161.1	175.8	116.8	•
3	•	•	•	130.5	•	141.1	160	137.9	•	•	•	208.4

maximum depreciation value of these methods is chosen in that period. The year that the method of depreciation is switched to another is called the switch point. The depreciation of the installed preprocessing centers of biomass, plant and warehouse for products in each year is indicated in Table 11. Also, the switch points are

Table 9 Flow of biofuel from plant k with size $p = 1$ to warehouse l in period t

Plant	Period											
	1	2	3	4	5	6	7	8	9	10	11	12
Warehouse 1 for products												
1	100	•	•	240	150	•	•	•	•	45.5	140	•
2	•	•	•	•	150	•	•	250	•	•	•	•
3	100	•	•	•	150	90	•	•	•	•	140	•
Warehouse 2 for products												
1	•	•	•	•	•	•	•	•	170	•	•	•
2	100	100	•	•	•	•	•	60	•	•	•	94
3	•	•	•	•	•	•	•	•	170	•	•	•
Warehouse 3 for products												
1	•	100	120	•	•	90	180	250	•	•	•	•
2	•	•	120	240	•	90	120	•	170	90	140	•
3	•	100	120	240	•	•	180	250	•	90	•	77

Table 10 Flow of biofuel from warehouse l to demand point w in period t

Warehouse for products	Period											
	1	2	3	4	5	6	7	8	9	10	11	12
Demand point 1												
1	12	•	•	•	•	•	•	26	19.33	•	40.5	•
2	•	•	•	•	•	•	20	•	1.667	•	93.67	13.33
3	.	13	28	23	26	21	•	•	•	•	311	6.67
Demand point 2												
1	25	•	•	•	•	20	•	•	20	•	31	•
2	•	•	•	•	27	•	•	•	•	•	•	•
3	•	27	19	25	•	•	21	19	•	•	•	19
Demand point 3												
1	22	•	•	•	10	237.67	•	18	.	•	37	•
2	•	•	•	•	•	39.67	•	•	18	•	•	18
3	•	21	34	20	•	149.67	139	•	•	•	•	•

highlighted in Table 11. According to the calculated depreciation in the model 1, the book value of each installed preprocessing centers of biomass, plant and warehouse for products at beginning of year are shown Table 12.

Sensitivity analysis

For more description of book value results, some figures are illustrated in "Appendix".

Lots of parameters exist in the proposed model which can change the objective function level. In this study the rate of return (ROR) of the project is calculated. The result from Fig. 3 shows ROR = 78 %. Unless interest rate is less than 78 % the project is not acceptable. Investment in this project would be done while the interest rate would be less than ROR. Figure 3 shows that while the interest rate is less than 78 % the objective function of the model is positive and the project is reasonable.

Conclusion

The depletion of the fossil fuels and major concerns about the security of energy in the future to produce fuels led to utilizing the renewable energies such as biofuels. This paper presented a general optimization model which enables the selection of preprocessing centers for the biomass, biofuel plants, and warehouses to store the biofuels. Two models are introduced to calculate the benefits of biofuel supply chain. At the first model, the depreciation costs of the installed centers are calculated by means of three methods (straight line, SOYD and DDB). The results of the first model indicated that at the preprocessing center, in the installed plants and the warehouses, method of

Fig. 2 The flow of biomass and biofuel from resources of biomass to demand points

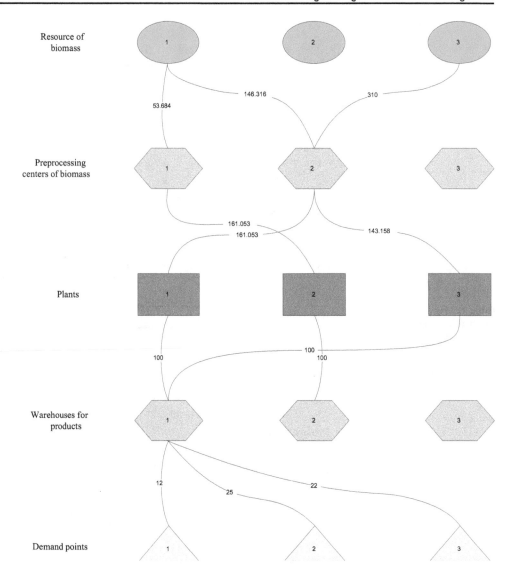

Table 11 The depreciation of the installed preprocessing centers of biomass, plant and warehouse for production each year

Year	Preprocessing centers of biomass			Plant			Warehouse for products		
	1	2	3	1	2	3	1	2	3
1	1301.8	1952.7	1353.8	13,017.8	11,716.0	14,319.5	1340.8	1314.8	1744.4
2	1101.5	1652.3	1145.6	11,015.0	9913.5	12,116.5	1134.5	1112.5	1476.0
3	932.0	1398.1	969.3	9320.4	8388.4	10252.4	960.0	941.4	1248.9
4	788.6	1183.0	820.2	7886.5	7097.8	8675.1	812.3	796.5	1056.8
5	667.3	1001.0	694.0	6673.2	6005.9	7340.5	687.3	674.0	894.2
6	564.7	847.0	587.2	5646.5	5081.9	6211.2	581.6	570.3	756.6
7	477.8	716.7	496.9	4777.8	**4787.5**	5255.6	492.1	482.6	640.2
8	**411.3**	**617.0**	420.5	**4113.4**	4787.5	4447.1	**438.7**	**420.5**	**554.5**
9	411.3	617.0	**355.8**	4113.4	4787.5	3762.9	438.7	420.5	554.5
10	411.3	617.0	352.6	4113.4	4787.5	3184.0	438.7	420.5	554.5
11	411.3	617.0	352.6	4113.4	4787.5	**2918.2**	438.7	420.5	554.5
12	411.3	617.0	352.6	4113.4	4787.5	2918.2	438.7	420.5	554.5

Table 12 The book value of each installed preprocessing centers of biomass, plant and warehouse for products at beginning of year

	Preprocessing center of biomass			Plant			Warehouse for products		
	1	2	3	1	2	3	1	2	3
Year									
0	10,000	15,000	10,400	100,000	90,000	110,000	10,300	10,100	13,400
1	8898.5	13,347.7	9254.4	88,985.0	80,086.5	97,883.5	9165.5	8987.5	11,924.0
2	7966.5	11,949.7	8285.1	79,664.6	71,698.1	87,631.0	8205.5	8046.1	10,675.1
3	7177.8	10,766.7	7464.9	71,778.1	64,600.3	78,955.9	7393.1	7249.6	9618.3
4	6510.5	9765.7	6770.9	65,104.9	58,594.4	71,615.4	6705.8	6575.6	8724.1
5	5945.8	8918.8	6183.7	59,458.4	53,512.5	65,404.2	6124.2	6005.3	7967.4
6	5468.1	8202.1	5686.8	54,680.5	48,725.0	60,148.6	5632.1	5522.7	7327.2
7	5056.7	7585.1	5266.3	50,567.1	43,937.5	55,701.5	5193.4	5102.3	6772.7
8	4645.4	6968.1	4910.6	46,453.7	39,150.0	51,938.6	4754.7	4681.8	6218.1
9	4234.0	6351.0	4557.9	42,340.3	34,362.5	48,754.6	4316.0	4261.4	5663.6
10	3822.7	5734.0	4205.3	38,226.8	29,575.0	45,836.4	3877.4	3840.9	5109.1
11	3411.3	5117.0	3852.6	34,113.4	24,787.5	42,918.2	3438.7	3420.5	4554.5

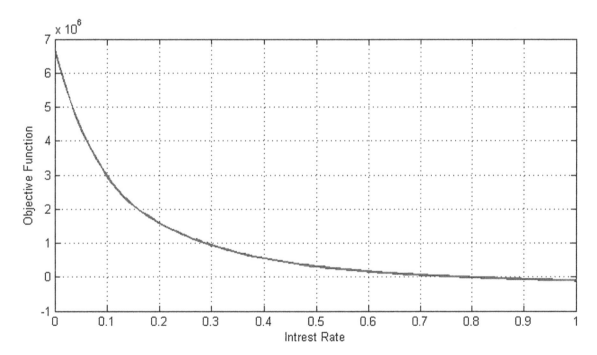

Fig. 3 Interest rate with objective function

calculating the depreciation changed from DDB to straight line in period ($t = 7, t = 7, t = 8$), ($t = 7, t = 6, t = 10$) and ($t = 7, t = 7, t = 7$), respectively. Therefore, for the installed centers the depreciation is utilized by the results of the first model. The results of the second model indicate that the small-size plants are purchased. Three preprocessing centers are purchased. One of the plants is rented and two are purchased and all of the warehouses of the biofuel are purchased. Also the results of the supply chain flows are indicated in the tables. Depreciation cost could be considered in designing all supply chains, such as automobile and oil. This new concept helps top managers decide well in designing supply chains and establishing plants and warehouses. Parameters in the model could be considered uncertain, for example, demand in the biofuel supply chain can be considered fuzzy.

Appendix

Switching point shows that the strategy of selecting depreciation method is changed from one to another. According to the fact that each organization desires to select the best depreciation method to reduce their cost, this model facilitates selecting depreciation cost by which they could be able to choose the best one with regards to the net present value of depreciation in each year. Switching points for each preprocessing centers, plants and warehouses are presented in Figs. 4, 5, 6, 7, 8, 9, 10, 11 and 12.

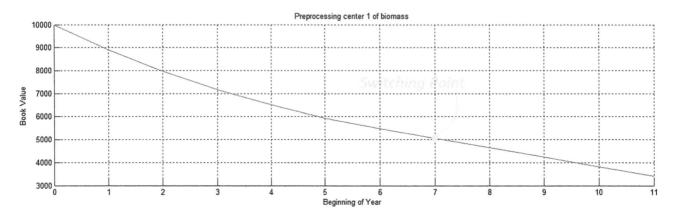

Fig. 4 Book value of preprocessing center 1 with regards to year

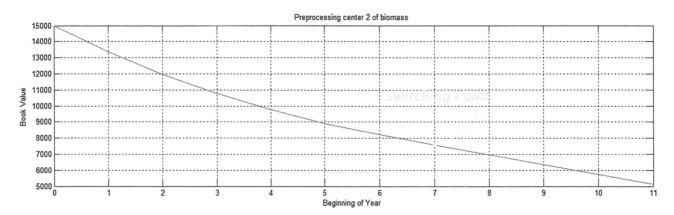

Fig. 5 Book value of preprocessing center 2 with regards to year

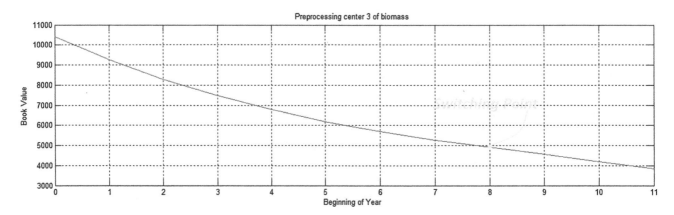

Fig. 6 Book value of preprocessing center 3 with regards to year

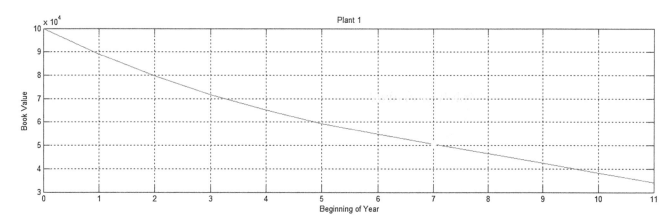

Fig. 7 Book value of plant 1 with regards to year

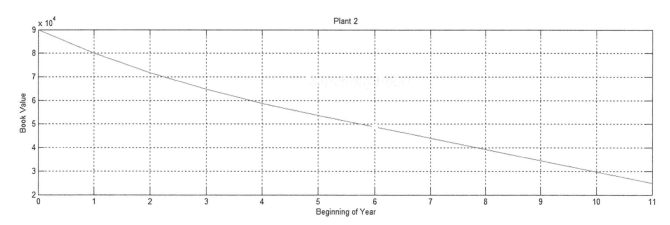

Fig. 8 Book value of plant 2 with regards to year

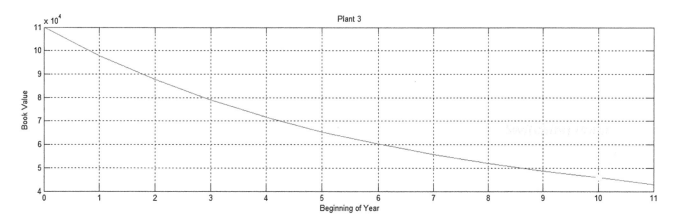

Fig. 9 Book value of plant 3 with regards to year

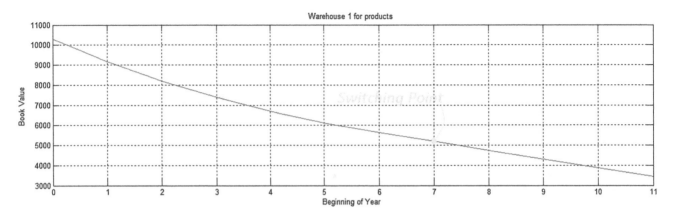

Fig. 10 Book value of warehouse 1 with regards to year

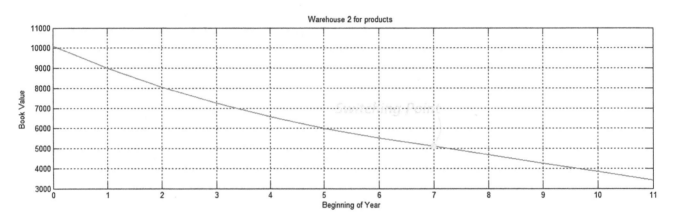

Fig. 11 Book value of warehouse 2 with regards to year

Fig. 12 Book value of warehouse 3 with regards to year

References

Akgul O, Zamboni A, Bezzo F, Shah N, Papageorgiou LG (2011) Optimization-based approaches for bioethanol supply chains. Ind Eng Chem Res 50(9):4927–4938. doi:10.1021/ie101392y

Akgul O, Shah N, Papageorgiou LG (2012) Economic optimisation of a UK advanced biofuel supply chain. Biomass Bioenergy 41:57–72. doi:10.1016/j.biombioe.2012.01.040

Alimardani M, Jolai F, Rafiei H (2013) Bi-product inventory planning in a three-echelon supply chain with backordering, Poisson demand, and limited warehouse space. J Ind Eng Int 9(1):1–13. doi:10.1186/2251-712X-9-22

AriaNezhad MG, Makuie A, Khayatmoghadam S (2013) Developing and solving two-echelon inventory system for perishable items in a supply chain: case study (Mashhad Behrouz Company). J Ind Eng Int 9(1):1–10. doi:10.1186/2251-712X-9-39

Ayoub N, Yuji N (2012) Demand-driven optimization approach for biomass utilization networks. Comput Chem Eng 36:129–139. doi:10.1016/j.compchemeng.2011.09.005

Ayoub N, Martins R, Wang K, Seki H, Naka Y (2007) Two levels decision system for efficient planning and implementation of bioenergy production. Energy Convers Manage 48(3):709–723. doi:10.1016/j.enconman.2006.09.012

Ayoub N, Elmoshi E, Seki H, Naka Y (2009) Evolutionary algorithms approach for integrated bioenergy supply chains optimization. Energy Convers Manag 50(12):2944–2955. doi:10.1016/j.enconman.2009.07.010

Chen C-W, Fan Y (2012) Bioethanol supply chain system planning under supply and demand uncertainties. Transp Res Part E Logist Transp Rev 48(1):150–164. doi:10.1016/j.tre.2011.08.004

Čuček L, Varbanov PS, Klemeš JJ, Kravanja Z (2012) Total footprints-based multi-criteria optimisation of regional biomass energy supply chains. Energy 44(1):135–145. doi:10.1016/j.energy.2012.01.040

Fazlollahi S, Maréchal F (2013) Multi-objective, multi-period optimization of biomass conversion technologies using evolutionary algorithms and mixed integer linear programming (MILP). Appl Therm Eng 50(2):1504–1513. doi:10.1016/j.applthermaleng.2011.11.035

Jackson SB, Liu XK, Cecchini M (2009) Economic consequences of firms' depreciation method choice: evidence from capital investments. J Acc Econ 48(1):54–68. doi:10.1016/j.jacceco.2009.06.001

Judd JD, Sarin SC, Cundiff JS (2012) Design, modeling, and analysis of a feedstock logistics system. Bioresour Technol 103(1):209–218. doi:10.1016/j.biortech.2011.09.111

Keirstead J, Samsatli N, Pantaleo AM, Shah N (2012) Evaluating biomass energy strategies for a UK eco-town with an MILP optimization model. Biomass Bioenergy 39:306–316. doi:10.1016/j.biombioe.2012.01.022

Kim J, Realff MJ, Lee JH, Whittaker C, Furtner L (2011) Design of biomass processing network for biofuel production using an MILP model. Biomass Bioenergy 35(2):853–871. doi:10.1016/j.biombioe.2010.11.008

Kostin AM, Guillén-Gosálbez G, Mele FD, Bagajewicz MJ, Jiménez L (2012) Design and planning of infrastructures for bioethanol and sugar production under demand uncertainty. Chem Eng Res Design 90(3):359–376. doi:10.1016/j.cherd.2011.07.013

Kumar R, Sharma AK, Tewari PC (2015) Cost analysis of a coal-fired power plant using the NPV method. J Ind Eng Int 11(4):495–504. doi:10.1007/s40092-015-0116-8

Leão RRDCC, Hamacher S, Oliveira F (2011) Optimization of biodiesel supply chains based on small farmers: a case study in Brazil. Bioresour Technol 102(19):8958–8963. doi:10.1016/j.biortech.2011.07.002

Leduc S, Schwab D, Dotzauer E, Schmid E, Obersteiner M (2008) Optimal location of wood gasification plants for methanol production with heat recovery. Int J Energy Res 32(12):1080–1091. doi:10.1002/er.1446

Mahmoudi R, Hafezalkotob A, Makui A (2014) Source selection problem of competitive power plants under government intervention: a game theory approach. J Ind Eng Int 10(3):59. doi:10.1007/s40092-014-0059-5

Meier A, Gremaud N, Steinfeld A (2005) Economic evaluation of the industrial solar production of lime. Energy Convers Manag 46(6):905–926. doi:10.1016/j.enconman.2004.06.005

Mele FD, Guillén-Gosálbez G, Jiménez L (2009) Optimal planning of supply chains for bioethanol and sugar production with economic and environmental concerns. Comput Aided Chem Eng 26:997–1002. doi:10.1016/S1570-7946(09)70166-X

Mobini M, Sowlati T, Sokhansanj S (2011) Forest biomass supply logistics for a power plant using the discrete-event simulation approach. Appl Energy 88(4):1241–1250. doi:10.1016/j.apenergy.2010.10.016

Mousavi SH, Nazemi A, Hafezalkotob A (2014) Using and comparing metaheuristic algorithms for optimizing bidding strategy viewpoint of profit maximization of generators. J Ind Eng Int 11(1):59–72. doi:10.1007/s40092-014-0094-2

Pérez-Fortes M, Laínez-Aguirre JM, Arranz-Piera P, Velo E, Puigjaner L (2012) Design of regional and sustainable bio-based networks for electricity generation using a multi-objective MILP approach. Energy 44(1):79–95. doi:10.1016/j.energy.2012.01.033

Petroleum B (2015) BP statistical review of world energy. British Petroleum, London. Available at http://www.bp.com/statisticalreview

Rentizelas AA, Tatsiopoulos IP (2010) Locating a bioenergy facility using a hybrid optimization method. Int J Prod Econ 123(1):196–209. doi:10.1016/j.ijpe.2009.08.013

Seifbarghy M, Kalani MM, Hemmati M (2015) A discrete particle swarm optimization algorithm with local search for a production-based two-echelon single-vendor multiple-buyer supply chain. J Ind Eng Int. doi:10.1007/s40092-015-0126-6

Tumuluru JS, Sokhansanj S, Wright CT, Boardman RD, Yancey NA (2011) A review on biomass classification and composition, co-firing issues and pretreatment methods. In: Proceedings of the American society of agricultural and biological engineers annual international meeting, pp 2053–2083. doi:10.13031/2013.37191

Velazquez-Marti B, Fernandez-Gonzalez E (2010) Mathematical algorithms to locate factories to transform biomass in bioenergy focused on logistic network construction. Renew Energy 35(9):2136–2142. doi:10.1016/j.renene.2010.02.011

Zamboni A, Bezzo F, Shah N (2009) Supply chain optimization for bioethanol production system in Northern Italy: environmentally conscious strategic design. Comput Aided Chem Eng 27:2037–2042. doi:10.1016/S1570-7946(09)70730-8

Zhang J, Osmani A, Awudu I, Gonela V (2013) An integrated optimization model for switchgrass-based bioethanol supply chain. Appl Energy 102:1205–1217. doi:10.1016/j.apenergy.2012.06.054

Zhu X, Yao Q (2011) Logistics system design for biomass-to-bioenergy industry with multiple types of feedstocks. Bioresour Technol 102(23):10936–10945. doi:10.1016/j.biortech.2011.08.121

12

A genetic algorithm for a bi-objective mathematical model for dynamic virtual cell formation problem

Mostafa Moradgholi[1] · Mohammad Mahdi Paydar[2] · Iraj Mahdavi[1] · Javid Jouzdani[3]

Abstract Nowadays, with the increasing pressure of the competitive business environment and demand for diverse products, manufacturers are force to seek for solutions that reduce production costs and rise product quality. Cellular manufacturing system (CMS), as a means to this end, has been a point of attraction to both researchers and practitioners. Limitations of cell formation problem (CFP), as one of important topics in CMS, have led to the introduction of virtual CMS (VCMS). This research addresses a bi-objective dynamic virtual cell formation problem (DVCFP) with the objective of finding the optimal formation of cells, considering the material handling costs, fixed machine installation costs and variable production costs of machines and workforce. Furthermore, we consider different skills on different machines in workforce assignment in a multi-period planning horizon. The bi-objective model is transformed to a single-objective fuzzy goal programming model and to show its performance; numerical examples are solved using the LINGO software. In addition, genetic algorithm (GA) is customized to tackle large-scale instances of the problems to show the performance of the solution method.

Keywords Virtual cell formation · Genetic algorithm · Workforce assignment · Bi-objective mathematical programming · Fuzzy goal programming

Introduction

In current competitive business environment, customers demand diverse products with higher quality at lower costs. Therefore, manufacturers tend to reduce investment on tools, parts and area and increase their flexibility. With more efficient overall control techniques, companies and businesses use effective approaches in supply, manufacturing and distribution. Production costs constitute a significant share in the total costs incurred by a company. Conventional manufacturing systems (e.g., workshop or flowshop) are not flexible enough to respond to changes. As a result, cellular manufacturing (CM) as technique, stem from group technology (GT) has emerged as a promising manufacturing system. CM is described as a manufacturing procedure which produces part families within a cell of machines serviced by operators and/or robots functioning only within the cell. CMSs have some advantages, such as reduction in lead times, work-in-process inventories, setup times, etc. (Heragu 1994; Wemmerlov and Hyer 1989). However, the performance of CMS depends significantly on the stability of demand reGArding the volume and mix.

Dynamic cellular manufacturing system (DCMS) is one of the methods proposed for increasing the applicability of CMS when the demand for products fluctuates. In DCMS, to meet the demand in each period, the configuration of cells can be changed from one period to another (Rheault et al. 1995). However, the actual reconfiguration of cells may be time-consuming and costly. Furthermore, if these

✉ Mohammad Mahdi Paydar
 paydar@nit.ac.ir

[1] Department of Industrial Engineering, Mazandaran University of Science and Technology, Babol, Iran

[2] Department of Industrial Engineering, Babol University of Technology, Babol, Iran

[3] Department of Industrial Engineering, Najafabad Branch, Islamic Azad University, Najafabad, Iran

changes occur very frequently with stationary machines, the implementation of these systems is burdensome if not impossible (Thomalla 2000). In VCMS, unlike traditional cellular manufacturing, machines are not physically grouped into cells nor actually moved from their positions. Hence, some costs like assigning machines to cells or relocation of machines are not incurred. In VCMS, better controlling and planning of production is obtained by grouping of machines into virtual cells. VCMSs are capable of responding to demand fluctuations in a reasonable amount of time due to their high flexibility.

When a product mix or part demand level changes from a period to another, the configuration of cells may not be optimal anymore. In other words, the cells are reconfigured in the beginning of a period leading to a change in machine groups and parts families and work teams. Dynamic virtual cell formation (DVCF), unlike conventional dynamic manufacturing systems, can be utilized in this reGArd while reducing some costs, such as actual machine relocation costs. Figure 1 depicts an example of cellular reconfiguration in a dynamic environment. It is supposed that there are nine machines which are stationary for two periods. It can be easily seen that the manufacturing cells are virtual.

Literature review

DCMS has been a point of attraction to both researchers and practitioners. Slomp et al. (2005) addressed the design of VCMS considering the limited availability of workers and worker skills. They presented a goal programming formulation, in which in the first stage, jobs and machines are grouped, and in the second step, workers are grouped to form a VCMS. Their objective was to assign the available capacity as efficiently as possible and also to make the VCMS as independent as possible. Nomden et al. (2006) reviewed the previous researches in the subject of VCMS.

They addressed several definitions of virtual cells, offered by several researchers, and presented the potential problems for future researches. Mak et al. (2007) proposed a methodology for designing VCMS considering CFP and production scheduling problems. Their methodology included (1) a mathematical model for minimizing the total materials/components travelling distance subject to constraints, such as delivery due dates of products, capacities of resources, and critical tool limitations, and (2) an ant colony optimization method for solving cell formation and production scheduling. Liang et al. (2011) surveyed manufacturing resource modelling methods with a focus on resource element approach. They presented a function-clustering-degree concept addressing the trade-off between the granularity and quantity of virtual cells to verify the reconfiguration of manufacturing systems for solving virtual cell formation problem (VCFP). Mahdavi et al. (2011a) proposed an FGP-based approach to bi-objective mathematical model of CFP and production planning in a DVCMS. The objective of their research was to minimize the exceptional elements (EEs), holding and backorder costs in a cubic space of machine–part–worker incidence matrix. Rezazadeh et al. (2011) presented a mathematical model for DVCFP in which product mix/demand is variant in each period. The assumptions of their model were (1) considering operation sequence for the variety of processes as alternatives, (2) considering machines time capacity, maximum cell size and work capacity for each virtual cell. The objective of their proposed model was finding optimum number of virtual cells to minimize production, material transportation, inventory and manufacturing costs in each period. Nikoofarid and Aalaei (2012) designed a mathematical model for production planning in a dynamic virtual cellular manufacturing (DVCM) considering demand and part mix variation, machine capacity and as machine and worker availability as the main constraints. Han et al. (2014) addressed the problem of virtual cellular multi-period dynamic reconfiguration. They developed a

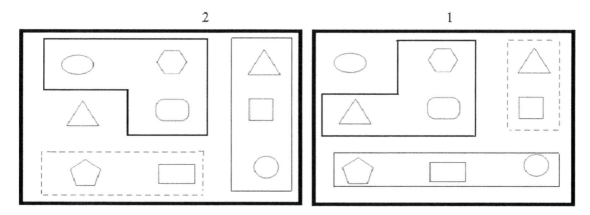

Fig. 1 Reconfiguration of virtual cells in a dynamic environment

model to incorporate the parameters of the problems, including product dynamic demand, machine capacity, operation sequence, balanced workload, alternative routings and batch setting. The objective of their mixed integer programming model is to minimize the total costs of operation, raw materials movements, inventory holding and process routes setup. Paydar and Saidi-Mehrabad (2015) developed a bi-objective possibilistic optimization mathematical model for formulating the integrated dynamic virtual cell formation and supply chain problem in a multi-echelon, multi-product and multi-period network. They developed a two-stage procedure, in which in the first stage, the proposed model is converted into an equivalent auxiliary crisp model, and in the second stage, a revised multi-choice goal programming approach is used for finding a compromise solution.

Although tremendous amount of research has already been conducted and published around CMS, the literature on DVCMS is still scarce. This paper is concentrated on DVCMS for processing multiple part types using multiple machine types and workers with different skills considering multiple candidate machine locations. We assume that several machines of any machine type are available for parts processing. In addition, we suppose that each worker is able to operate more than one machine in one cell.

In this paper, a DVCMS with several part types assigned to virtual cells to be processed by machines with different potential locations and cross-functional workers in a multi-period planning horizon is studied. In addition, more than one machine of each type may be available for part processing; i.e., duplicate machines are also considered. In this paper, unlike previous researches, we consider virtual cell of workers, machines (with candidate locations) and parts, simultaneously. The cost (number) of transportations between cells, as a major issue in DVCMS, is minimized. Furthermore, the number of exceptional elements is minimized as an objective function of the proposed model. The addressed problem is obviously NP-hard and considering dynamic conditions makes it even harder. Therefore, deterministic approaches may fail to efficiently solve real-world instances of the problem. Hence, metaheuristic approaches should be applied to obtain a satisfying solution in a reasonable amount of time. According to the literature and previously published researches (Mahdavi et al. 2009; Paydar and Saidi-Mehrabad 2013; Bootaki et al. 2014), genetic algorithm (GA) is capable of finding efficient solutions in the cell formation problems, and therefore, this algorithm is utilized for solving the proposed DVCMS mathematical problem. The main contributions of this paper to the literature on DVCMS are as follows:

1. Considering duplicate machines;
2. Simultaneous grouping of workers, machines and parts into virtual cells;
3. Proposing a bi-objective model that optimizes both:

    (a) The exceptional and void elements, and;
    (b) The total cost consisted of the fixed setup costs, the variable machine operation costs and the worker salary costs;

4. Developing a GA algorithm as a solution approach to the proposed model.

Problem description and formulation

In this section, the proposed mathematical model is formulated using a 4D machine–part–worker-location incidence matrix.

Sets

i index for part type ($i = 1, 2 \ldots , P$);
m index for machine type ($m = 1, 2 \ldots , M$);
w index for worker type ($w = 1, 2, \ldots , W$);
k index for cell ($k = 1, 2, \ldots , C$);
l index for location ($l = 1, 2, \ldots , L$);
t index for time period ($t = 1, 2, \ldots , T$).

Input parameters

r_{mw} 1 if worker type w is capable of operating machine type m and 0 otherwise;
a_{im} 1 if part type i can be processed on machine type m and 0 otherwise;
RW_{wt} available time for worker w in period t;
RM_{mt} available time for machine m in period t;
t_{imw} processing time of part i on machine type m with worker type w;
D_{it} demand of part i in period t;
SW_{wt} salary cost of worker type w in period t;
C_m fixed investment cost of machine type m;
α_m variable cost of machine type m.

Decision variables

X_{imwklt} 1 if part type i is to be processed on machine type m in location l with worker type w in cell k in period t and 0 otherwise;
NW_{wkt} number of workers of type w allotted to cell k in period t;

Y_{ml} 1 if machine m is located in location l and 0 otherwise;

F_{lkt} 1 if location l is assigned to cell k in period t and 0 otherwise;

Z_{ikt} 1 if part i is processed in cell k in period t and 0 otherwise;

W_{wkt} 1 if worker type w assigned to cell k in period t and 0 otherwise.

Mathematical model

Min Z_1

$$= \sum_{t=1}^{T} \sum_{k=1}^{C} \sum_{l=1}^{L} \left(\sum_{i=1}^{P} \sum_{m=1}^{M} \sum_{w=1}^{W} Y_{ml} \times F_{lkt} \times Z_{ikt} \times W_{wkt} \right.$$

$$\left. - \sum_{i=1}^{P} \sum_{m=1}^{M} \sum_{w=1}^{W} Y_{ml} \times F_{lkt} \times Z_{ikt} \times W_{wkt} \times a_{im} \times r_{mw} \right) \tag{1-1}$$

$$+ \sum_{t=1}^{T} \sum_{i=1}^{P} \sum_{m=1}^{M} \sum_{l=1}^{L} \sum_{k=1}^{C} \sum_{w=1}^{W} a_{im} \tag{1-2}$$

$$\times r_{mw} (1 - Y_{ml} \times F_{lkt} \times Z_{ikt} \times W_{wkt}).$$

Min Z_2

$$= \sum_{m=1}^{M} \sum_{l=1}^{L} C_m \times Y_{ml} \tag{2-1}$$

$$+ \sum_{t=1}^{T} \sum_{k=1}^{C} \sum_{m=1}^{M} \alpha_m \times D_{it} \times t_{imw} \times X_{imwklt} \tag{2-2}$$

$$+ \sum_{t=1}^{T} \sum_{k=1}^{C} \sum_{w=1}^{W} SW_{wt} \times NW_{wkt}. \tag{2-3}$$

Subject to

$$X_{imwklt} \le Y_{ml} \quad \forall i, m, w, k, l, t. \tag{3}$$

$$X_{imwklt} \le F_{lkt} \quad \forall i, m, w, k, l, t. \tag{4}$$

$$X_{imwklt} \le W_{wkt} \quad \forall i, m, w, k, l, t. \tag{5}$$

$$X_{imwklt} \le Z_{ikt} \quad \forall i, m, w, k, l, t. \tag{6}$$

$$\sum_{m=1}^{M} Y_{ml} \le 1 \quad \forall l. \tag{7}$$

$$\sum_{k=1}^{C} F_{lkt} = 1 \quad \forall l, t. \tag{8}$$

$$\sum_{k=1}^{C} Z_{ikt} = 1 \quad \forall i, t. \tag{9}$$

$$NW_{wkt} \le A \times W_{wkt} \quad \forall k, w, t. \tag{10}$$

$$\sum_{i=1}^{P} \sum_{m=1}^{M} \sum_{l=1}^{L} D_{it} \times t_{imw} \times X_{imwklt} \le NW_{wkt} \times RW_{wt} \tag{11}$$
$$\forall w, k, t.$$

$$\sum_{i=1}^{P} \sum_{w=1}^{W} D_{it} \times t_{imw} \times X_{imwklt} \le Y_{ml} \times RM_{mt} \quad \forall m, l, k, t. \tag{12}$$

$$\sum_{k=1}^{C} \sum_{w=1}^{W} \sum_{l=1}^{L} X_{imwklt} = a_{im} \quad \forall i, m, t. \tag{13}$$

$$\sum_{k=1}^{K} X_{imwklt} \le a_{im} \times r_{mw} \quad \forall i, m, w, l, t. \tag{14}$$

$$Y_{ml} \in \{0, 1\} \quad \forall m, l. \tag{15}$$

$$F_{lkt} \in \{0, 1\} \quad \forall l, k, t. \tag{16}$$

$$W_{wkt} \in \{0, 1\} \quad \forall w, k, t. \tag{17}$$

$$Z_{ikt} \in \{0, 1\} \quad \forall i, k, t. \tag{18}$$

$$X_{imwklt} \in \{0, 1\} \quad \forall i, m, w, l, k, t. \tag{19}$$

$$NW_{wkt} \ge 0 \text{ and integer} \quad \forall w, k, t. \tag{20}$$

The model has two objectives: in the first objective in (1-1), the goal is minimizing the number of exceptional elements, and in (1-2), the goal is to minimize the total number of voids; in the second objective, in (2-1), we minimize the fixed cost associated with machine investment and installation, and in (2-2), the variable cost of machines is minimized and (2-3) is to minimize the workers' salary cost.

Constraint (3) ensures that if machine m is not assigned to location l, then certainly X_{imwklt} equals zero. Constraint (4) is to ensure that if location l is not assigned to cell k in period t, then certainly X_{imwklt} is equal to zero. Constraint (5) guarantees that if worker type w is not assigned to cell k in period t, then certainly X_{imwklt} is zero. Constraint (6) is for ensuring that if part type i is not assigned to cell k in period t, then certainly X_{imwklt} equals zero. Obviously, only one machine can be location in each location l. This is considered by Constraint (7). Each location l should be assigned to one cell k in each period; this fact is modeled by Constraint (8). Constraint (9) ensures that each part i is assigned to only one location l in each cell in the tth period. Constraint (10) determines the number of workers type w in all cells type k in the tth period where A is a large positive number.

Constraint (11) ensures that the sum of assigned time for workforce should not be more than available time. It is noticeable in this constraint that operations are performed only in cells to which the corresponding workers are assigned. This is because if $NW_{wkt} = 0$, then no operations in that cell with worker type w can be performed and the left side of the constraint is equal to zero. Obviously, total assigned time to the machine m in the cell k in any period should not exceed the available time for machine m in cell k in each period. This fact is modeled by Constraint (12). Constraint (13) guarantees that each part is assigned to be processed on a machine in a period t with a worker w at a location l and in a cell k. Constraint (14) expresses the fact that each part could be manufactured in only one cell and by only one worker. Constraints (15)–(20) specify the allowed intervals and types of decision variables.

Linearization of the proposed model

The proposed model is obviously non-linear in the first part and the second part of the first objective function. Fortunately, most of the software have the ability to solve complex non-linear models; however, experiences show that solving such problems is usually time-consuming and results in local optima. Therefore, a linear model is practically more preferable and more convenient and efficient to solve. In addition, solving the model in small sizes utilizing fuzzy goal programming that the genetic algorithm method may be used for model verification. To linearize the model, some auxiliary variables and constraints are required to be defined and added to the original model. To linearize Eqs. (1-1) and (1-2), an auxiliary variable, Q_{imwklt}, is introduced as follows:

$$Q_{imwklt} = Y_{ml} \times F_{lkt} \times Z_{ikt} \times W_{wkt}.$$

Regarding Q_{imwklt}, the following constraints are also added to the model. It is easy to check that depending on the values of the binary variables, Y_{ml}, F_{lkt}, Z_{ikt} and W_{wkt}, the defined auxiliary variable, Q_{imwklt}, acts as the multiplication of the four binary variables.

$$Q_{imwklt} - Y_{ml} - F_{lkt} - Z_{ikt} - W_{wkt} + 3.5 \geq 0 \\ \forall\, i,m,w,l,k,t. \tag{21}$$

$$3.5 \times Q_{imwklt} - Y_{ml} - F_{lkt} - Z_{ikt} - W_{wkt} \leq 0 \\ \forall\, i,m,w,l,k,t. \tag{22}$$

$$Q_{imwklt} \in \{0,1\} \quad \forall\, i,m,w,l,k,t. \tag{23}$$

Fuzzy goal programming-based approach

One of the most important differences between one-objective and multi-objective optimization is multi-objective optimization can solve multi-dimensional objective problems. One of the famous methods for solving multi-objective problems is goal programming (GP). However, the application of GP in real-world problems may face two important difficulties: the first is the mathematical expression of the decision maker's imprecise aspiration levels for the goals and the second is the need to optimize all goals simultaneously. Fuzzy goal programming (FGP) is a mathematical decision-making mechanism to incorporate uncertainty and imprecision into the formulation. In practice, a high degree of fuzziness and uncertainty is included in the data set (Mahdavi et al. 2011b). The FGP has been tackled through different methods, such as probability distribution, penalty function, fuzzy numbers, preemptive FGP, interpolated membership function and the weighted additive model. Zimmermann first proposed fuzzy programming for solving the multi-objective linear programming problems (Zimmermann 1978). A number of researchers have extended the fuzzy set theory to the field of goal programming proposed by Narasimhan (1980). The fuzzy model of a generalized multi-objective multi-constrained optimization problem (Yang et al. 1991) can be expressed in what follows. Consider a problem with the following minimization objectives:

$$Z_l(X) \leq g_l \quad l = 1,2,\ldots,b \tag{24}$$

and subject to constraints:

$$d_j(X) \leq D_j \quad j = 1,2,\ldots,m, \tag{25}$$

where l is the index of goals, b represents the number of fuzzy-minimum goal constraints, g_l is the goal value (target value) for objective l given by the decision maker (DM), X is a k-dimensional decision vector, goal constraints are represented by $Z_l(X)$ and finally, $G = \{X|d_j(x) \leq D_j, j = 1,\ldots,m\}$ is the set of system constraints and defines the feasible space in which m represents the number of system constraints.

Let p_l denote the maximum tolerance limit for g_l determined by the DM. Thus, using the concept of fuzzy sets, the membership function of the objective functions can be defined as follows (Zimmermann 1978):

$$\begin{cases} \mu_{Z_l}(X) = \begin{cases} 1 & \text{if } Z_l(X) < g_l \\ 1 - \dfrac{Z_l(X) - g_l}{p_l} & \text{if } g_l \le Z_l(X) \le g_l + p_l \\ 0 & \text{if } Z_l(X) > g_l + p_l \end{cases} \end{cases} \tag{26}$$

The term $\mu_{zl}(x)$ indicates the desirability of solution X in terms of the objective l. The corresponding graph of Eq. (26) is shown in Fig. 2.

The α-level sets Z_l^λ, $\forall l \in \{1, 2, \ldots, b\}$, $\forall \lambda \in [0, 1]$ are defined as:

$$Z_l^\lambda = \{X | \mu_{Z_l}(X) \ge \lambda, 0 \le \lambda \le 1\}, \quad l = 1, 2, \ldots b.$$

Then, the decision space is defined by intersecting the system constraints with the intersection of λ-level sets as follows:

$$Z^* = \left\{ \bigcap_{l=1}^{b} Z_l^\lambda \right\} \bigcap G, \quad 0 \le \lambda \le 1.$$

According to the extension principle, the membership function of Z is defined as follows:

$$\mu_Z(X) = \min_{l=1,2,\ldots m} \{\mu_{Z_l}(X)\}.$$

Finally, the optimal solution, Z^* X^*, must maximize $\mu_Z(X)$ by solving the following mathematical programming.

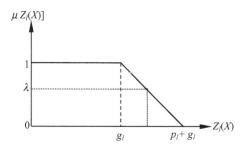

Fig. 2 Membership function related to objectives

2	1	4	1	3	3

Fig. 3 Example of chromosome Y_{mi}

1	3	2	1
2	2	1	3
2	1	3	2
3	1	3	1

Fig. 4 Chromosome demonstration for variable Z_{ikt}

2	3	2	1
3	2	1	3
2	1	3	2
3	1	3	1

Fig. 5 Chromosome representation for variable F_{lkt}

3	3	2	1
2	2	1	3
2	1	3	2
3	1	3	1
1	3	1	2
2	1	1	1

Fig. 6 Chromosome representation for variable w_{wkt}

$$\text{Max } Z^* = \lambda$$
$$\text{s.t.}$$
$$\lambda \le \mu_{Z_l}(X), \quad l = 1, 2, \ldots, b \tag{27}$$
$$d_j(X) \le D_j, \quad j = 1, 2, \ldots, m$$
$$0 \le \lambda \le 1$$

The GA approach

GA is inspired by Darwin's theory of evolution and genetic knowledge and is based on elitism. It simulates the genetic evolution of orGAnisms, and its generic usage is as an optimization method. The excellent books by Davis (1991)] and Goldberg (1989) described many possible variants of GAs. GA is based on an analogy to the phenomenon of natural selection in biology. First, a chromosome structure is defined to represent the solutions to the problem. An initial solution population is generated either randomly or using a heuristic. Each chromosome is then improved through a selection/elitism mechanism. More specifically, members of the population are selected based on an evaluation function, called "fitness", which associates a value to each member according to its objective function. The higher a member's fitness value, the more likely it is to be selected. Thus, the less fit individuals are replaced by those with higher value. Genetic operators are then applied to the selected members to produce a new generation. This

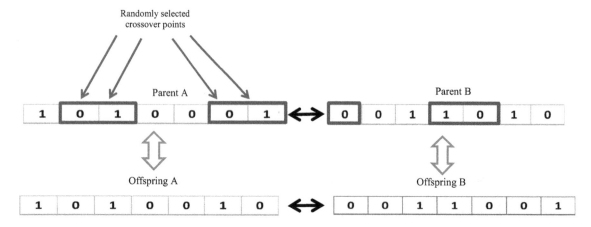

Fig. 7 Uniform crossover

Fig. 8 Proposed mutation

process is repeated until some stopping criteria are reached (Mahdavi et al. 2009). The main components of GA for implementation are:

1. The scheme for coding.
2. The initial population.
3. Adaptation function for evaluating the fitness of each member of the population.
4. Selection procedure.
5. The genetic operators used for combining the solution features for producing a new generation.
6. Certain control parameter values (e.g., population size, number of iterations, genetic operator probabilities, etc.).

The scheme for coding

For any implementation of GA, the first stage is to map solution characteristics into the format of a chromosome string. Each chromosome is made up of a sequence of genes from a certain alphabet. The alphabet can be a set of binary numbers, real numbers, integers, symbols, or matrices Goldberg (1989).

In genetic algorithms, each chromosome represents a feasible solution in the search space and is formed by a fixed number of genes. Usually, genes are represented using binary codes. In this paper, a four-component chromosome is used for solution representation.

The first component of the chromosome, formed according to the decision variable Y_{ml}, is a row vector in which the number of column represents the location and the value within

Table 1 Operation process times in example 1

	Part 1			Part 2			Part 3		
	W1	W2	W3	W1	W2	W3	W1	W2	W3
M1	0.3	0	0	0	0	0	0	0.3	0.3
M2	0	0	0	0.2	0.4	0	0	0	0
M3	0	0.4	0.2	0	0.3	0.1	0	0	0

the column determines the type of machine. For example, in Fig. 3, the first element of the vector is 2 meaning that a machine of type 2 is in located in the location 1.

The next component models the variable, Z_{ikt}, and is a matrix with P rows and T columns. The element in row p and column t of the matrix shows the number of the cell in which part p is processed in period t. For instance, in Fig. 4, the matrix with four rows and four columns determines the cell numbers for four parts in four periods. The number 3 in the first row and second column means that part number 1 in period 2 is processed in cell 3.

The variables F_{lkt} and w_{wkt} are represented using a similar matrix utilized to code the variable Z_{ikt}. The only difference is that rows in the matrices for F_{lkt} and w_{wkt} represent locations and workers, respectively. Figure 5 illustrates an example of the matrix for F_{lkt} with the element in the first row and the second column being equal to 3; showing that location 1 in period 2 is assigned to the cell number 2. Similarly, in Fig. 6, the matrix for variable w_{wkt} is formed by six workforces and four periods. The element in the first row and the first column specifics that worker 1 in period 1 is assigned to the cell 3.

Table 2 Machines available times, machines fixed costs and machines variable costs in example 1

	M1	M2	M3
RM	70	70	70
C	500	400	350
α	20	30	25

Table 3 Machines available times, workers hiring and number of available workers in example 1

	W1	W2	W3
RW	60	60	60
SW	400	500	450
AW	4	4	4

The initial population

Another aspect of GA implementation is generating a set of initial solutions known as the initial population. The number of initial solutions to be included in the population is called population size. The population size is a key factor

Table 5 Output information related to assign each machine to the locations using LINGO in example 1

	L1	L2	L3	L4
M1	0	0	1	1
M2	1	0	0	0
M3	0	1	0	0

in a successful GA implementation. A small population size increases the speed of the algorithm; however, it may prevent the algorithm from converging to satisfying solutions. On the other hand, although a large population usually results in better solutions, it may significantly increase CPU time (Back et al. 1997)

Fitness function

In GA implementation, a fitness function is used to evaluate the chromosomes for reproduction. The purpose of the fitness function is to measure the quality of the candidate solutions in the population with respect to the objective and constraint functions of the model. The fitness function is

Table 4 Output information of the proposed model related to assign machines and workers to each cell in each period using LINGO in example 1

	Parts assigned to			Machines in			Worker assigned to			Location assigned to		
	Cell 1	Cell 2	Cell 3	Cell 1	Cell 2	Cell 3	Cell 1	Cell 2	Cell 3	Cell 1	Cell 2	Cell 3
Period 1	3	–	1, 2	1	–	1, 3, 2,	1	–	1, 3	3	–	4, 2, 1
Period 2	3	–	1, 2	1	–	1, 3, 2,	1	–	1, 3	4	–	3, 2, 1

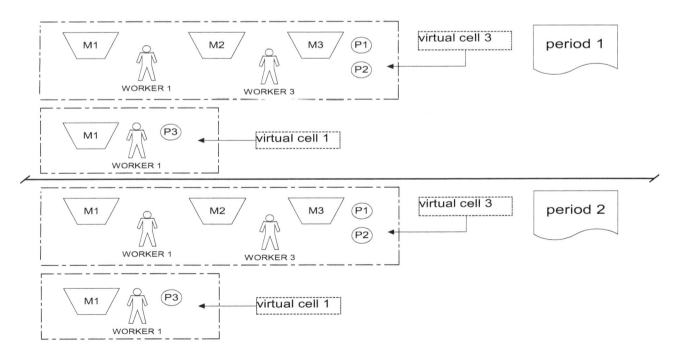

Fig. 9 Cell reconfiguration schema in each period for example 1

Table 6 Obtained values for FGP variables

Variable	λ	Z_1	Z_2
Value	0.812501	198	161,400

Table 7 In the interpretation of the results obtained from the proposed GA in example 1

	L1	L2	L3	L4
M1	1	1	0	0
M2	0	0	0	1
M3	0	0	1	0

calculated according to Eq. (27), where λ is the same $*$, and a penalty function is used in fitness function (Z_X^*) to satisfy the constraints.

$$Z_X^* = \min\left(\frac{Z^*}{1 + \text{penalty function}}\right).$$

Selection rule

The roulette wheel selection procedure, as proposed by Goldberg (1989), is the selection strategy used in the proposed algorithm. The goal of the selection strategy is to allow the "fittest" individuals to be considered more often to reproduce children for the next generation. Each individual is assigned a probability of being selected based on its fitness value. Although individuals with higher fitness value have a higher selection probability, all individuals in the population should be given a chance to be selected. Hence, after ranking the individuals, the parents are selected randomly based on their fitness.

Genetic operators

Reproduction is carried out by applying crossover and mutation operators on the selected parents to produce offspring. The crossover and mutation operators for the proposed algorithm are discussed in what follows.

Uniform crossover

For every pair of randomly selected parents, a small proportion of randomly selected genes is exchanged. The crossover process is illustrated in Fig. 7. Individuals parent A and parent B produce offspring A and offspring B after applying the crossover. We define parent A as the direct parent of offspring A, and parent B as the direct parent of offspring B.

Mutation operator

The conventional mutation operator randomly alters the value of the genes according to a small probability of mutation; thus, it is merely a random walk and does not guarantee a positive direction toward the optimal solution. The proposed heuristic mutation remedies this deficiency. In this scheme, an individual is randomly chosen from the population (Fig. 8).

Parameters

The parameters required to run the algorithm are population size, number of generations, number of iterations, crossover and mutation probabilities. These parameters have a crucial role in the performance of the GAs. The number of generations is a function of the size of the

Fig. 10 Convergence of λ in example 1

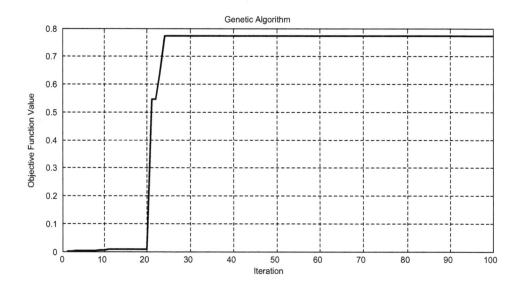

Table 8 Output information of the proposed model related to assign the machines and workers to each cell in each period using the genetic algorithm in example 1

	Parts assigned to			Machine in			Worker assigned to			Location assigned to		
	Cell 1	Cell 2	Cell 3	Cell 1	Cell 2	Cell 3	Cell 1	Cell 2	Cell 3	Cell 1	Cell 2	Cell 3
Period 1	3	1, 2	–	1	1, 2, 3	–	1	1, 2	–	2	1, 3, 4	–
Period 2	–	–	1, 2, 3	–	–	1, 2, 3	–	–	1, 3	–	–	1, 3, 4

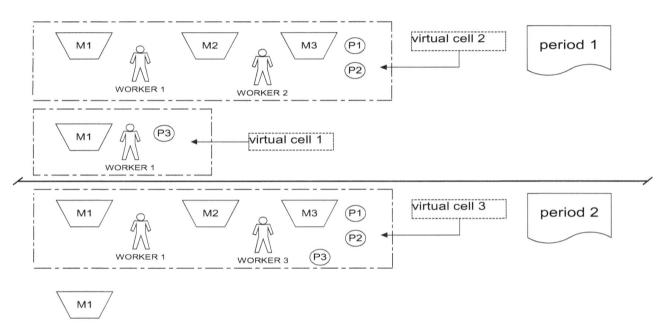

Fig. 11 Cell reconfiguration schema in each period for example 1 in the MATLAB software

problem at hand. As the solution space extends, the GA requires a larger number of generations to reach a satisfying convergence point. Population size may vary depending on the application. The number of iterations must be adjusted to allow the GA to complete the convergence process. The crossover operator has a significant effect on the performance of GA, and therefore, usually, a relatively large probability value is considered for this parameter. Mutation operator is basically used to maintain diversity in the population and is performed with a low probability.

Computational results

Solving the model with LINGO software

In this section, we present an example for which the branch and bound in the LINGO software and the genetic algorithm are utilized as solution methods. In addition, to

Table 9 GAP of between LINGO and GA

Algorithm	Objective function values		
	λ	Z_1	Z_2
LINGO	0.812501	198	161,400
The proposed GA	0.77273	201	163,200
GAP (%)	4.8	1.5	1.1

evaluate the performance of the proposed model, a comparison of the outcomes is provided.

This example includes three cells, three parts, three machines, three workers, four locations and two periods in which all the presented hypothesis in "Problem description and formulation" are valid. Our goal is determining machine locations and cells and worker assignments.

Our data for the model include a_{im} which is a 2D variable to determine part–machine relations and is shown by the following matrix:

Fig. 12 Convergence of λ in example 2

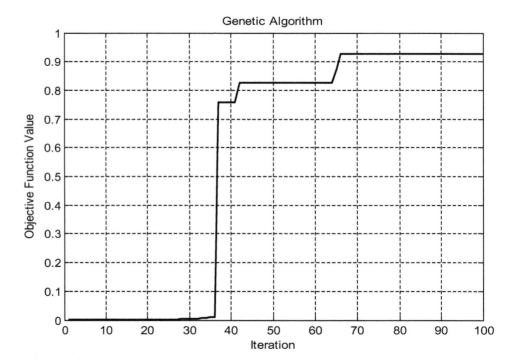

Table 10 Output information related to assign each machine to the locations used genetic algorithm in example 2

	L1	L2	L3	L4	L5	L6
M1	1	0	0	0	0	1
M2	0	1	0	0	1	0
M3	0	0	1	1	0	0

$$a_{im} = \begin{matrix} 1 & 0 & 1 \\ 0 & 1 & 1 \\ 1 & 0 & 0 \end{matrix}$$

Another input is r_{mw} that determines the worker–machine assignments as shown as follows:

$$r_{mw} = \begin{matrix} 1 & 0 & 0 \\ 1 & 1 & 0 \\ 0 & 1 & 1 \end{matrix}$$

Table 12 Obtained values for GA variables in example 2

Variable	λ	Z_1	Z_2
Value	0.92646	880	486,000

Demand volume for each part in each period is given by:

$$D_{it} = \begin{matrix} 30 & 100 \\ 70 & 80 \\ 90 & 60 \end{matrix}$$

The time spent for processing each part on each machine by each worker is represented by a 3D matrix as follows (Table 1).

The other input parameters for machines and workforces are depicted in Tables 2 and 3, respectively.

FGP model of the problem is coded and solved using the LINGO Software run on a desktop PC equipped with an Intel® Core™ i3 @ 3200 GHz and 4 GBs of RAM running

Table 11 Output information of the proposed model related to the machines and workers to each cell in each period using the genetic algorithm in example 2

	Parts assigned to			Machine in			Worker assigned to			Location assigned to		
	Cell 1	Cell 2	Cell 3	Cell 1	Cell 2	Cell 3	Cell 1	Cell 2	Cell 3	Cell 1	Cell 2	Cell 3
Period 1	–	1, 2	3, 4, 5	–	1, 2, 3	1, 2	–	2, 4	1	–	2, 4, 6	1, 5
Period 2	2, 3, 4	1, 5	–	1, 2, 2, 3	1, 3	–	1, 4	4	–	2, 4, 5, 6	1, 3	–
Period 3	5	–	1, 2, 3, 4	1	–	1, 2, 3	4	–	1, 3	1	–	3, 5, 6

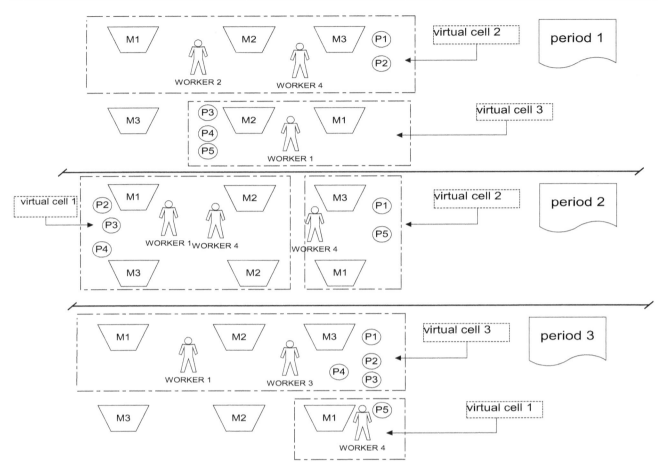

Fig. 13 Cell reconfiguration schema in each period for example 2 in the MATLAB software

Fig. 14 Convergence of λ in example 3

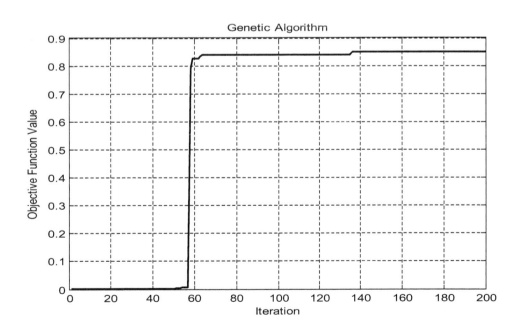

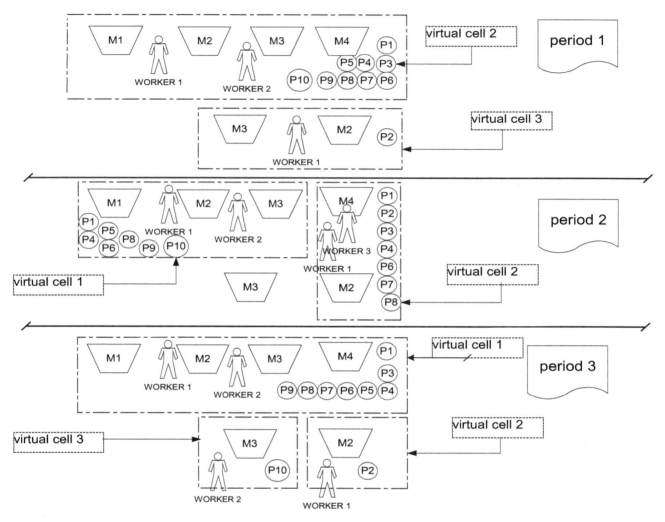

Fig. 15 Cell reconfiguration schema in each period for example 3 in the MATLAB software

Microsoft Windows^{TM} 7 Ultimate. The results were obtained after 13 h and 54 min. In Table 4, parts, machines and locations assignments to cells in each period are presented. In the results, for example, in cell 3 in the first period, two machines of type 3, one machine of type 1 and one machine of type 2 are assigned.

In Table 5, assignments of machines to locations are presented. It should be noted that the locations of machines are determined in the first period and are fixed for all periods. For instance, the machine, M2, is assigned to location 1 for all periods.

For more understanding of the results, the reconfigurations of this example are depicted in Fig. 9.

In Table 6, the obtained values for the FGP variables are presented. The optimized values of the first and second objective functions (Z_1 and Z_2) are 198 and 161,400, respectively.

Table 13 Output information related to assign each machine to the locations used genetic algorithm in example 3

	L1	L2	L3	L4	L5	L6
M1	1	0	0	0	0	0
M2	0	0	1	0	0	1
M3	0	0	0	1	1	0
M4	0	1	0	0	0	0

Solving example 1 using genetic algorithm (GA)

We coded the proposed GA solution method by MATLAB 2010 and applied the program to the same numerical example solved by the LINGO Software using the same desktop PC mentioned above and discussed in "Solving the model with LINGO software". In the results, obtained after 330 s, λ was 0.77273 for which Fig. 10 illustrates the

Table 14 Output information of the proposed model related to assign machines and workers to each cell in each period using the genetic algorithm

	Parts assigned to			Machine in			Worker assigned to			Location assigned to		
	Cell 1	Cell 2	Cell 3	Cell 1	Cell 2	Cell 3	Cell 1	Cell 2	Cell 3	Cell 1	Cell 2	Cell 3
Period 1		1, 3, 4, 5, 6, 8, 7, 9, 10	2		1, 2, 3, 4	2		1, 2	1		1, 2, 4, 6	3
Period 2	1, 4, 5, 8, 6, 9, 10	1, 2, 3, 4, 6, 7, 8		1, 2, 3	2, 4		1, 2	1, 3		1, 3, 5	2, 6	
Period 3	1, 3, 4, 5, 6, 7, 8, 9	2	10	1, 2, 3, 4	2	3	1, 2	1	2	1, 2, 3, 4	6	5

For more understanding, reconfigurations of this example are given in Fig. 15

Table 15 Obtained values for GA variables in example 3

Variable	λ	Z_1	Z_2
Value	0.84897	1405	961,200

convergence in 24 iterations. In this figure, vertical and horizontal axes show λ and the number of iterations, respectively.

Therefore, the obtained values translate into the matrix depicted in Table 7.

For the parts, machines and workers, the best solution is derived from X_{imwklt}. The corresponding values are presented in Table 8.

For more understanding of the solution, the reconfigurations of this example are given in Fig. 11.

To illustrate the performance of the proposed method, the difference percentage GAP between the proposed GA and LINGO results is calculated which is shown in Table 9.

The performance GAP shows that the proposed GA outperforms LINGO Solver with a significant GAP.

Consideration of some examples with greater dimensions

For further investiGAtion, we considered the second example with the greater dimensions. It has three periods, five types of parts, three types of machines, four types of workers, three cells and eight locations. After entering input data to the model, the results were obtained as follows:

Table 16 Comparison of the proposed GA and LINGO for small-sized examples

No. of example	Example size						Objective value (λ)		CPU time		GAP (%)
	No. of part types	No. of machine types	No. of worker types	No. of cells	No. of locations	No. of periods	GA	LINGO	GA	LINGO	
1	2	2	2	2	3	2	0.93325	0.93325	0:02:55	1:08:04	0
2	2	3	3	2	4	2	0.83241	0.86241	0:03:46	6:10:36	3
3	3	3	3	3	4	2	0.77273	0.812501	0:05:30	13:54:00	4.8

Table 17 Comparison of the proposed GA and LINGO for medium-sized test examples

No. of example	Example size						Objective value (λ)		CPU time		GAP (%)
	No. of part types	No. of machine types	No. of worker types	No. of cells	No. of locations	No. of periods	GA	LINGO (Z^{best})	GA	LINGO (Z^{best})	
1	4	3	3	2	6	2	0.88157	0.92573	0:14:35	15:00:00	4.7
2	5	3	4	3	8	3	0.92646	0.97332	0:22:07	15:00:00	4.8
3	6	4	4	3	8	3	0.87427	0.93241	0:38:24	15:00:00	6.2

Table 18 Comparison of the proposed GA and LINGO for large-sized test examples

No. of example	Example size						Objective value (λ)		CPU time		GAP (%)
	No. of part types	No. of machine types	No. of worker types	No. of cells	No. of locations	No. of periods	GA	LINGO (Z^{best})	GA	LINGO (Z^{best})	
1	7	3	3	2	8	2	0.86731	0.82367	1:43:36	15:00:00	5
2	8	3	3	3	6	3	0.85294	–	2:50:14	–	–
3	9	4	3	2	8	3	0.8245	–	4:12:47	–	–
4	10	4	3	3	6	3	0.84897	–	5:59:20	–	–

Table 19 Results of the test

Problem category	Objective function		Computational time	
	$\mu_{GA} - \mu_{LINGO}$	P value	$\mu'_{GA} - \mu'_{LINGO}$	P value
Small	−0.0233	0.096	−0.292	0.099
Medium	−0.04972	0.004	−0.60762	0.000
Large	Not applicable			

The model is solved in about 22 min (1327 s) and achieved the solution. λ is converged to 0.92646 and Fig. 12 depicts this convergence. The vertical axis depicts the value for λ and the horizontal axis represents the number of iterations. After 65 iterations, λ was converged to the aforementioned value.

The obtained values for the variables are presented as follows (Tables 10, 11, 12).

For more understanding, reconfigurations of this example are given in Fig. 13.

Example 3: In this example, we considered a problem with ten parts, three workers, four machines, three cells, three periods and six locations. We ran the model, and after 6 h (21,560 s), the solution was found. At last, λ value was converged to 0.84897 after 138 iterations. Figure 14 depicts these values for iteration larger than 64. Vertical axis shows λ value, and horizontal axis depicts the iterations.

Output values for the variables are as follows (Tables 13, 14, 15).

Computational results for the proposed GA

To show the effectiveness of the proposed GA in solving the proposed model, first three small-sized examples which can be solved optimally using LINGO are presented. The results of solving these examples using the LINGO and the proposed GA are compared in Table 16. The relative differences between the objective values, achieved by the two methods, are shown as GAP in Table 16. According to this table, in the worst case,

solution GAP between the proposed GA and the global optimum found by the LINGO software is 4.8 %. This shows that the proposed GA is capable of obtaining the near optimal solutions in a reasonable computational time. As shown in Table 16, the LINGO software has solved the third example in a time equal to 13:54:00 which is not a reasonable computational time for solving a small-sized example.

To demonstrate the performance of the proposed GA in solving medium-sized examples, three test problems, in which LINGO fails to solve optimally, are designed and solved using the proposed GA. In a maximization problem, the LINGO software found a possible interval for optimum value of objective function (Z^*) that is limited by the Z^{bound} and Z^{best} values, where $Z^{best} \leq Z^* \leq Z^{bound}$. Z^{best} shows the best known feasible objective function value, and Z^{bound} represents the upper bound of the objective function. Approaching to the current values for the best known solution and the bound, Z^{best} is either the optimal solution, or very close to it. At such a point, the solver can be interrupted and report the current best solution with the aim of shortening additional computation time. Therefore, we limit runtime to 15 h to save computational effort and report the best solution obtained after 15 h. To validate the results found by GA, the solutions achieved by GA for three examples are compared with Z^{best} obtained by LINGO after 15 h. These results are shown in Table 17. The GAP between the objective value of the proposed GA and Z^{best} in the worst case is 6.2 %. This confirms that the proposed GA is able to obtain solutions to those examples effectively. In addition, computational time for those examples by GA is less than 40 min showing its superior efficiency. To evaluate the performance of the proposed GA in solving the large-sized examples, four random instances are solved by GA and LINGO, and the obtained results are shown in Table 18. LINGO solver is not capable of even finding feasible solutions for the last three test problems even after 15 h. However, GA has solved these examples in less than 6 h. Moreover, in the first example of Table 18, the solution obtained by GA is better than Z^{best}, achieved by LINGO after 15 h.

To further clarify the superiority of the proposed algorithm, a statistical test is conducted for each group of problems. More specifically, a paired t test is utilized to compare the objective function value and computational time of the proposed algorithm and those of LINGO solver. The results of the test are presented in Table 19. In this Table, μ_{LINGO} and μ_{GA} are the mean objective function value for LINGO and GA, respectively, and μ'_{LINGO} and μ'_{GA} are the mean computational time for LINGO and GA, respectively.

Conclusions

In this paper, a bi-objective mathematical model for dynamic virtual cellular manufacturing system is developed which have several advantages toward the previous researches in the literature. One of the majors to another researches is introducing method to the assignment of workers alongside assignment of machines to locations in DVCMS which this mater concludes to complexity of the model. However, considering these features simultaneously causes to better planning and close to real-life situations. The most important features of this paper is as follows:

1. Developing the dimensions of cellular manufacturing problem, including machines, parts, workers and locations leading to a more realistic model.
2. Assigning workers, location to machines and parts to cells to operation processing in dynamic virtual cellular manufacturing system simultaneously.
3. Calculating the number of inter-cell transportations cost and exceptional elements in dynamic virtual cellular manufacturing system.
4. Calculating machine variable and fixed costs in each periods and also workforce hiring cost.
5. Forming virtual cells in each period.

Further researches on the proposed model may be attempted in future studies by incorporating the following issues:

- Developing a mathematical model with considering uncertainty with fuzzy parameters;
- Operation scheduling can be considered in virtual cellular manufacturing problem;
- Distances between each machine and transportation of materials can be added to the mathematical model.

References

Back T, Fogel D, Michalawecz Z (1997) Handbook of evolutionary computation. Oxford University Press, Oxford

Bootaki B, Mahdavi I, Paydar MM (2014) A hybrid GA-AUGMECON method to solve a cubic cell formation problem considering different worker skills. Comput Ind Eng 75:31–40

Davis L (1991) Handbook of genetic algorithms. Van Nostrand, New York

Goldberg DE (1989) Genetic algorithms: search, optimization and machine learning. Addison-Wesley Inc, Boston

Han W, Wang F, Lv J (2014) Virtual cellular multi-period formation under the dynamic environment. IERI Procedia 10:98–104

Heragu SS (1994) Group technology and cellular manufacturing. IEEE Trans Syst Man Cybern 24(2):203–214

Liang F, Fung RYK, Jiang Z (2011) Modelling approach and behavior analysis of manufacturing resources in virtual cellular manufacturing systems using resource element concept. Int J Comput Integr Manuf 24(12):1168–1182

Mahdavi I, Paydar MM, Solimanpur M, Heidarzade A (2009) Genetic algorithm approach for solving a cell formation problem in cellular manufacturing. Expert Syst Appl 36:6598–6604

Mahdav I, Aalaei A, Paydar MM, Solimanpur M (2011) Multi-objective cell formation and production planning in dynamic virtual cellular manufacturing systems. Int J Prod Res 49(21):6517–6537

Mahdavi I, Paydar MM, Solimanpur M (2011) Solving a new mathematical model for cellular manufacturing system: a fuzzy goal programming approach. Iran J Oper Res 2(2):35–47

Mak KL, Peng XX, Lau TL (2007) An ant colony optimization algorithm for scheduling virtual cellular manufacturing systems. Int J Comput Integr Manuf 20(6):524–537

Narasimhan R (1980) Goal programming in a fuzzy environment. Decis Sci 11:325–326

Nikoofarid E, Aalaei A (2012) Production planning and worker assignment in a dynamic virtual cellular manufacturing system. Int J Manag Sci Eng Manag 7(2):89–95

Nomden G, Slomp J, Suresh NC (2006) Virtual manufacturing cells: a taxonomy of past research and identification of future research. Int J Flex Manuf Syst 17:71–92

Paydar MM, Saidi-Mehrabad M (2013) A hybrid genetic-variable neighborhood search algorithm for the cell formation problem based on grouping efficacy. Comput Oper Res 40:980–990

Paydar MM, Saidi-Mehrabad M (2015) Revised multi-choice goal programming for integrated supply chain design and dynamic virtual cell formation with fuzzy parameters. Int J Comput Integr Manuf 28(3):251–265

Rezazadeh H, Mahini R, Zarei M (2011) Solving a dynamic virtual cell formation problem by linear programming embedded particle swarm optimization algorithm. Appl Soft Comput 11:3160–3169

Rheault M, Drolet J, Abdulnour G (1995) Physically reconfigurable virtual cells: a dynamic model for a highly dynamic environment. Comput Ind Eng 29(1–4):221–225

Slomp J, Chowdary BV, Suresh NC (2005) Design of virtual manufacturing cells: a mathematical programming approach. Robot Comput Integr Manuf 21(3):273–288

Thomalla CS (2000) Formation of virtual cells in manufacturing systems. In: Proceedings of group technology/cellular manufacturing world symposium, San Juan, pp 13–16

Wemmerlov U, Hyer NL (1989) Cellular manufacturing in the US industry: a survey of users. Int J Prod Res 27(9):1511–1530

Yang T, Ignizio JP, Kism HJ (1991) Fuzzy programming with nonlinear membership functions: piecewise linear approximation. Fuzzy Sets Syst 11:39–53

Zimmermann HJ (1978) Fuzzy programming and linear programming with several objective functions. Fuzzy Sets Syst 1:45–55

13

Multi-objective optimization of discrete time–cost tradeoff problem in project networks using non-dominated sorting genetic algorithm

Mohammadreza Shahriari[1]

Abstract The time–cost tradeoff problem is one of the most important and applicable problems in project scheduling area. There are many factors that force the mangers to crash the time. This factor could be early utilization, early commissioning and operation, improving the project cash flow, avoiding unfavorable weather conditions, compensating the delays, and so on. Since there is a need to allocate extra resources to short the finishing time of project and the project managers are intended to spend the lowest possible amount of money and achieve the maximum crashing time, as a result, both direct and indirect costs will be influenced in the project, and here, we are facing into the time value of money. It means that when we crash the starting activities in a project, the extra investment will be tied in until the end date of the project; however, when we crash the final activities, the extra investment will be tied in for a much shorter period. This study is presenting a two-objective mathematical model for balancing compressing the project time with activities delay to prepare a suitable tool for decision makers caught in available facilities and due to the time of projects. Also drawing the scheduling problem to real world conditions by considering nonlinear objective function and the time value of money are considered. The presented problem was solved using NSGA-II, and the effect of time compressing reports on the non-dominant set.

Keywords Time–cost tradeoff · Time value of money · Crashing · NSGA-II · Multi-objective problem · AOA network

Introduction

Discrete time–cost tradeoff problem (DTCTP) has many applications, and a lot of research has been conducted on this area. In these studies, customer needs to get services in shorter time periods and necessity of reducing the project cost were considered together, and this approach raised the importance of DTCTP for business owners and researchers. In 1991, Hindelang and Muth presented the DTCTP for the first time (Hindelang and Muth 1979). Prabuddha et al. (1997) and Vladimir et al. (2001) showed that this problem belongs to NP-hard problem. One of the basic assumptions in this problem is that the activities cost is a function of the activities duration, which is a decision variable. The lower limit of duration is crash duration, and the upper limit of duration is normal duration. Kelley and Walker (1959), Fulkerson (1961), Kelly (1961), Ford and Fulkerson (1962), Siemens (1971), Goyal (1975), and Elmaghraby and Salem (1981) presented the linear mathematical models and Moder et al. (1983) considered continuous activities cost. DTCTP was solved using many different exact methods, such as dynamic programing (Hindelang and Muth 1979), enumeration algorithm (Patterson and Harvey 1979), and branch and bound (Demeulemeester et al. 1996, 1998; Erenguc et al. 2001), but none of the methods could solve the DTCTP in large scale, so many researchers decided to use heuristics and metaheuristic algorithms to solve this problem. Akkan (1998) used a heuristic algorithm based on Lagrange released, while Liu et al. (2000) used genetic algorithm (GA) for solving DTCTP and Elmaghraby and

✉ Mohammadreza Shahriari
  shahriari@iau.ae; Shahriari.mr@gmail.com

[1] Faculty of Management, South Tehran Branch, Islamic Azad University, Tehran, Iran

Kamburowski (1992) considered reward and penalty on objective function. Ann and Erenguc (1998) introduced activities compression, and Van Slyke was the first researcher who used the Monte Carlo simulation method for compressing activities. It has many advantages, such as longer projects time, flexibility of selecting distribution functions of activities time, and ability to calculate the value of the path critically (Van Slyke 1963), and DTCTP was considered with time switch in it (Vanhoucke et al. 2002). All the presented researchers have only one objective function, and the aim of all is decreasing the project cost or time. The researchers found that for determining the duration of activities, many things must be considered. For example, one may wish to finish the project in shortest time with minimum cost. To this effect, two finishing time and project cost objective functions can be considered in a scalar unique objective function (Sasaki and Gen 2003a, b), and can solve this multi-objective problem using the LP-metric method (Pasandideh et al. 2011). Mohammadreza Shahriari et al. introduced a mathematical model for time–cost tradeoff problem with budget limitation based on a mixed-integer programing, considering the time value of money (Mohammadreza Shahriari et al. 2010). Jaeho Son et al. (2013) presented a new formulation technique is introduced to merge the two independent scenarios mathematically, and it guarantees the optimal solution. Öncü - Hazır (2014) presented a mathematical model to support project managers from a wide range of industries in scheduling activities to minimize deviations from project goals. Tiwari and Johari (2015) introduced an approach, which was experimented on several case studies that proved its usefulness. The intertwined approach using simple and popular Microsoft Office tools (Excel and MSP) is logical, fast, and provides a set of feasible project schedule meeting the deadline and that do not violate resource limits. Mario Vanhoucke extend the standard electromagnetic metaheuristic with problem specific features and investigate the influence of various EM parameters on the solution quality (Vanhoucke 2014).

Kaveh et al. (2015) showed that two new metaheuristic algorithms, charged system search (CSS) and colliding body optimization (CBO), are utilized for solving this problem. The results show that both of these algorithms find reasonable solutions; however, CBO could find the result in a less computational time having a better quality.

Ke et al. proposed a model to deal with an intelligent algorithm combining stochastic simulations and GA, where a stochastic simulation technique is employed to estimate random functions and GA is designed to search optimal schedules under different decision-making criteria (Ke et al. 2012).

In this study, we used NSGA-II for solving this two-objective problem.

The assumption of activities compressing in time–cost tradeoff problem was presented to achieve the specific solutions. In this solution, with spending money, the project time will decrease. Generally, delay in finishing the projects is inevitable. The World Commission on Dams (WCD) made a research on 99 large projects and reported that only 50 of these projects finished in due time, 30 % of them have 1–2 years delay, and 20 % have more than 2 years delay (4 projects have more than 10 years delay), and the main reasons of delays are financial problems, inefficiency of contractors and operation management, unreal scheduling, and employers dissatisfaction (WCD 2000). Unlike the studies in this area, we consider the "maximum duration" for the upper level of activities duration, whatever moving from normal time to compressed time caused more compressing cost, and whatever moving from normal time to presented upper bound of the time caused more saving of money. The final front of the solutions in NSGA-II presents a variety of solutions for decision makers, and they have opportunity to select a proper solution based on the available budget and appropriate time for the project. Another problem considered in this study is the time value of money. In large projects, it is important to know the proper time of spending or saving the money due to the time value of money. This study has five parts. The second part presents a mathematical model. Third part deals with solving algorithm. In part 4, a numerical example is presented, and the final part belongs to conclusion and further studies.

Mathematical model

Nomenclatures

t_i	Happening time of event i
$D_{f(ij)}$	Minimum allowed time of activity $i-j$
$D_{n(ij)}$	Normal time for activity $i-j$
$D_{m(ij)}$	Maximum allowed time of activity $i-j$
d_{ij}	Scheduled (actual) time of activity $i-j$
H	Indirect (overhead) project cost
K_n	Direct project cost
$C_{f(ij)}$	Compressing cost of activity $i-j$
$C_{n(ij)}$	Normal cost of activity $i-j$
$C_{m(ij)}$	The cost of delaying in activity $i-j$
C_{ij}	Compressing cost rate of activity $i-j$
C'_{ij}	Saving rate of delaying for activity $i-j$
t_{Max}	Maximum allowed time for finishing the project
C_{Max}	Maximum available budget
I_0	Interest rate

$$y_{ij} = \begin{cases} 1 & \text{If the activity } i-j \text{ compressed} \\ 0 & \text{Otherwise} \end{cases}$$

$$y'_{ij} = \begin{cases} 1 & \text{If the activity } i-j \text{ has delay} \\ 0 & \text{Otherwise} \end{cases}$$

$$y''_{ij} = \begin{cases} 1 & \text{If the activity } i-j \text{ has been done in normal time} \\ 0 & \text{Otherwise} \end{cases}$$

Problem definitions

When the beginning activities of a project compressed, the compressing budget would be involved until the finishing project time, and this value of money would be involved for shorter time. This effect is obvious for the money that is involved with delaying in activities time, so the time value of money is an effective factor in this area. In the presented model, we used the time value of money to calculate the best time for compressing or delaying the activities. We added the compressing cost to the total cost function and subtract the venue of saving money of delaying on the activities from the total cost function, so the Pert network make better tradeoff between time and cost, considering the time value of money. The cost function contains direct cost, overhead cost, compressing cost, and delaying cost as follows:

$$Z_1 = H(t_n - t_1) + K_n + \sum_i \sum_j C_{ij}.y_{ij}\{D_{n(ij)} - d_{ij}\}$$
$$- \sum_i \sum_j C'_{ij}.y'_{ij}\{D_{n(ij)} - d_{ij}\} \tag{1}$$

For interpolation of the equations that contain the time value of money for compressing, we act as follows. Consider some of the activities have been compressed, and for each time unit of compression, C_{ij} is the unit of money spend, and therefore, $\sum_i \sum_j C_{ij}(t_n - t_i)$ is the total money that spends for compression involved from day t_i to day t_n (finishing time of project). For calculating the saving money of delaying, we act the same, so we have:

$$Z_1 = H(t_n - t_1) + K_n$$
$$+ \sum_i \sum_j C_{ij}.y_{ij}\{\{D_{n(ij)} - d_{ij}\} + (t_n - t_i)I_0\}$$
$$- \sum_i \sum_j C'_{ij}.y'_{ij}\{\{D_{n(ij)} - d_{ij}\} + (t_n - t_i)I_0\} \tag{2}$$

Much research has been conducted on the effects of the project cost and the cost of compression. Ameen (1987) presented the definition of cost gradient and offered a new technique, "CAPERTSIM", for decision making under uncertainty in time–cost tradeoff in compressing and the relation between them. In this research, a cost gradient index is defined as the ratio of money spend to compressing

value and considering the time duration of each activity as a probabilistic variable presents a simple simulation-based model for time compressing.

In this study, we present a nonlinear relation between activities durations and C_{ij} and C'_{ij}, as shown in Fig. 1:

According to Fig. 1, if the relation between compressing time and compressing cost was an exponential distribution, then we would have:

$$C(d_{ij}) = \alpha e^{-\beta \cdot d_{ij}} \tag{3}$$

If we had the coordination of D_n and D_f, the values of α and β could be calculated as follows:

$$\beta = \frac{Ln\left(\frac{C_n}{C_f}\right)}{D_f - D_n} \tag{4}$$

$$\alpha = e^{\{Ln(C_n) + \beta \cdot D_n\}} \tag{5}$$

Using the same approach for the saving coefficient, we have:

$$C'(d_{ij}) = \alpha' e^{-\beta' \cdot d_{ij}} \tag{6}$$

$$\beta' = \frac{Ln\left(\frac{C_m}{C_n}\right)}{D_n - D_m} \tag{7}$$

$$\alpha' = e^{\{Ln(C_m) + \beta' \cdot D_m\}} \tag{8}$$

The second objective function is considered as the finishing time of the project as follows:

$$Z_2 = t_n \tag{9}$$

Mathematical model

$$\text{Min } Z_1 = H(t_n - t_1) + K_n + \sum_i \sum_j y_{ij} \cdot \alpha_{ij} e^{-\beta_{ij} \cdot d_{ij}}$$
$$\{1 + (t_n - t_i)I_0\}$$
$$- \sum_i \sum_j y'_{ij} \cdot \alpha'_{ij} e^{-\beta'_{ij} \cdot d_{ij}}\{1 + (t_n - t_i)I_0\} \tag{10}$$

$$\text{Min } Z_2 = t_n \tag{11}$$

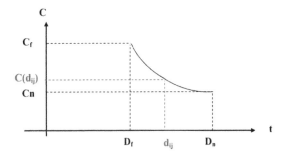

Fig. 1 Relation between activities durations and C_{ij} (C'_{ij})

$S.t:$

$$t_j - t_i \geq d_{ij} \tag{12}$$

$$y_{ij} \cdot D_{f(ij)} \leq y_{ij} \cdot d_{ij} \leq y_{ij} \cdot D_{n(ij)}; \quad \forall i,j \tag{13}$$

$$y'_{ij} \cdot D_{f(ij)} \leq y'_{ij} \cdot d_{ij} \leq y'_{ij} \cdot D_{n(ij)}; \quad \forall i,j \tag{14}$$

$$y''_{ij} \cdot d_{ij} = D_{n(ij)} \cdot y''_{ij}; \quad \forall i,j \tag{15}$$

$$y_{ij} + y'_{ij} + y''_{ij} = 1; \quad \forall i,j \tag{16}$$

$$t_n \leq t_{\text{Max}} \tag{17}$$

$$H(t_n - t_1) + \sum_i \sum_j y_{ij} \cdot \alpha_{ij} e^{-\beta_{ij} \cdot d_{ij}} \{1 + (t_n - t_i)I_0\}$$
$$- \sum_i \sum_j y'_{ij} \cdot \alpha'_{ij} e^{-\beta'_{ij} \cdot d_{ij}} \{1 + (t_n - t_i)I_0\} \leq C_{\text{Max}} \tag{18}$$

$$t_i \geq 0 \tag{19}$$

Equation (10) is the first objective function that deals with minimizing project cost, considering the time value of money. Equation 11 is the second objective function that minimizes the finishing time of project. Considering this two objective function together moves the problem to balance the compression and delaying of the project activities.

Solving algorithm

Because the presented problem belongs to NP-hard problem, the metaheuristic algorithms were used to solve the problem. One of these algorithms is GA. Sou-Sen Leu (2000) used GA based on fuzzy theory and considered the effects of uncertainty of the parameters in time–cost tradeoff. Li and Cao (1999) created "MLGAS" technique by combining GA and Learning machines method and claimed that when the relation between activities and cost is nonlinear, the presented technique has better solutions. Heng et al. (Burns 1994) presented a new algorithm based on GA and prepared a computational program to evaluate the efficiency of the presented algorithm. This research is the most complete research in this area that used GA for time–cost tradeoff. Chau and Chan (1997) claimed that exact methods, such as DP and LP, have very long solving time and are not suitable for solving time–cost tradeoff, so he developed the GA for this problem and considered the resource constraints for each activity.

The single objective optimization algorithm could find the better solution for one objective function, and if more than one objective considered for a problem, these algorithms have no efficiency. When the problem has more than one objective function, the results can be shown as a Pareto front of non-dominant solutions. This Pareto front contains the solutions that are

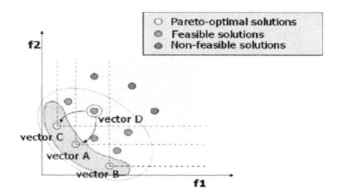

Fig. 2 Pareto sets in multi-objective problems (Deb 2001)

acceptable operation for all objective functions. When we have Pareto front of solutions, none of the solutions in Pareto frons has better result for all objective function comparing with other solutions in Pareto front, and we have not a single optimal solution (Tavakkoli-Moghaddam et al. 2008). Figure 2 shows the Pareto sets in multi-objective problems (Deb 2001).

NSGA-II algorithm

Non-dominated sorting genetic algorithm (NSGA-II) is one of the most efficient and famous multi-objective algorithms, which was presented by Deb (2001) and Zade et al. (2014) and proved its usefulness in multi-objective problems (Deb et al. 2002). The NSGA-II can convergent with Pareto sets of solutions, and the results could spread to all sets. NSGA-II uses non-dominant sorting for convergent confidence and also crowding distance for cutting the bad solutions for earning better solutions (Gen and Cheng 1997; Amiri and Khajeh 2016). Totally, its higher

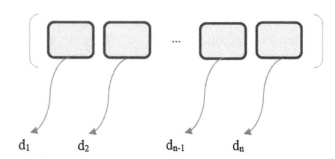

Fig. 3 Problem chromosome structure

Table 1 NSGA-II parameters value		
	$npop$	50
	P_C	0.7
	P_M	0.3
	nIt	100

Fig. 4 Procedure on non-dominant sorting

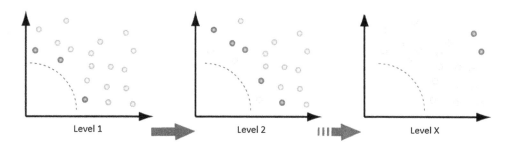

competence makes this algorithm a good selection for multi-objective problems.

The steps of NSGA-II are as follows.

Initialization

The structure of the chromosome that used in this study is shown in Fig. 3. This chromosome is a $1 \times n$ matrix, and each genome in this chromosome is duration of activities.

In addition, the initial information of this algorithm is:

npop The population size represents the number of chromosomes in each iteration of algorithm.

P_C Probability of crossover operator that represents the number of parents in the mating pool.

P_M Probability of mutation operator that represents the number of chromosomes mutating in each iteration of algorithm.

nIt Maximum algorithm iterations.

The values of these parameters are shown if Table 1.

Fast non-dominant sorting and crowding distance

In this step, all chromosomes ranked using fast non-dominant sorting and crowding distance concepts. In fast non-dominant, sorting the population is sorted based on domination concept. Each solution in this step is compared with all other chromosomes and determines which one is dominant or non-dominant. Finally, we have a set of non-dominant solutions that forms the first boundary of solutions. For determining the second boundary, the solutions that located in first boundary ignore and the procedure is repeated again. This procedure runs until all solutions are located in solution boundaries. In this procedure, the worst situation happened when each boundary contains one solution. In this situation, the complexity of algorithm is

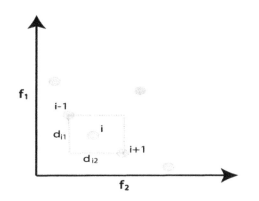

Fig. 5 Crowding distance of a specific solution (Deb 2001)

Parent 1	3	5	4	7	10	8	13	15	11	16
Parent 2	2	7	3	9	11	9	15	14	17	19
Randomly generated	0	0	1	1	0	1	0	0	1	0
Offspring 1	2	7	4	7	11	8	15	14	11	19
Offspring 2	3	5	3	9	10	9	13	15	17	16

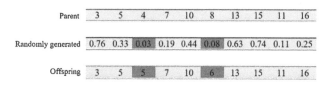

Fig. 6 Crossover operator

Parent	3	5	4	7	10	8	13	15	11	16
Randomly generated	0.76	0.33	0.03	0.19	0.44	0.08	0.63	0.74	0.11	0.25
Offspring	3	5	5	7	10	6	13	15	11	16

Fig. 7 Mutation operator

Fig. 8 Mechanism of NSGA-II
evolution

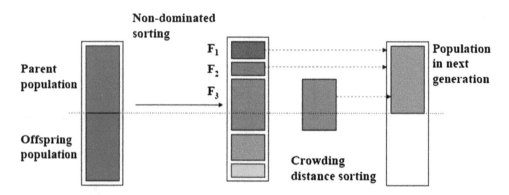

Table 2 Value of example
parameters

Act	(D_f)	(D_n)	(D_m)	(C_f)	(C_n)	(C_m)	Act	(D_f)	(D_n)	(D_m)	(C_f)	(C_n)	(C_m)
A1	3	8	10	2000	500	375	A22	4	9	11	4050	1900	1425
A2	3	6	8	2050	850	637	A23	6	9	11	6020	500	375
A3	3	5	6	2480	480	360	A24	6	8	10	1900	360	270
A4	10	15	19	1830	380	285	A25	2	5	6	2070	870	652
A5	4	8	10	2100	300	225	A26	5	7	9	4000	500	375
A6	11	12	15	1890	690	517	A27	2	6	8	15,020	1180	885
A7	5	9	11	320	200	150	A28	6	15	19	64,600	7000	5250
A8	4	11	14	25,000	1200	900	A29	5	12	15	3710	700	525
A9	5	6	8	770	230	172	A30	2	4	5	470	370	277
A10	6	8	10	615	175	131	A31	4	8	10	7720	920	690
A11	3	7	9	6420	2300	1725	A32	14	19	24	16,800	5000	3750
A12	12	19	24	19,980	5700	4275	A33	3	5	6	1110	940	705
A13	3	5	6	3150	550	412	A34	7	12	15	27,550	5300	3975
A14	2	4	5	525	125	94	A35	5	7	9	5600	3400	2550
A15	11	13	16	3520	660	495	A36	9	11	14	3580	300	225
A16	6	9	11	2535	885	664	A37	4	8	10	8690	770	577
A17	4	7	9	12,660	3000	2250	A38	6	11	14	2190	690	517
A18	6	9	11	610	250	187	A39	1	5	6	6520	3000	2250
A19	10	13	16	9980	980	735	A40	6	8	10	2940	2200	1650
A20	5	7	9	3440	1440	1080	A41	4	9	11	12,600	1100	825
A21	8	11	14	3970	280	210	A42	3	8	10	3850	600	450

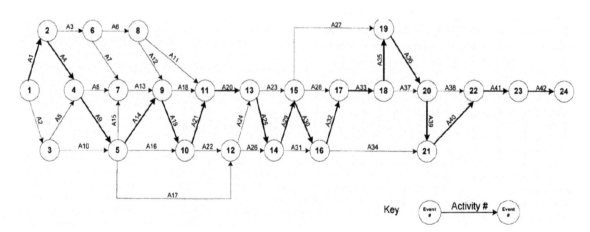

Fig. 9 AOA network of presented example

$O(MN^2)$, where M is the number of objectives and N is the population size. The procedure on non-dominant sorting is shown in Fig. 4.

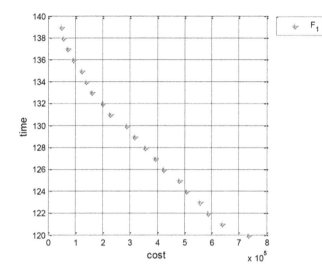

Fig. 10 Pareto front of the results for presented example

For determining the density of the solutions around a specific solution, the average distance between the specific solutions and two adjacent solutions is calculated and considered as a crowding distance. In other words, if we draw a rectangle that two adjacent solutions located on its vertex, sum of its length and width would be the crowding distance of the specific solution, as shown in Fig. 5. A specific solution with less crowding distance means less solution density around the specific solution, so for selecting the solutions for the next generation, the more crowding distance is better than less crowding distance (Deb 2001).

Parents

After non-dominant sorting and calculating crowding distance of each solution, the parents are ready for crossover and mutation operators based on selecting strategy.

Selecting strategy   Parents are selected for crossover and mutation operators based on crowded tournament selection operator. In this operator, two solutions are compared with

Table 3 Result of the first and second objective functions for 50 runs of NSGA-II and the values of compressing and delaying on activities

No.	Time	Cost	Crash	Delay	No.	Time	Cost	Crash	Delay
1	120	735,183	65	20	26	129	315,744	50	24
2	120	735,183	65	20	27	130	284,427	48	24
3	121	637,017	61	22	28	130	284,427	48	24
4	122	585,702	59	24	29	130	284,427	48	24
5	123	555,798	59	23	30	131	225,226	48	25
6	124	504,895	54	23	31	131	225,226	48	25
7	124	504,895	54	23	32	132	197,302	46	24
8	124	504,895	54	23	33	132	197,302	46	24
9	124	504,895	54	23	34	132	197,302	46	24
10	125	479,169	53	23	35	132	197,302	46	24
11	126	422,887	52	23	36	133	158,326	46	26
12	126	422,887	52	23	37	133	158,326	46	26
13	126	422,887	52	23	38	133	158,326	46	26
14	126	422,887	52	23	39	133	158,326	46	26
15	127	392,340	50	23	40	134	138,559	44	27
16	127	392,340	50	23	41	135	121,757	46	28
17	127	392,340	50	23	42	136	92,339	44	27
18	127	392,340	50	23	43	136	92,339	44	27
19	128	355,261	50	25	44	136	92,339	44	27
20	128	355,261	50	25	45	136	92,339	44	27
21	128	355,261	50	25	46	137	71,646	42	28
22	128	355,261	50	25	47	137	71,646	42	28
23	129	315,744	50	24	48	137	71,646	42	28
24	129	315,744	50	24	49	138	53,281	42	29
25	129	315,744	50	24	50	139	47,021	40	28

each other, and winner is selected. The ith solution has two properties:

1. Has a rank or degree on non-domination that shows by r_i,
2. Has a crowding distance that shows by d_i,

The ith solution is the winner of the competition comparing with jth solution if and only if, one of two below conditions is established (Deb 2001):

1. The ith solution has better rank in a non-dominant sorting procedure ($r_i < r_j$) that means this solution has

Table 4 Duration of activities in solutions 1 to 25

Solution	1	2	3	4	5	6	7	8	9	10	11	12	13	14	15	16	17	18	19	20	21	22	23	24	25
A1	3	3	3	3	3	3	3	3	3	3	3	3	3	3	3	3	3	3	3	3	3	3	3	3	3
A2	7	7	7	7	7	7	7	7	7	7	7	7	7	7	7	7	7	7	7	7	7	7	7	7	7
A3	4	4	6	4	5	4	4	6	6	6	5	4	4	4	6	4	5	6	5	6	4	5	4	5	6
A4	13	13	13	13	13	13	13	13	13	13	14	13	13	13	13	13	14	13	14	13	13	14	13	14	13
A5	9	9	9	9	9	9	9	9	9	9	9	9	9	9	9	9	9	9	9	9	9	9	9	9	9
A6	11	11	12	12	12	12	12	12	12	12	12	12	12	12	12	12	12	12	12	12	12	12	12	12	12
A7	8	8	10	10	10	10	10	10	10	10	10	10	10	10	10	10	10	10	10	10	10	10	10	10	10
A8	14	14	14	14	14	14	14	14	14	14	14	14	14	14	14	14	14	14	14	14	14	14	14	14	14
A9	5	5	5	5	5	5	5	5	5	5	5	5	5	5	5	5	5	5	5	5	5	5	5	5	5
A10	8	8	8	8	8	8	8	8	8	8	8	9	8	8	8	8	8	8	8	8	8	8	8	8	8
A11	8	8	8	8	8	8	8	8	8	8	8	8	8	8	8	8	8	8	8	8	8	8	8	8	8
A12	20	20	20	20	20	20	20	20	20	20	20	20	20	20	20	20	20	20	20	20	20	20	20	20	20
A13	4	4	6	4	4	4	4	6	6	6	4	4	4	4	6	4	4	6	4	6	4	4	4	4	6
A14	4	4	5	4	5	5	5	5	5	5	5	5	5	5	5	5	5	5	5	5	5	5	5	5	5
A15	12	12	14	14	14	14	14	14	14	14	14	14	14	14	14	14	14	14	14	14	14	14	14	14	14
A16	10	10	10	10	10	10	10	10	10	10	10	10	10	10	10	10	10	10	10	10	10	10	10	10	10
A17	8	8	8	8	8	8	8	8	8	8	8	8	8	8	8	8	8	8	8	8	8	8	8	8	8
A18	10	10	10	10	10	10	10	10	10	10	10	10	10	10	10	10	10	10	10	10	10	10	10	10	10
A19	11	11	13	11	11	11	13	14	13	14	11	13	13	11	13	11	13	13	13	13	11	11	13	13	13
A20	5	5	7	5	7	5	5	8	7	7	7	7	7	7	7	7	7	7	7	7	7	7	7	7	7
A21	8	8	9	8	9	9	8	9	9	9	9	9	9	9	9	9	9	9	9	9	9	9	9	9	9
A22	5	5	9	5	9	5	5	9	9	9	9	9	9	9	9	9	9	9	9	9	9	9	9	9	9
A23	9	9	10	9	9	9	9	9	10	10	9	9	9	9	9	9	9	9	9	9	9	9	9	9	9
A24	7	7	7	7	7	9	7	7	7	7	7	7	7	7	7	7	7	7	7	7	7	7	7	7	7
A25	2	2	3	2	2	2	2	2	2	2	2	2	2	2	2	2	2	2	2	2	2	2	2	2	2
A26	9	9	9	9	9	9	9	9	9	9	9	9	9	9	9	9	9	9	9	9	9	9	9	9	9
A27	8	8	8	8	8	8	8	8	8	8	8	8	8	8	8	8	8	8	8	8	8	8	8	8	8
A28	17	17	17	17	17	17	17	17	17	17	17	17	17	17	17	17	17	17	17	17	17	17	17	17	17
A29	6	6	8	6	6	6	6	6	8	8	6	6	6	6	6	6	6	6	6	6	6	6	6	6	6
A30	2	2	2	2	2	2	2	2	2	2	2	2	2	2	2	2	2	2	2	2	2	2	2	2	2
A31	9	9	9	9	9	9	9	9	9	9	9	9	9	9	9	9	9	9	9	9	9	9	9	9	9
A32	15	15	20	15	15	15	15	20	20	20	20	20	15	15	15	20	20	20	15	15	20	20	15	15	20
A33	3	3	3	3	3	3	3	3	3	3	3	3	3	3	3	3	3	3	3	3	3	3	3	3	3
A34	13	13	13	13	13	13	13	13	13	13	13	13	13	13	13	13	13	13	13	13	13	13	13	13	13
A35	5	5	8	5	5	5	5	5	5	8	5	5	5	5	5	5	5	5	5	5	5	5	5	5	8
A36	9	9	10	9	9	9	9	9	9	9	9	9	9	9	9	9	9	9	9	9	9	9	9	9	9
A37	10	10	10	10	10	10	10	10	10	10	10	10	10	10	10	10	10	10	10	10	10	10	10	10	10
A38	7	7	8	8	8	8	8	8	8	8	8	8	8	8	8	8	8	8	8	8	8	8	8	8	8
A39	2	2	2	2	2	2	2	2	2	2	2	2	2	2	2	2	2	2	2	2	2	2	2	2	2
A40	6	6	6	6	6	6	6	6	6	6	6	6	6	6	6	6	6	6	6	6	6	6	6	6	6
A41	4	4	4	4	4	4	4	4	4	4	4	4	4	4	4	4	4	4	4	4	4	4	4	4	4
A42	3	3	3	3	3	3	3	3	3	3	3	3	3	3	3	3	3	3	3	3	3	3	3	3	3

better non-domination degree comparing with jth solution,

2. If ith solution has more crowding distance comparing with jth solution $(d_i > d_j)$, when the rank of both solutions are equal.

Crossover operator

For crossover operator, two parents were randomly selected, and two offspring produced using a uniform crossover operator. In this operator, for each genome of the parent's

Table 5 Duration of activities in solutions 26–50

Solution	26	27	28	29	30	31	32	33	34	35	36	37	38	39	40	41	42	43	44	45	46	47	48	49	50
A1	3	3	3	3	3	3	3	3	3	3	3	3	3	3	3	3	3	3	3	3	3	3	3	3	3
A2	7	7	7	7	7	7	7	7	7	7	7	7	7	7	7	7	7	7	7	7	7	7	7	7	7
A3	4	5	4	6	6	6	6	6	4	5	4	5	5	4	6	6	6	4	4	5	6	4	5	6	6
A4	13	14	13	13	13	13	13	13	13	14	13	14	14	13	13	13	13	13	13	14	13	13	14	13	13
A5	9	9	9	9	9	9	9	9	9	9	9	9	9	9	9	9	9	9	9	9	9	9	9	9	9
A6	12	12	12	12	12	12	12	12	12	12	12	12	12	12	12	12	12	12	12	12	12	12	12	12	12
A7	10	10	10	10	10	10	10	10	10	10	10	10	10	10	10	10	10	10	10	10	10	10	10	10	10
A8	14	14	14	14	14	14	14	14	14	14	14	14	14	14	14	14	14	14	14	14	14	14	14	14	14
A9	5	5	5	5	5	5	5	5	5	5	5	5	5	5	5	5	5	5	5	5	5	5	5	5	5
A10	9	8	8	8	8	8	8	8	8	8	8	8	8	8	8	8	8	8	8	8	8	8	8	8	8
A11	8	8	8	8	8	8	8	8	8	8	8	8	8	8	8	8	8	8	8	8	8	8	8	8	8
A12	20	20	20	20	20	20	20	20	20	20	20	20	20	20	20	20	20	20	20	20	20	20	20	20	20
A13	4	4	4	6	6	6	6	6	4	4	4	4	4	4	6	6	6	4	4	4	6	4	4	6	6
A14	5	5	5	5	5	5	5	5	5	5	5	5	5	5	5	5	5	5	5	5	5	5	5	5	5
A15	14	14	14	14	14	14	14	14	14	14	14	14	14	14	14	14	14	14	14	14	14	14	14	14	14
A16	10	10	10	10	10	10	10	10	10	10	10	10	10	10	10	10	10	10	10	10	10	10	10	10	10
A17	8	8	8	8	8	8	8	8	8	8	8	8	8	8	8	8	8	8	8	8	8	8	8	8	8
A18	10	10	10	10	10	10	10	10	10	10	10	10	10	10	10	10	10	10	10	10	10	10	10	10	10
A19	13	13	11	13	13	13	13	13	11	11	13	13	13	11	13	13	13	11	13	13	13	11	13	13	13
A20	7	7	7	7	7	7	7	7	7	7	7	7	7	7	7	7	7	7	7	7	7	7	7	7	7
A21	9	9	9	9	9	9	9	9	9	9	9	9	9	9	9	9	9	9	9	9	9	9	9	9	9
A22	9	9	9	9	9	9	9	9	9	9	9	9	9	9	9	9	9	9	9	9	9	9	9	9	9
A23	9	9	9	9	10	9	10	9	9	9	9	9	9	9	9	9	10	9	9	9	9	9	9	9	9
A24	7	7	7	7	7	7	7	7	7	7	7	7	7	7	7	7	7	7	7	7	7	7	7	7	7
A25	2	2	2	2	2	2	2	2	2	2	2	2	2	2	2	2	2	2	2	2	2	2	2	2	2
A26	9	9	9	9	9	9	9	9	9	9	9	9	9	9	9	9	9	9	9	9	9	9	9	9	9
A27	8	8	8	8	8	8	8	8	8	8	8	8	8	8	8	8	8	8	8	8	8	8	8	8	8
A28	17	17	17	17	17	17	17	17	17	17	17	17	17	17	17	17	17	17	17	17	17	17	17	17	17
A29	6	6	6	6	8	6	8	6	6	6	6	6	6	6	6	6	8	6	6	6	6	6	6	6	6
A30	2	2	2	2	2	2	2	2	2	2	2	2	2	2	2	2	2	2	2	2	2	2	2	2	2
A31	9	9	9	9	9	9	9	9	9	9	9	9	9	9	9	9	9	9	9	9	9	9	9	9	9
A32	20	20	15	20	20	20	20	15	20	20	15	15	20	15	20	20	20	15	15	15	15	20	20	20	20
A33	3	3	3	3	3	3	3	3	3	3	3	3	3	3	3	3	3	3	3	3	3	3	3	3	3
A34	13	13	13	13	13	13	13	13	13	13	13	13	13	13	13	13	13	13	13	13	13	13	13	13	13
A35	5	5	5	8	8	5	8	5	5	5	5	5	5	5	8	5	8	5	5	5	5	5	5	5	8
A36	9	9	9	9	9	9	9	9	9	9	9	9	9	9	9	9	9	9	9	9	9	9	9	9	9
A37	10	10	10	10	10	10	10	10	10	10	10	10	10	10	10	10	10	10	10	10	10	10	10	10	10
A38	8	8	8	8	8	8	8	8	8	8	8	8	8	8	8	8	8	8	8	8	8	8	8	8	8
A39	2	2	2	2	2	2	2	2	2	2	2	2	2	2	2	2	2	2	2	2	2	2	2	2	2
A40	6	6	6	6	6	6	6	6	6	6	6	6	6	6	6	6	6	6	6	6	6	6	6	6	6
A41	4	4	4	4	4	4	4	4	4	4	4	4	4	4	4	4	4	4	4	4	4	4	4	4	4
A42	3	3	3	3	3	3	3	3	3	3	3	3	3	3	3	3	3	3	3	3	3	3	3	3	3

chromosome, a binary random variable is produced. If this variable is 1, the genome of the parents is changed with each other, and if this variable is 0, the genome stays in its place (Chau and Chan 1997; Deb et al. 2000). The crossover operator is shown in Fig. 6.

Mutation operator

The mutation operator is done in all four matrices of solution chromosome. For each genome, a uniform random variable between 0 and 1 is produced. If the value of this random variable is less than mutation rate, the genome mutates randomly, and if the value of the random variable is greater than mutation rate, the genome does not change (Chau and Chan 1997). The mutation operator is shown in Fig. 7.

Offsprings evaluation and combining with parents

In this section, we evaluate the offsprings that are created with crossover and mutation operators and assign a fitness quantity to each offspring. Then, we combine the parents and offsprings, and create a new population. This combination keeps the better solutions in new population. In multi-objective optimization problems, elithism has ambiguity. In these cases, we use a non-domination rank for each solution, so that each solution is rated based on non-domination.

After combination of the parents and offsprings, each solution is ranked based on the fast non-dominant sorting and crowding distances.

The mechanism of NSGA-II evolution is shown in Fig. 8.

Stop condition

The last step of NSGA-II is checking the stop condition. In multi-objective metaheuristic algorithms, there is no standard stop condition, so we consider a predefined algorithm iteration.

Numerical example

For illustrating the steps of presented algorithm, we used the numerical example of the paper entitled "Crashing PERT network using mathematical Programming" published in "International Journal of Project Management" in 2001. This example has been used in many studies as an authentic example for time compressing. In this example, all activities are assumed to be done in normal or compressed time. We extend this example for considering the delay in project activities. The values of example parameters are presented in Table 2.

The other parameters are $K_n = 100,000$, $H = 2000$, $I_0 = 0.1$, $C_{\text{Max}} = 1000,000$, and $t_{\text{Max}} = 140$.

The AOA network of presented example is shown in Fig. 9.

The results of solving the example with NSGA-II are presented as a Pareto front in Fig. 10.

As it is shown in Fig. 10, for faster finishing of projects, we need to pay more money, and in this situation, we must compress the activities more than normal situations. On the other hand, if we want to finish the projects in maximum acceptable due time, we have more money saving. The result of the first and second objective functions for 50 runs of NSGA-II and the values of compressing and delaying on activities are presented in Table 3.

As it is shown in Table 3, only 20 results are unique in 50 obtained ones, and the other results are repetitive. In addition, from the result no. 1 to no. 50, the time of finishing project increases, the project cost increases, and the compression and delaying activities show increasing trends. The duration of activities in each solution presented in Tables 4 and 5 consequently.

Conclusion and further studies

In this study, we showed that adding some assumptions to DTCTP can draw the problem nearer to real-world situations. One of these assumptions is adding the time value of money, because in many projects scheduling, the time value of money has a very important effect on making decision about compressing of the activities. In addition, adding the ability of delaying on project activities is another important factor appended to time–cost tradeoff problems. Moreover, presenting a Pareto front of results to decision makers gives them the opportunity to select the better solution due to project limitations and make the decision making procedure more flexible.

References

Akkan C (1998) A Lagrangian heuristic for the discrete time–cost tradeoff problem for activity-on-arc project networks. Working Paper, Koc University, Istanbul

Ameen (1987) A computer assisted pert simulation. J Syst Manage

Amiri Maghsoud, Khajeh Mostafa (2016) Developing a bi-objective optimization model for solving the availability allocation problem in repairable series–parallel systems by NSGA II. J Indus Eng Int 12(1):61–69

Ann T, Erenguc SS (1998) The resource constrained project scheduling problem with multiple crashable modes: a heuristic procedure. Eur J Operat Res 107(2):250–259

Burns SA (1994) The LP/IP hybrid method for construction time–cost trade off analysis. Construct Manage Econ J

Chau DKH, Chan WT, Govindan K (1997) A time-cost trade-off model with resource consideration using genetic algorithm. Civil Eng Syst 14:291–311

Deb K (2001) Multi-objective optimization using evolutionary algorithms. Wiley, Chichester

Deb K, Agrawal S, Pratap A, Meyarivan T (2000) A fast elitist non-dominated sorting genetic algorithm for multi-objective optimization: NSGA-II. In: Proceedings of the parallel problem solving from nature VI (PPSN-VI) conference, pp 849–858

Deb K, Pratap A, Agarwal S, Meyarivan T (2002) A fast and elitist multiobjective genetic algorithm: NSGA-II. IEEE Trans Evol Comput 6(2):182–197

Deineko VG, Woeginger GJ (2001) Hardness of approximation of the discrete time–cost tradeoff problem. Operat Res Lett 29(5):207–210

Demeulemeester E, Herroelen W, Elmaghraby SE (1996) Optimal procedures for the discrete time/cost trade-off problem in project networks. Eur J Operat Res 88(1):50–68

Demeulemeester E, De Reyck B, Foubert B et al (1998) New computational results on the discrete time/cost trade-off problem in project networks. J Operat Res Soc 49(6):1153–1163

Elmaghraby SE, Kamburowski J (1992) The analysis of activity network under generalized precedence relations. Manage Sci 38(9):1245–1263

Elmaghraby SE, Salem A (1981) Optimal linear approximation in project compression. OR Technical Report 171, North Carolina State University at Raleigh

Erenguc SS, Ahn T, Conway DG (2001) The resource constrained project scheduling problem with multiple crashable modes: an exact solution method. Naval Res Log 48(2):107–127

Ford LR, Fulkerson DR (1962) Flows in networks. Princeton University Press, Princeton

Fulkerson DR (1961) A network flow computation for project cost curves. Manage Sci 7:167–178

Gen M, Cheng R (1997) Genetic algorithms and engineering design. Wiley, New York

Goyal SK (1975) A note on the paper: a simple CPM time/cost trade-off algorithm. Manage Sci 21:718–722

Hazır O, Haouari M, Erel E (2014) Robust optimization for the discrete time-cost tradeoff problem with cost uncertainty. Handbook on project management and scheduling, vol 2, Part of the series International Handbooks on Information Systems pp 865–874

Hindelang TJ, Muth JF (1979) A dynamic programming algorithm for decision CPM networks. Operat Res 27(2):225–241

Kaveh A, Mahdavi VR (2015) Resource allocation and time-cost trade-off using colliding bodies optimization. Colliding Bodies optimization, pp 261–277

Ke H, Ma W, Chen X (2012) Modeling stochastic project time–cost trade-offs with time-dependent activity durations. Appl Math Comput 218:9462–9469

Kelley JE (1961) Critical path planning and scheduling: mathematical basis. Oper Res 9:296–320

Kelley JE, Walker MR (1959) Critical path planning and scheduling: an introduction. Mauchly Associates, Ambler

Leu SS (2000) A GA-based fuzzy optimal model for construction time- cost trade off. Int J Project Manage

Li H, Cao JN (1999) Using machine learning and GA to solve time-cost trade-off problems. J Project Manage

Liu SX, Wang MG, Tang LX et al (2000) Genetic algorithm for the discrete time/cost trade-off problem in project network. J North-east Univ China 21(3):257–259

Moder JJ, Phillips CR, Davis EW (1983) Project management with CPM, PERT and precedence diagramming, 3rd edn. Van Nostrand Reinhold Company, New York

Pasandideh SHR, Niaki STA, Hajipour V (2011) A multi-objective facility location model with batch arrivals: two parameter-tuned meta-heuristic algorithms. J Intell Manuf. doi:10.1007/s10845-011-0592-7

Patterson JH, Harvey RT (1979) An implicit enumeration algorithm for the time/cost tradeoff problem in project network analysis. Found Control Eng 4(2):107–117

Prabuddha DE, Dunne EJ, Ghosh JB et al (1997) Complexity of the discrete time–cost tradeoff problem for project networks. Operat Res 45(2):302–306

Sasaki M, Gen M (2003a) A method of fuzzy multi-objective nonlinear programming with GUB structure by hybrid genetic algorithm. Int J Smart Eng Des 5(4):281–288

Sasaki M, Gen M (2003b) Fuzzy multiple objective optimal system design by hybrid genetic algorithm. Appl Softw Comput 2(3):189–196

Shahriari M et al (2010) A new mathematical model for time cost trade-off problem with budget limitation based on time value of money. Appl Math Sci 4(63):3107–3119

Siemens N (1971) A simple CPM time/cost trade-off algorithm. Manage Sci 17:B-354–B-363

Son J, Hong T, Lee S (2013) A mixed (continuous + discrete) time-cost trade-off model considering four different relationships with lag time. KSCE J Civil Eng 17(2):281–291

Tavakkoli-Moghaddam R, Safari J, Sassani F (2008) Reliability optimization of series-parallel systems with a choice of redundancy strategies using a genetic algorithm. Reliab Eng Syst Safe 93:550–556

The report of the world commission on Dams (WCD) (2000) Dams and development. Earthscan publications L + d, London and sterling

Tiwari S, Johari S (2015) Project scheduling by integration of time cost trade-off and constrained resource scheduling. J Inst Eng (India) Ser 96(1):37–46

Van Slyke RM (1963) Monte Carlo methods and the PERT problem. Operat Res. 33:141–143

Vanhoucke M (2014) Generalized Discrete Time-Cost Tradeoff Problems. Handbook on project management and scheduling, vol. 1 Part of the series International Handbooks on Information Systems pp 639–658

Vanhoucke M, Demeulemeester E, Herroelen W (2002) Discrete time/cost trade-offs in project scheduling with time-switch constraints. J Operat Res Soc 53(7):741–751

Zade AE, Sadegheih A, Lotfi M (2014) A modified NSGA-II solution for a new multi-objective hub maximal covering problem under uncertain shipments. J Indus Eng Int 10(4):185–197

JIT single machine scheduling problem with periodic preventive maintenance

Mohammadreza Shahriari[1] · Naghi Shoja[2] · Amir Ebrahimi Zade[3] ·
Sasan Barak[4] · Mani Sharifi[5]

Abstract This article investigates a JIT single machine scheduling problem with a periodic preventive maintenance. Also to maintain the quality of the products, there is a limitation on the maximum number of allowable jobs in each period. The proposed bi-objective mixed integer model minimizes total earliness-tardiness and makespan simultaneously. Due to the computational complexity of the problem, multi-objective particle swarm optimization (MOPSO) algorithm is implemented. Also, as well as MOPSO, two other optimization algorithms are used for comparing the results. Eventually, Taguchi method with metrics analysis is presented to tune the algorithms' parameters and a multiple criterion decision making technique based on the technique for order of preference by similarity to ideal solution is applied to choose the best algorithm. Comparison results confirmed the supremacy of MOPSO to the other algorithms.

Keywords Scheduling · Single machine · Periodic maintenance · Total earliness-tardiness · Multi-objective optimization · MCDM

✉ Mohammadreza Shahriari
  shahriari@iau.ae

[1] Faculty of Management, South Tehran Branch, Islamic Azad University, Tehran, Iran

[2] Department of Industrial Engineering, Firoozkooh Branch, Islamic Azad University, Firoozkooh, Iran

[3] Department of Industrial Engineering, Amirkabir University of Technology, Tehran, Iran

[4] Faculty of Economics, Technical University of Ostrava, Ostrava, Czech Republic

[5] Faculty of Industrial and Mechanical Engineering, Qazvin Branch, Islamic Azad University, Qazvin, Iran

Introduction

Most of the manufacturing organizations try to implement some of the ideas adopted by Just in Time (JIT) philosophy, like on time delivery or minimum inventory (Salameh and Ghattas 2001). In this paper we introduce and formulate a JIT single machine scheduling problem with a periodic preventive maintenance on the machine. In most of the scheduling problems it is assumed that the machine is available interruptedly while, in practice, it may be unavailable due to causes like breakdown or preventive maintenance. According to the British Standard Institute (BSI), "Maintenance is a combination of any actions to retain an item in, or restore it to an acceptable condition" (BSI 1984). Periodic preventive maintenance, which is a fundamental part of JIT production, consists of regular preventive measures to increase machine reliability and to decrease breakdown probability during manufacturing process. In some of the manufacturing processes, overuse of the tool might decrease quality of the work piece. Therefore, when the work piece is expensive or when the accuracy is necessary, we change the tool before it is amortized. A well-known example is the printed circuit board manufacturing process in which the drilling machine is one of the most important devices. Thus not only should the machine stop to maintain after a period of processing time, but also the machine should stop to change the micro-drill after fixed times of using. Accordingly, the proposed model holds a limitation on the maximum number of processed jobs during each period. The objective of the proposed bi-objective mixed integer model is minimizing total earliness-tardiness costs and makespan simultaneously.

Machine unavailability problem has been investigated in the literature due to causes like machine breakdown, tool change or preventive maintenance. Machine breakdown or product quality loss is probable when a machine continues to work unceasingly for a long time. To avoid this situation, preventive maintenance is conducted on the machine which may be periodic or flexible (Xu et al. 2015). In a flexible maintenance the earliest and latest start time are determined and the maintenance process is allowed to start during this period. Yang et al. were the first to study a single machine scheduling problem with a flexible maintenance (Yang et al. 2002). They investigated the problem to minimize makespan and provided a proof for NP-Hardness of the problem. Qi et al. (1999) studied a problem in which multiple maintenance processes and jobs are to be scheduled on a single machine. They proposed heuristics as well as Branch and Bound based approaches to determine the sequence of jobs and maintenance start times while total completion time is minimized. Furthermore, Chen proposed a mixed binary integer programming and a heuristic to minimize mean flow time (Chen 2006). Also, Wan (Wan 2014) investigates on minimizing total earliness and tardiness in a single machine scheduling problem with a common due date for all jobs and a flexible maintenance. Luo et al. proposed polynomial algorithms for a single machine scheduling with flexible maintenance and various objective functions (Luo et al. 2015). Low et al. studied a single machine scheduling with flexible maintenance under two strategies, the first one was the flexible maintenance and the latter was changing the tool, after a predetermined number of jobs were conducted on the machine (Low et al. 2010). Their model was aimed at minimizing the makespan.

In addition, there are various scheduling researches addressing a periodic maintenance process. For example (Liao and Chen 2003) proposes a branch and bound based algorithm to minimize maximum tardiness or (Chen 2006) proposes a heuristic to minimize mean flow time in a problem with periodic maintenance. Benmansour et al. investigated on a JIT single machine scheduling problem with periodic maintenance in which the objective was to minimize maximum tardiness and maximum earliness (Benmansour et al. 2014), they also proposed a heuristic to solve the problem efficiently. On the other hand, in some cases like (Liao et al. 2007; Yang et al. 2008) it is assumed that the machine must stop for maintenance after a fixed number of processes. Hsu et al. studied a single machine scheduling problem, with a makespan minimizing objective, under two strategies; namely periodic maintenance and limited number of operations during each period (Hsu et al. 2010). They proposed a two stage binary integer programming and two efficient heuristics, best fit butterfly (BBF) and best fit decreasing (DBF), for solving large scale problems. Also Ebrahimyzade and Fakhrzad proposed a new mathematical model and dynamic genetic algorihtm (GA) to solve this problem (Ebrahimy Zade and Fakhrzad

2013). Computational results revealed that the solutions from the proposed dynamic GA had a better quality than the BBF and DBF.

On time delivery and minimum inventory costs are amongst the most important targets of a JIT manufactoring system. However, minimizing earliness and tardiness costs does not necessarily imply minimum inventory. Therefore, in some cases like Behnamian (2014), Gao et al. (2014), Behnamian and Fatemi Ghomi (2014), Gao (2010), and Eren (2007) total earliness–tardiness and makespan minimization are considered simultaneously, however, none of them mentioned the preventive maintenance, despite its substantial role in JIT philosophy. Table 1 delineates some properties of the proposed model with the most related articles in the literature.

Considering the above literature review, main contributions of the article are as follows:

1. Proposing a mathematical model for JIT single machine scheduling problem with periodic maintenance.
2. Simultaneous minimization of makespan, earliness and tardiness in a JIT scheduling problem with periodic maintenance.
3. Proposing three different multi-objective optimization algorithms to solve the problem.

The rest of this paper is organized as follows. The proposed mixed integer model is presented in the next section and considering computational complexity of the problem we propose three multi-objective optimization algorithms for solving the problems in the following section. Parameters tuning for each algorithms are described next, followed by section on computational results and finally conclusions and further research directions are provided.

Proposed model

In this section, we define the discussed problem formally. The problem is composed of n nonresumable jobs available at time zero to be scheduled on a single machine. It is assumed that the machine does not have ready time and the jobs will be delivered to customers immediately after they are completed. Each job has a unique due date. When a job is submitted to the customer before the due date it is called an early job and if it is delivered after the due date, it is a tardy job. It is assumed that both earliness and tardiness are costly, although when earliness is desired from the customer's point of view, a negative cost may be applied in the model. Preventive maintenance is conducted on the machine after a fixed period T. Moreover, to maintain the quality of the products, there is a limitation on the

Table 1 Some extensions to the scheduling problem with maintenance process

Article	Maintenance	Tool change	Minimizing	Solution approach
Yang et al. (2002)	Flexible	–	Makespan	Exact
Qi et al. (1999)	Flexible	–	Total completion time	Heuristics and branch and bound
Chen (2006)	Periodic/flexible	–	Mean flow time	Heuristic
Wan (2014)	Flexible	–	Total tardiness-earliness	Exact
Luo et al. (2015)	Flexible	✔	Makespan	Exact
Low et al. (2010)	Flexible	✔	Makespan	Heuristic
Liao et al. (2007)	–	✔	Makespan	Branch and bound
Yang et al. (2008)	–	✔	Makespan	Heuristic
Liao and Chen (2003)	Periodic	–	Maximum tardiness	Heuristic
Benmansour et al. (2014)	Periodic	–	Maximum earliness-tardiness	Heuristic
Hsu et al. (2010)	Periodic	✔	Makespan	Heuristic
Ebrahimy Zade and Fakhrzad (2013)	Periodic	✔	Makespan	GA
This article	Periodic	✔	(1) Total tardiness-earliness and (2) Makespan	MOPSO

maximum number of allowable jobs during each working period and after that, during the maintenance period, the tool will be changed.

The first objective in the proposed model minimizes total earliness-tardiness. On the other hand, machine/operator idleness and unnecessary work in process (WIP) are costly to the manufacturing system but the first objective does not encompass them. For example assume a problem in which the fixed processing period is 5, fixed maintenance duration is 2, and maximum number of doable jobs during each processing period are 3. Process time and due date for each jobs are presented in Table 2. Two different sequences, namely A and B, are demonstrated in Fig. 1. As evident, total earliness-tardiness for both sequences is 5, thus they are equivalent from the first objective's point of view. However, in sequence B we have less idle time and subsequently total working time for the machine and operator are less than sequence A which results in a less expensive manufacturing system. For this purpose, the second objective in the proposed model minimizes the makespan.

The set of problem parameters and indices in the proposed model are as follows:

i — Index for the jobs; $i = \{1, 2,..., n\}$.
j — Index for the position of a job in a period; $j = \{1,2,..., k\}$.
s — Index for the periods.
p_i — Process time for job i.
t — Duration of a working period (between two consecutive periods).
m — Fixed maintenance duration.
d_i — Due date for job i.
α_i — Earliness penalty coefficient for job i.
β_i — Tardiness penalty coefficient for job i.

And the set of decision variables are as follows:

x_{ijs} — Is a binary variable which is 1 if job i is in the jth position of period s; otherwise it is 0.
C_{ijs} — Completion time for job i in the jth position of period s.
T_i — Tardiness for job i.
E_i — Earliness for job i.

Considering the fact that neither the number of periods nor the number of jobs in each period are predetermined, initially $\left(l = \max\left\{n, \left[\frac{\max d_i}{t+m}\right] + 1\right\}\right)$ maximum number of

Table 2 Process time and due dates

Job number	Process time	Due date	Completion time (A)	Completion time (B)
1	4	4	4	4
2	3	17	17	12
3	1	5	10	5
4	2	9	9	9

Fig. 1 Two different sequences

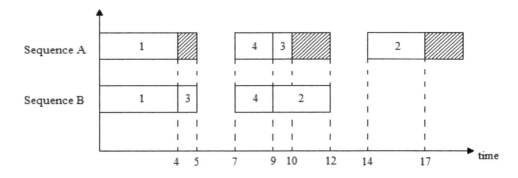

Fig. 2 A hypothetical solution

possible periods are assumed, each of which containing k positions for the jobs. Thereafter considering the duration of each period, maximum number of applicable jobs in each period, and the objective functions; the jobs are arranged in the periods. Figure 2 presents a possible solution for a problem with $n = 7$, $k = 4$.

The proposed mixed integer nonlinear model is as follows:

$$\min \sum_i \alpha_i E_i + \beta_i T_i \tag{1}$$

$$\min \max_{i,j,k}\{c_{ijs}\} \tag{2}$$

Subjects to:

$$\sum_j \sum_s x_{ijs} = 1 \quad \forall i \tag{3}$$

$$\sum_i x_{ijs} \leq 1 \; \forall j, s \tag{4}$$

$$\sum_j \sum_i p_i x_{ijs} \leq t \quad \forall s \tag{5}$$

$$\sum_i \sum_j x_{ijs} \leq k \quad \forall s \tag{6}$$

$$c_{ijs} = (t + m)(k - 1)x_{ijs} + \left(\sum_{i'=1}^{n}\sum_{j'=1}^{j} p_{i'}x_{i'j's}\right)x_{ijs} \quad \forall i, j, s \tag{7}$$

$$T_i \geq \sum_j \sum_s c_{ijs} - d_i \quad \forall i \tag{8}$$

$$E_i \geq d_i - \sum_j \sum_s c_{ijs} \quad \forall i \tag{9}$$

$$x_{ijs} \in \{0, 1\}, \quad c_{ijs}, T_i, E_i \geq 0 \tag{10}$$

In the above equations, Eq. (1) is the first objective of the proposed model that minimizes total earliness-tardiness and Eq. (2) is the second objective that minimizes the makespan. Equation (3) guarantees that each job appears in a unique position of a period. According to constraints Eq. (4), it is guaranteed that at most one job will be located in each position of a period. Constraints Eq. (5) ensure that the sum of processing time for the jobs being in the same period is less than t. According to Eq. (6), maximum number of doable jobs in each period is less than k. Equation (7) calculates the completion time for job i in the jth position of period s. Right side of this equation totalizes duration of the previous periods with the process time of the jobs prior to i (including i) in the jth period. For a given job i, tardiness and earliness are $\max\{0, C_i - d_i\}$ and $\max\{0, d_i - C_i\}$ respectively. Constraints Eqs. (8) and (9) provide lower bounds for tardiness (T_i) and earliness (E_i) respectively. According to Eq. (10), x_{ijs} is a binary variable and C_{ijs}, T_i, E_i are nonnegative variables.

Proposed solving algorithm

Due to the computational complexity of the problem, multi-objective particle swarm optimization (MOPSO) algorithm is implemented. In addition to the MOPSO, two other multi-objective optimization algorithms are used for comparing the results.

MOPSO algorithm

Optimization algorithm of particle swarm which was first introduced in 1995 (Yousefi et al. 2013), is a population based stochastic optimization technique inspired by social

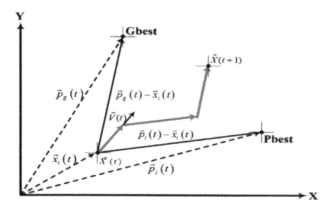

Fig. 3 Decoding and encoding solution

behavior of the swarm members such as bird flocking, fish training, etc. PSO's capability of storing the previous solution helps to reduce memory usage and speed up the CPU. In addition, a few parameters which should be adjusted in PSO and obtaining remarkable results in literature make PSO a useful method to apply. A brief description of how the algorithm works is as follows: Initially, a particle is identified as the best particle in a neighborhood of particles, based on its fitness. All particles are then accelerated in the direction of this particle as well as the direction of their own best solutions that they have covered previously.

Consider a group of N particles that are searching a global optimal solution in a D dimensions space. Vectors $X_i = (x_{1i}, \; x_{2i},..., \; x_{Di})$, $V_i = (v_{1i}, \; v_{2i},..., \; v_{Di})$ and $Pbest = (Pbest_{i1}, \; Pbest_{i2}, \; ..., \; Pbest_{iD})$ Shows position, velocity, and the best personal position of each particle respectively. $Gbest$ is also a *leader* position visited by the whole particle swarm population. In any iteration of PSO, the position and velocity of particles are updated according to Eqs. (11) and (12) and Fig. 3.

$$v_{ij}^{t+1} = w \times v_{ij}^t + C_1 \times r_1 \times (Pbest_{ij}^t - x_{ij}^t) + C_2 \times r_2 \\ \times (leader_{ij}^t - x_{ij}^t) \qquad (11)$$

$$x^{t+1}ij = x^t ij + v^{t+1}ij \qquad (12)$$

Based on Eqs. (11) and (12), v_j^t and v_{ij}^{t+1} are the velocity of particle i in jth dimension and iteration t and $t + 1$, x_{ij}^t and x_{ij}^{t+1} are the position of particle i in jth dimension and iteration t and $t + 1$, $Pbest_{ij}^t$ is the best previous position of particle i in jth dimension and iteration t, and $leader_{ij}^t$ (best global position) is the leader position for particle i in jth dimension and iteration t. C_1 and C_2 are the personal learning coefficient and the social learning factor respectively. r_1 and r_2 are random numbers between 0 and 1 and ω is an inertia factor usually in the range [0.8, 1.2].

Multi-objective PSO (MOPSO) is an extension of PSO proposed by (Coello et al. 2004) for multi-objective optimization problems. MOPSO stores the non-dominated solutions in 'repository', an external archive for solutions. The members of repository are not dominating each other and they provide an approximation of real Pareto frontier of the optimization problem. Repository members are updated by region based selection (grid index). Furthermore, in MOPSO, each particle chooses a solution in the repository as its leader with region based selection, instead of a unique global best for all particles. The Pseudo Code of MOPSO is presented as follows:

```
Start
Input nParticle, C1, C2, W, Max iteration;
Generate initial particle;
Evaluation fitness value of initial particle;
Calculate best personal memory;
Create best leader memory;
Create grid index for solution dimension;
Determine repository member;
Determine grid for repository members;
        For it=1to Max iteration
                For j=1to nParticle
                        Update particle position;
                        Evaluate fitness value of
                particle;
                        Update best personal memory;
                        Update best leader memory;
                End for
                Determine repository members;
                Combine new repository members with
                old repository members rep = {rep∪rep_new};
                Update repository members by non-
                dominance sorting algorithm;
                Update grid index for solution
                dimension;
                Find grid for repository members;
                Delete extra repository members;
        End for
Output: Extract the repository front;
End
```

For decoding process of MOPSO algorithm, N random numbers between [0, 1] is used, where N is the number of works. Encoding process is implemented by ordering these numbers increasingly while the minimum and maximum numbers are given to the first and the last works, respectively. Figure 4 illustrates the process.

Fig. 4 Decoding and encoding solution

Non-dominated sorting genetic algorithm II (NSGA-II)

NSGA-II algorithm was proposed by Deb, Pratap, Agarwal and Meyarivan (Deb et al. 2002). It uses a ranking scheme called the fast non-dominated sorting approach which requires a computational complexity of at most to rank the individuals, where M is the number of objectives and N is population size.

The Pseudo Code of NSGA-II is presented as follows:

```
Begin
Input nPop, Pc, Pm, Max iteration;
Generate initial population;
Evaluation fitness value of initial population;
Assign rank base on Pareto non-dominance sort;
For i=1 to Max iteration do
        For j=1 to 2×round((P_c×nPop)/2) ;
                Select parent by binary tournament selection;
                Apply crossover;
        End for
        Combine offspring and population;
        For j=1 to round(P_m×nPop);
                Select Chromosome by random selection;
                Apply mutation;
        End for
        Combine mutation members and population;
        Assign rank based on Pareto non-dominance sorting
        algorithm;
        Calculated the crowding distance of individuals in each
        front;
        Select the best nPop individual base on rank and
        crowded distance;
        End for
Output: Extract the best Pareto front;
End
```

where P_m is percentage of mutation and P_c is percentage of cross over.

Non-dominated ranking genetic algorithms (NRGA)

Al Jadaan et al. (2008) presented NRGA by exchanging the selection strategy of NSGA-II from the tournament selection to the roulette wheel. The Pseudo Code of NRGA is presented as follows:

```
Begin
Input nPop, Pc, Pm, Max iteration;
Generate initial population;
Evaluation fitness value of initial population;
Assign rank base on Pareto non-dominance sort;
For i=1 to Max iteration do
        For j=1 to 2×round((P_c×nPop)/2) ;
                Select parents by Pareto roulette wheel selection;
                Apply tournament crossover;
        End for
        Combine offspring and population;
        For j=1 to round(P_m×nPop) ;
                Select Chromosome by random selection;
                Apply mutation;
        End for
        Combine mutation members and population;
        Assign rank based on Pareto non-dominance sorting
        algorithm;
        Calculated the crowding distance of individuals in each
        front;
        Select the best nPop individual base on rank and
        crowded distance;
        End for
Output: Extract the best Pareto front;
End
```

PSO algorithm was developed for continuous searching space optimization problems. Since we encode the searching space of this problem continuously, therefore, PSO algorithm is used appropriately in this paper. Moreover, regarding to population based property of the MOPSO, we use two well-known population based algorithms namely NRGA and NSGAII to verify and validate the MOPSO results.

Parameter tuning

Undoubtedly, the results of a MOPSO, NRGA, and NSGA II algorithms to attain better fitness function value significantly depends on their parameters. The main parameters of a PSO algorithm that should be calibrated at the best level are: cognitive factor (C1), social factor (C2), swarm size (N-Particle), number of iterations (N-It), number of repository (N Rep), number of grid (N Grid), and inertia factor (w). Also, the main parameters of NRGA and NSGA

Tables 3 S/N for three repetitions in for MOPSO with orthogonal array L^{27}

C1	C2	W	Max It	N Particle	N Rep	N Grid	MOCV$_1$	MOCV$_2$	MOCV$_3$	S/N
1	1	1	1	1	1	1	1.003	1.154	2.044	−3.36837
1	1	1	1	2	2	2	1.354	1.837	1.593	−4.1193
1	1	1	1	3	3	3	2.364	1.015	3.427	−7.86824
1	2	2	2	1	1	1	2.781	1.141	1.014	−5.25651
1	2	2	2	2	2	2	1.019	1.011	1.205	−0.68496
1	2	2	2	3	3	3	1.223	2.826	4.379	−9.80119
1	3	3	3	1	1	1	1.014	1.338	1.345	−1.88222
1	3	3	3	2	2	2	1.027	2.011	1.339	−3.6121
1	3	3	3	3	3	3	1.009	1.019	1.145	−0.50182
2	1	2	3	1	2	3	1.622	5.844	1.417	−11.1161
2	1	2	3	2	3	1	3.494	1.884	1.852	−8.05895
2	1	2	3	3	1	2	2.750	5.420	7.217	−14.7239
2	2	3	1	1	2	3	1.013	1.034	1.009	−0.16115
2	2	3	1	2	3	1	1.004	1.024	1.027	−0.15824
2	2	3	1	3	1	2	1.019	1.026	1.028	−0.20889
2	3	1	2	1	2	3	1.025	1.030	1.009	−0.18368
2	3	1	2	2	3	1	1.017	1.012	1.018	−0.13505
2	3	1	2	3	1	2	1.006	1.012	1.026	−0.12676
3	1	3	2	1	3	2	1.184	1.018	1.015	−0.63009
3	1	3	2	2	1	3	1.261	1.614	1.024	−2.42516
3	1	3	2	3	2	1	2.211	1.016	1.029	−3.6671
3	2	1	3	1	3	2	1.034	1.018	1.219	−0.78147
3	2	1	3	2	1	3	1.018	1.029	1.228	−0.79561
3	2	1	3	3	2	1	1.158	1.014	1.180	−0.98247
3	3	2	1	1	3	2	1.014	1.021	1.020	−0.15784
3	3	2	1	2	1	3	1.012	1.011	1.025	−0.13805
3	3	2	1	3	2	1	1.014	1.021	1.021	−0.16069

II algorithms are: percentage of crossover (P_c), percentage of mutation (P_m), maximum iteration (Max It), and population size (N pop).

Therefore, in this section, the parameters of all algorithms are calibrated by using Taguchi method (Fazel Zarandi et al. 2013). Instead of the Fisher factorial designs, Taguchi developed fractional factorial experiments (FFEs) to reduce complexity of experiments in the full factorial designs. Taguchi method analyses the results in two ways: (1) analysis of variance for experiments with a single replicate, (2) the signal to noise ratio (S/N) for experiments with multiple replications where (N) is noise factor and (S) is controllable factors. Since the one with multiple replications has a better performance, the S/N is applied in this research to analyze the solutions. For more information regarding the Taguchi method see Taguchi et al. (2005).

For tuning of the proposed algorithm's parameters with Taguchi model properly, a new response which is representing different quality of a solution is considered for the experiments. In Pareto based algorithms, two main goals are interesting (1) convergence and (2) diversity. Mean ideal distance (MID) is the one that measure the convergence rate of the algorithms. MID measures the convergence rate of Pareto frontier members to a certain point (0, 0). Also, diversity measures the extension of the Pareto frontier. Spacing is the one that measure the diversity rate of the algorithms. It measures the standard deviation of the distances among solutions of the Pareto frontier. The Diversity and the MID metrics, as representatives of the multi-objective goals, are used to define multi-objective coefficient of variation (MOCV) in Eq. (13) as follows:

$$\text{MOCV} = \frac{\text{MID}}{\text{Diversity}} \tag{13}$$

By this definition, the two goals of the Pareto-based algorithms are considered simultaneously as a single response and more reliable outputs can be expected.

To conduct the Taguchi method more comprehensively, we have implemented a three stages orthogonal array experiments with MOCV. The purpose of conducting these

Fig. 5 The mean S/N plot for different levels of the MOPSO parameters

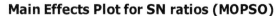

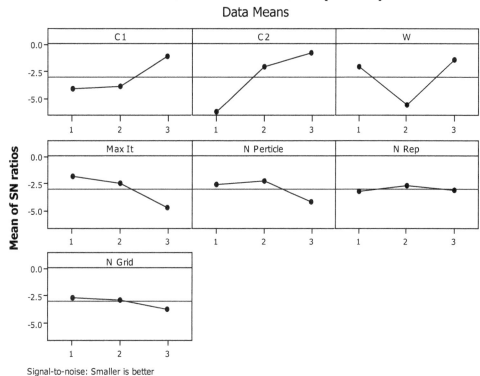

Table 4 Optimal parameters for MOPSO

Parameter	MOPSO			
	Level 1	Level 2	Level 3	Optimal value
Personal learning coefficient (C1)	1	1.4962	2	2
Global learning coefficient (C2)	1	1.4962	2	2
Inertia weight (W)	0.6	0.7298	0.9	0.9
Maximum iteration (Max It)	$5 \times n$	$10 \times n$	$15 \times n$	$5 \times n$
Particle size (N particle)	100	150	200	150
Repository size (N Rep)	75	100	100	100
Number of grids (N Grid)	5	8	10	5

arrays is to define the optimum level for each controllable parameter which cooperates to get the better fitness function value. From orthogonal arrays, each problem is run three times and since a solution with the highest MOCV is desired, the aim is to find minimize *S/N* calculated by Eq. 14 (Sadeghi et al. 2014).

$$S/N = -10 \ \log\left(\frac{1}{n}\sum_{i=1}^{n} MOCV_i^2\right) \qquad (14)$$

where, $MOCV_i$, $i = 1, 2, 3$ is the solution in ith replication of the Taguchi method and $n = 3$ is the number of replications in experiments.

Table 3 shows the experimental results of MOPSO with L^{27} orthogonal array under different scenarios of the parameters combinations, respectively where "1", "2",

and "3" refer to the first, the second, and the third level of the parameters. Regarding Eq. (14), these tables present S/Ns as well. In addition, it can be seen from Fig. 5 that the highest mean of S/N is the best. Therefore, Table 4 contains the optimal parameter values of the MOPSO algorithm. The same calculation is done for NRGA and NSGA II algorithms and the optimal parameters along with their levels are presented in Table 5.

Results comparison and discussion

In this section, we study the ability of algorithms (MOPSO, NRGA, and NSGAII) on test problems which is implemented in Matlab Software R2013a. It should be noted

Table 5 Optimal parameters for NSGA-II and NRGA

Parameter	NSGA-II				NRGA-II			
	Level 1	Level 2	Level 3	Optimal value	Level 1	Level 2	Level 3	Optimal value
Percentage of crossover (Pc)	0.7	0.8	0.9	0.7	0.7	0.8	0.9	0.9
Percentage of mutation (Pc)	0.2	0.15	0.1	0.15	0.2	0.15	0.1	0.1
Maximum iteration (Max It)	$5 \times n$	$10 \times n$	$15 \times n$	$5 \times n$	$5 \times n$	$10 \times n$	$15 \times n$	$15 \times n$
Population size (N pop)	100	150	200	100	100	150	200	100

Table 6 The result metric for NSGA-II

	MID	Max spread	SNS	NPS	RAS	Spacing
Problem 1	8,188,090	81,755	65,535	1	0	0
Problem 2	3,115,196	61,820	4,316,702	2	0.0173	0
Problem 3	1,514,438	105,448	1,520,510	7	0.0170	472
Problem 4	477,649	169,720	343,502	5	0.0064	250
Problem 5	922,707	212,320	868,632	3	0.0098	0
Problem 6	365,532	271,606	99,730	6	0.0289	4504
Problem 7	1,234,825	289,389	1,087,554	4	0.0148	2831
Problem 8	888,024	351,010	583,274	6	0.0147	2992
Problem 9	2,405,870	425,846	2,131,479	7	0.0203	3172
Problem 10	889,989	489,836	441,929	5	0.0150	3155
Problem 11	902,871	540,905	442,327	3	0.0013	0
Problem 12	835,618	621,949	256,050	3	0.0096	986
Problem 13	1,554,134	662,422	1,084,839	3	0.0092	3872
Problem 14	1,699,642	726,679	1,189,341	3	0.0020	762
Problem 15	5,726,711	771,869	5,521,994	5	0.0208	10,168
Problem 16	1,495,514	888,459	735,868	3	0.0107	7939
Problem 17	1,872,152	932,262	1,147,962	3	0.0033	2370
Problem 18	101,626,317	1,015,117	65,535	1	0	0
Problem 19	3,054,732	1,158,626	2,016,031	8	0.0131	6669
Problem 20	1,185,397	1,168,536	22,055	2	0.0136	0

that, since there is no executive library for this problem, all data in this paper have been generated randomly. However, these data are chosen in such a way which mirrors the real condition of the company and can serve as proxy for real cases. Therefore, 20 samples are generated and six well-known metrics including: MID, Max Spread, SNS, NPS, RAS, and Spacing (see Coello et al. 2007) are calculated for each algorithm and sample (see Tables 6, 7, 8). After obtaining the results of metrics, we evaluate and rank the algorithms with MCDM model. Technique for Order of Preference by Similarity to Ideal Solution (TOPSIS) method is one of well-known multi-criteria decision making modeling (Akhavan et al. 2015).

In this research, the metrics are used as criteria and algorithms are considered as alternatives. The goal is to prioritize alternatives based on criteria and select the algorithm which has the best performance.

In this model, first of all, the average of six metrics (criteria) in all problems is calculated then this matrix is normalized. After calculating the normalized decision matrix, the Euclidean distance of alternatives from positive and negative ideal solutions (d_i^+, d_i^-) are calculated using Eqs. (15–16).

$$d_i^+ = \sqrt{\sum_{j=1}^n \left(v_{ij} - v_j^+\right)^2} \quad i = 1, 2, \ldots, m \quad (15)$$

$$d_i^- = \sqrt{\sum_{j=1}^n \left(v_{ij} - v_j^-\right)^2} \quad i = 1, 2, \ldots, m \quad (16)$$

Finally, relative closeness value for each alternative calculated using Eq. 17. The alternative which has larger relative closeness value should be selected as the best one.

$$Cl_i = \frac{d_i^-}{d_i^- + d_i^+} \quad i = 1, 2, \ldots, m \quad (17)$$

Table 7 The result metric for NRGA

	MID	Max spread	SNS	NPS	RAS	Spacing
Problem 1	8,188,090	81,755	65,535	1	0	0
Problem 2	3,141,348	61,820	4,353,686	2	0.017	0
Problem 3	2,654,742	104,720	2,943,465	4	0.019	61
Problem 4	1,908,975	170,280	1,842,959	9	0.009	922
Problem 5	5,337,288	211,018	5,916,110	4	0.016	1640
Problem 6	8,707,682	260,742	10,344,769	3	0.007	152
Problem 7	7,372,236	292,630	8,172,759	4	0.011	557
Problem 8	11,490,043	343,836	13,649,436	3	0.006	766
Problem 9	8,468,223	416,458	8,997,759	5	0.009	2104
Problem 10	9,741,373	485,164	10,342,254	5	0.017	2362
Problem 11	18,018,400	529,442	21,413,375	3	0.011	6501
Problem 12	14,945,207	596,653	16,567,010	4	0.001	249
Problem 13	31,992,291	638,213	44,339,500	2	0.001	0
Problem 14	18,291,967	725,449	20,279,436	4	0.006	2840
Problem 15	19,658,874	770,875	21,796,666	4	0.015	7084
Problem 16	13,764,059	822,929	14,172,228	6	0.005	4625
Problem 17	30,127,022	899,187	35,791,283	3	0.005	403
Problem 18	9,870,254	964,115	9,748,874	6	0.010	3563
Problem 19	36,327,763	1,073,741	43,165,207	3	0.012	0
Problem 20	22,699,998	1,129,400	24,111,918	5	0.005	1651

Table 8 The result metric for MOPSO

	MID	Max spread	SNS	NPS	RAS	Spacing
Problem 1	8,188,090	81,755	65,535	1	0	0
Problem 2	3,141,348	61,820	4,353,686	2	0.017	0
Problem 3	2,654,742	104,720	2,943,465	4	0.019	61
Problem 4	1,908,975	170,280	1,842,959	9	0.009	922
Problem 5	5,337,288	211,018	5,916,110	4	0.016	1640
Problem 6	8,707,682	260,742	10,344,769	3	0.007	152
Problem 7	7,372,236	292,630	8,172,759	4	0.011	557
Problem 8	11,490,043	343,836	13,649,436	3	0.006	766
Problem 9	8,468,223	416,458	8,997,759	5	0.009	2104
Problem 10	9,741,373	485,164	10,342,254	5	0.017	2362
Problem 11	18,018,400	529,442	21,413,375	3	0.011	6501
Problem 12	14,945,207	596,653	16,567,010	4	0.001	249
Problem 13	31,992,291	638,213	44,339,500	2	0.001	0
Problem 14	18,291,967	725,449	20,279,436	4	0.006	2840
Problem 15	19,658,874	770,875	21,796,666	4	0.015	7084
Problem 16	13,764,059	822,929	14,172,228	6	0.005	4625
Problem 17	30,127,022	899,187	35,791,283	3	0.005	403
Problem 18	9,870,254	964,115	9,748,874	6	0.010	3563
Problem 19	36,327,763	1,073,741	43,165,207	3	0.012	0
Problem 20	22,699,998	1,129,400	24,111,918	5	0.005	1651

The average decision matrix, normalized weighted decision matrix, Euclidean distance of alternatives, and relative closeness values of alternatives for MOPSO, NRGA, and NSGAII are shown in Table 9. The results show that MOPSO algorithm's performance in solving problems is better than others.

Table 9 The result of TOPSIS method for problem

	Average decision matrix						
	MID	Max spread	SNS	NPS	RAS	Spacing	MID
NSGA-II	6,997,770.5	547,278.8	1,197,042.5	4	0.01139	2507.1	0.032
NRGA	14,135,291.8	528,921.5	15,900,711.5	4	0.009131	1774.0	0.016
MOPSO	1,037,890.2	622,772.9	12,296.3	4.15	0.024065	2206.4	0.213

	Normalize decision matrix				
	Max spread	SNS	NPS	RAS	Spacing
NSGA-II	0.055	0.015	0.015	0.028	0.067
NRGA	0.054	0.193	0.015	0.035	0.095
MOPSO	0.063	0.000	0.016	0.013	0.076

	d_i^+	d_i^-	CL	Rank
NSGA-II	0.180	0.184	0.494	2
NRGA	0.011	0.276	0.038	3
MOPSO	0.276	0.030	0.902	1

Conclusion

This paper presented a Bi-objective model for scheduling a JIT single machine with a periodic preventive maintenance while total earliness-tardiness and makespan are minimized simultaneously. Furthermore, the proposed mixed integer model takes the maximum number of allowable jobs in each period which helps to maintain quality of the products.

To solve the model, we propose NSGA-II, NRGA, and MOPSO algorithms. The parameters of these algorithms are tuned by Taguchi method, and finally, six performance metrics are used to analyze the diversity and convergence of proposed algorithms. Based on MADM analysis of these metrics, we have shown that MOPSO has better metric performance to other algorithms and has better uniformity within the solutions of their Pareto curves.

References

Akhavan P, Barak S, Maghsoudlou H, Antuchevičienė J (2015) FQSPM-SWOT for strategic alliance planning and partner selection; case study in a holding car manufacturer company. Technol Econ Dev Econ 21(2):165–185

Al Jadaan O, Rajamani L, Rao C (2008) Non-dominated ranked genetic algorithm for solving multiobjective optimization problems. J Theor Appl Inf Technol 15(5):60–67

Behnamian J (2014) A parallel competitive colonial algorithm for JIT flowshop scheduling. J Comput Sci 5(5):777–783

Behnamian J, Fatemi Ghomi SMT (2014) Multi-objective fuzzy multiprocessor flowshop scheduling. Appl Soft Comput 21:139–148

Benmansour R, Allaoui H, Artiba A, Hanafi S (2014) Minimizing the weighted sum of maximum earliness and maximum tardiness costs on a single machine with periodic preventive maintenance. Comput Oper Res 47:106–113

BSI (1984) Glossary of maintenance management terms in terotechnology, BS 3811

Chen JS (2006) Single-machine scheduling with flexible and periodic maintenance. J Oper Res Soc 57(6):703–710

Coello CAC, Pulido GT, Lechuga MS (2004) Handling multiple objectives with particle swarm optimization. Evolut Comput IEEE Trans 8(3):256–279

Coello CC, Lamont GB, Van Veldhuizen DA (2007) Evolutionary algorithms for solving multi-objective problems. Springer, Berlin

Deb K, Pratap A, Agarwal S, Meyarivan T (2002) A fast and elitist multiobjective genetic algorithm: NSGA-II. Evolut Comput IEEE Trans 6(2):182–197

Ebrahimy Zade A, Fakhrzad MB (2013) A dynamic genetic algorithm for solving a single machine scheduling problem with periodic maintenance. ISRN Ind Eng 2013:11

Eren T (2007) A multicriteria flowshop scheduling problem with setup times. J Mater Process Technol 186(1–3):60–65

Fazel Zarandi MH, Mosadegh H, Fattahi M (2013) Two-machine robotic cell scheduling problem with sequence-dependent setup times. Comput Oper Res 40(5):1420–1434

Gao J (2010) A novel artificial immune system for solving multiobjective scheduling problems subject to special process constraint. Comput Ind Eng 58(4):602–609

Gao KZ, Suganthan PN, Pan QK, Chua TJ, Cai TX, Chong CS (2014) Pareto-based grouping discrete harmony search algorithm for multi-objective flexible job shop scheduling. Inf Sci 289:76–90

Hsu C-J, Low C, Su C-T (2010) A single-machine scheduling problem with maintenance activities to minimize makespan. Appl Math Comput 215(11):3929–3935

Liao CJ, Chen WJ (2003) Single-machine scheduling with periodic maintenance and nonresumable jobs. Comput Oper Res 30(9):1335–1347

Liao CJ, Chen CM, Lin CH (2007) Minimizing makespan for two parallel machines with job limit on each availability interval. J Oper Res Soc 58(7):938–947

Low C, Ji M, Hsu C-J, Su C-T (2010) Minimizing the makespan in a single machine scheduling problems with flexible and periodic maintenance. Appl Math Model 34(2):334–342

Luo W, Cheng TCE, Ji M (2015) Single-machine scheduling with a variable maintenance activity. Comput Ind Eng 79:168–174

Qi X, Chen T, Tu F (1999) Scheduling the maintenance on a single machine. J Oper Res Soc 50(10):1071–1078

Sadeghi J, Sadeghi S, Niaki STA (2014) A hybrid vendor managed inventory and redundancy allocation optimization problem in supply chain management: an NSGA-II with tuned parameters. Comput Oper Res 41:53–64

Salameh MK, Ghattas RE (2001) Optimal just-in-time buffer inventory for regular preventive maintenance. Int J Prod Econ 74(1–3):157–161

Taguchi G, Chowdhury S, Wu Y (2005) Taguchi's quality engineering handbook. Wiley, New York

Wan L (2014) Scheduling jobs and a variable maintenance on a single machine with common due-date assignment. Sci World J 2014:5

Xu D, Wan L, Liu A, Yang D-L (2015) Single machine total completion time scheduling problem with workload-dependent maintenance duration. Omega 52:101–106

Yang D-L, Hung C-L, Hsu C-J, Chern M-S (2002) Minimizing the makespan in a single machine scheduling problem with a flexible maintenance. J Chin Inst Ind Eng 19(1):63–66

Yang D-L, Hsu C-J, Kuo W-H (2008) A two-machine flowshop scheduling problem with a separated maintenance constraint. Comput Oper Res 35(3):876–883

Yousefi M, Omid M, Rafiee S, Ghaderi S (2013) Strategic planning for minimizing CO_2 emissions using LP model based on forecasted energy demand by PSO Algorithm and ANN. J Homepage www. IJEE.IEEFoundation.org 4(6):1041–1052

15

Multi-choice stochastic bi-level programming problem in cooperative nature via fuzzy programming approach

Sumit Kumar Maiti[1] · Sankar Kumar Roy[2]

Abstract In this paper, a Multi-Choice Stochastic Bi-Level Programming Problem (MCSBLPP) is considered where all the parameters of constraints are followed by normal distribution. The cost coefficients of the objective functions are multi-choice types. At first, all the probabilistic constraints are transformed into deterministic constraints using stochastic programming approach. Further, a general transformation technique with the help of binary variables is used to transform the multi-choice type cost coefficients of the objective functions of Decision Makers(DMs). Then the transformed problem is considered as a deterministic multi-choice bi-level programming problem. Finally, a numerical example is presented to illustrate the usefulness of the paper.

Keywords Bi-level programming · Stochastic programming · Multi-choice programming · Fuzzy programming · Non-linear programming

Introduction and literature review

Bi-level programming problem under cooperative environment

Real-life decision-making problems in which there are multiple Decision Makers(DMs), who make decisions successively, are often formulated as a multi-level

programming problems. Assuming that each DM makes a decision without any communication with some other DMs, as a solution concept to the problems, Stackelberg solution is employed. However, for decision-making problems in decentralized firms, it is quite natural to assume that there exist communication and some cooperative relationship among the DMs.

Anandalingam (1988) considered a mathematical programming model of decentralized multi-level systems and discussed the solution procedure. Anandalingam and Apprey (1991) proposed and discussed the multi-level programming with conflict resolution. Lai (1996) discussed hierarchical optimization and obtained a satisfactory solution for this multi-level programming. Sinha and Sinha (2004) considered linear multi-level programming under fuzzy environment. Dempe and Starostina (2007) considered a fuzzy bi-level programming problem and described the solution procedure with the help of multi-criteria optimization technique. In 2001, Roy (2001) proposed an approach to multi-objective bi-matrix games for Nash equilibrium solution. In 2006, Roy (2006) presented a fuzzy programming techniques for Stackelberg game. He used in his paper a fuzzy programming technique to solve Stackelberg game and compared the solution with the Kuhn–Tucker transformation technique. In 2007, Roy (2007) solved two-person multi-criteria bi-matrix games using fuzzy programming technique. Dey et al. (2014) presented a technique for order preference by similarity to ideal solution (TOPSIS) algorithm to linear fractional bi-level multi-objective decision-making problem in 2014. In 2012, Lachhwani and Poonia (2012) suggested for solving multi-level fractional programming problems in a large hierarchical decentralized organization using fuzzy goal programming approach. Zheng et al. (2011) discussed a class of bi-level multi-objective programming problem under fuzzy environment.

✉ Sankar Kumar Roy
sankroy2006@gmail.com

[1] School of Applied Sciences and Humanities, Haldia Institute of Technology, Purba Midnapore, 721 157 Haldia, India

[2] Department of Applied Mathematics with Oceanology and Computer Programming, Vidyasagar University, 721 102 Midnapore, India

Shih et al. (1996) proposed the multi-level programming problem with fuzzy approach and also discussed the solution concepts assuming cooperative communication among the DMs. Their methods were based on the idea that the DM at the lower level optimizes his or her objective function, taking a goal or preference of the DM at the upper level into consideration. The DM identifies the membership functions of fuzzy goals for their objective functions, and especially, the DM at the upper level also specifies those of fuzzy goals for decision variables. The DM at the lower level solves a fuzzy programming problem with constraints on a satisfactory degree of the DM at the upper level.

In this paper, we consider the multi-choice stochastic bi-level programming problem in cooperative environment and also assume that the DMs at the upper level and at the lower level have own fuzzy goals with respect to their objective functions.

The mathematical model of bi-level programming problem is as follows:

Model 1

$$\max_{\text{for DM}_{11}} Z_{11}(x) = \sum_{j=1}^{n} c_{11j} x_j$$

$$\max_{\text{for DM}_{2f}} Z_{2f}(x) = \sum_{j=1}^{n} c_{2fj} x_j$$

$$\text{subject to} \sum_{j=1}^{n} a_{ij} x_j \leq b_i \quad i = 1, 2, \ldots, m,$$

$$x_j \geq 0 \quad j, \quad f = 1, 2, \ldots, n.$$

Stochastic programming

In most of the real-life decision-making problems in mathematical programming, the parameters are considered as random variables. The branch of mathematical programming which deals with the theory and methods for the solution of conditional extremum problems under incomplete information about the random parameters is called stochastic programming. Many problems in applied mathematics may be considered as belonging to any one of the following classes:

1. Descriptive problems, in which, with the help of mathematical methods, information is processed about the investigated event, some laws of the event being induced by others.
2. Optimization problems in which from a set of feasible solutions, an optimal solution is chosen.

Besides the above division of applied mathematical problems, these problems may be further classified as deterministic and stochastic problems. In the process of solution of the stochastic problem, several mathematical models

have been developed. However, probabilistic methods were for a long time applied exclusively to the solution of the descriptive type of problems. Research on the theoretical development of stochastic programming has been going on for last four decades and to solve the several real-life problems in management science, it has been applied successfully. The chance constrained programming was first developed by Charnes and Cooper (1978).

Multi-choice programming

Multi-choice programming is a mathematical programming problem, in which DM is allowed to set multiple number of choices for a parameter. The situation of multiple choices for a parameter exists in many managerial decision-making problems. The multi-choice programming cannot only avoid the wastage of resources but also decide on the appropriate resource from multiple resources. A method for modeling the multi-choice programming problem, using binary variables was presented by Chang (2007). He has also proposed a revised method for multi-choice goal programming model which does not involve multiplicative terms of binary variables to model the multiple aspiration levels Chang (2008). Acharya and Acharya (2013) presented the generalized transformation technique for a multi-choice linear programming problems in which constraints are associated with some multi-choice parameters. Recently, Mahapatra et al. (2013) and Roy (2006) discussed the multi-choice stochastic transportation problem involving extreme value distribution and exponential distribution in which the multi-choice concept involved only in the cost parameters. In 2014, Maity and Roy (2014) presented also multi-choice multi-objective transportation problem using utility function approach. Recently, Maity and Roy (2015) studied a mathematical model for a transportation problem consisting of a multi-objective environment with non-linear cost and multi-choice demand. Roy (2015) discussed the solution procedure for multi-choice transportation problem using Langrange's interpolating polynomial approach. Roy (2014) presented multi-choice stochastic transportation problem involving Weibull distribution.

In this paper, we consider a generalized transformation technique to transform a multi-choice stochastic bi-level programming problem to an equivalent mathematical programming model. Using the transformation technique, the transformed model can be derived. Applying fuzzy programming technique, optimal solution of the proposed model is obtained.

The organization of the paper is as follows: following the introduction and literature review in Sect. 1, mathematical model of multi-choice stochastic bi-level programming problem (MCSBLPP) is presented in Sect. 2. Mathematical formulation is presented in Sect. 3 and solution procedure in Sect. 4. To verify the proposed methodology of the paper, a numerical example is presented in Sect. 5. In Sect. 6, the results of the given problems have been discussed here. Section 7 presents sensitivity analysis with our proposed problem. Finally, conclusion is presented in Sect. 8.

Mathematical model of MCSBLPP

In the mathematical model of Sect. 1, considering the cost coefficients of the objective functions for both DMs are multi-choice types and also assume that all the parameters of the constraints are random variables. Then the corresponding mathematical model of bi-level programming problem is to be treated as multi-choice stochastic bi-level programming problem (**MCSBLPP**) and is stated as below:

Model 2

$$\max_{\text{for DM}_{11}} Z_{11}(x) = \sum_{j=1}^{n} \left(c_{11j}^{(1)}, c_{11j}^{(2)}, \ldots, c_{11j}^{(k_j)} \right) x_j,$$

$$\max_{\text{for DM}_{2f}} Z_{2f}(x) = \sum_{j=1}^{n} \left(c_{2fj}^{(1)}, c_{2fj}^{(2)}, \ldots, c_{2fj}^{(k_j)} \right) x_j,$$

$$(1)$$

$$\text{subject to } P_r \left[\sum_{j=1}^{n} a_{ij}x_j \leq b_i \right] \geq p_i \quad i = 1, 2, \ldots, m,$$

where $x_j \geq 0 \quad f = 1, 2, \ldots, n \quad 0 \leq p_i \leq 1; \quad \forall i, j.$ and p_i is the pre-specified level of probability.

Model formulation

Assuming that a_{ij} $(i = 1, 2, \ldots, m; j = 1, 2, \ldots, n)$ and b_i $(i = 1, 2, \ldots, m)$ are normal random variables, $c_{11j} = (c_{11j}^{(1)}, c_{11j}^{(2)}, \ldots, c_{11j}^{(k_j)}) \forall j$ and $c_{2fj} = (c_{2fj}^{(1)}, c_{2fj}^{(2)}, \ldots, c_{2fj}^{(k_j)}) \forall j$ are multi-choice parameters.

The following cases are to be considered:

1. Only a_{ij} $(i = 1, 2, \ldots, m; j = 1, 2, \ldots, n)$ follows normal distribution.
2. Only b_i $(i = 1, 2, \ldots, m)$ follows normal distribution.
3. Both a_{ij} $(i = 1, 2, \ldots, m; j = 1, 2, \ldots, n)$ and b_i $(i = 1, 2, \ldots, m)$ follow normal distributions.

Conversion of probabilistic constraints to deterministic constraints

Only a_{ij} $(i = 1, 2, \ldots, m; j = 1, 2, \ldots, n)$ follows Normal distribution.

Assuming that $\bar{a}_{ij}$ and $Var(a_{ij}) = \sigma_{a_{ij}}^2$ are the mean and variance of a_{ij}. Considering that the multivariate distribution of a_{ij} is also known along with the covariance, $cov(a_{ij}, a_{kl})$ between the random variables a_{ij} and a_{kl}. We consider f_i as

$$f_i = \sum_{j=1}^{n} a_{ij}x_j \quad i = 1, 2, \ldots, m.$$

As x_j s are constants (not yet known), let the mean $\bar{f}_i$ and variance $Var(f_i)$ are defined as follows: $\bar{f}_i = \sum_{j=1}^{n} \bar{a}_{ij}x_j$ $(i = 1, 2, \ldots, m)$ and $Var(f_i) = X^T V_i X$ where V_i is ith covariance matrix which is defined as follows:

$$V_i = \begin{pmatrix} Var(a_{i1}) & cov(a_{i1}, a_{i2}) & \ldots & cov(a_{i1}, a_{in}) \\ cov(a_{i2}, a_{i1}) & Var(a_{i2}) & \ldots & cov(a_{i2}, a_{in}) \\ \ldots & \ldots & \ldots & \ldots \\ \ldots & \ldots & \ldots & \ldots \\ cov(a_{in}, a_{i1}) & cov(a_{in}, a_{i2}) & \ldots & Var(a_{in}) \end{pmatrix}$$

The constraints of Eq. 1 can be rewritten as:

$$P_r[f_i \leq b_i] \geq p_i \quad i = 1, 2, \ldots, m,$$

$$i.e., P_r \left[\frac{f_i - \bar{f}_i}{\sqrt{Var(f_i)}} \leq \frac{b_i - \bar{f}_i}{\sqrt{Var(f_i)}} \right] \geq p_i \quad i = 1, 2, \ldots, m$$

Therefore, $P_r[f_i \leq b_i] = \phi \left[\dfrac{b_i - \bar{f}_i}{\sqrt{Var(f_i)}} \right].$ (2)

where $\phi(x)$ represents the cumulative distribution function of the standard normal distribution evaluated at x. Defining e_i as $\phi(e_i) = p_i$.

Then the constraints in Eq. 2 can be stated as

$$\phi \left[\frac{b_i - \bar{f}_i}{\sqrt{Var(f_i)}} \right] \geq \phi(e_i) \quad i = 1, 2, \ldots, m.$$

These inequalities will be satisfied only if

$$\frac{b_i - \bar{f}_i}{\sqrt{Var(f_i)}} \geq e_i \quad i = 1, 2, \ldots, m,$$

$$i.e., \bar{f}_i + e_i \sqrt{Var(f_i)} - b_i \leq 0 \quad i = 1, 2, \ldots, m.$$

Thus, finally, the probabilistic constraints (1) can be transformed into deterministic constraints as:

$$\sum_{j=1}^{n} \bar{a}_{ij}x_j + e_i \sqrt{X^T V_i X} - b_i \leq 0 \quad i = 1, 2, \ldots, m.$$

Thus, we obtain a multi-choice deterministic model (**Model 3**) as follows:

Model 3

$$\max_{\text{for DM}_{11}} Z_{11}(x) = \sum_{j=1}^{n} \left(c_{11j}^{(1)}, c_{11j}^{(2)}, \ldots, c_{11j}^{(k_j)} \right) x_j,$$

$$\max_{\text{for DM}_{2f}} Z_{2f}(x) = \sum_{j=1}^{n} \left(c_{2fj}^{(1)}, c_{2fj}^{(2)}, \ldots, c_{2fj}^{(k_j)} \right) x_j,$$

$$\text{subject to } \sum_{j=1}^{n} \bar{a}_{ij}x_j + e_i \sqrt{X^T V_i X} - b_i \leq 0 \quad i = 1, 2, \ldots, m,$$

where $x_j \geq 0 \quad \forall j, f.$ (3)

Only b_i $(i = 1, 2, \ldots, m)$ follows normal distribution

Considering that $\bar{b}_i$ and $Var(b_i)$ are the mean and variance of b_i, the constraints of Eq. 1 can be rewritten as

$$P_r\left[\sum_{j=1}^{n} a_{ij}x_j \leq b_i\right] = P_r\left[\frac{\sum_{j=1}^{n} a_{ij}x_j - \bar{b}_i}{\sqrt{Var(b_i)}} \leq \frac{b_i - \bar{b}_i}{\sqrt{Var(b_i)}}\right]$$

$$= P_r\left[\frac{b_i - \bar{b}_i}{\sqrt{Var(b_i)}} \geq \frac{\sum_{j=1}^{n} a_{ij}x_j - \bar{b}_i}{\sqrt{Var(b_i)}}\right] \geq p_i$$

$$i.e., 1 - P_r\left[\frac{b_i - \bar{b}_i}{\sqrt{Var(b_i)}} \leq \frac{\sum_{j=1}^{n} a_{ij}x_j - \bar{b}_i}{\sqrt{Var(b_i)}}\right] \geq p_i,$$

$$i.e., P_r\left[\frac{b_i - \bar{b}_i}{\sqrt{Var(b_i)}} \leq \frac{\sum_{j=1}^{n} a_{ij}x_j - \bar{b}_i}{\sqrt{Var(b_i)}}\right] \leq 1 - p_i. \tag{4}$$

Defining e_i as $\phi(e_i) = 1 - p_i$, the constraints in Eq. 4 can be stated as

$$\phi\left[\frac{\sum_{j=1}^{n} a_{ij}x_j - \bar{b}_i}{\sqrt{Var(b_i)}}\right] \leq \phi(e_i) \quad i = 1, 2, \ldots, m.$$

This inequality will be satisfied only if

$$\frac{\sum_{j=1}^{n} a_{ij}x_j - \bar{b}_i}{\sqrt{Var(b_i)}} \leq e_i \quad i = 1, 2, \ldots, m.$$

Thus, finally, the probabilistic constraints (1) can be transformed into deterministic constraints as:

$$\sum_{j=1}^{n} a_{ij}x_j - \bar{b}_i - e_i\sqrt{Var(b_i)} \leq 0 \quad i = 1, 2, \ldots, m.$$

Thus, we have obtained a multi-choice deterministic model (**Model 4**) as follows:

Model 4

$$\max_{\text{for DM}_{11}} Z_{11}(x) = \sum_{j=1}^{n}\left(c_{11j}^{(1)}, c_{11j}^{(2)}, \ldots, c_{11j}^{(k_j)}\right)x_j,$$

$$\max_{\text{for DM}_{2f}} Z_{2f}(x) = \sum_{j=1}^{n}\left(c_{2fj}^{(1)}, c_{2fj}^{(2)}, \ldots, c_{2fj}^{(k_j)}\right)x_j,$$

$$\text{subject to } \sum_{j=1}^{n} a_{ij}x_j - \bar{b}_i - e_i\sqrt{Var(b_i)} \leq 0 \quad i = 1, 2, \ldots, m,$$

where $x_j \geq 0 \quad \forall j, f.$ \hfill (5)

Both a_{ij} $(i = 1, 2, \ldots, m; \ j = 1, 2, \ldots, n)$ and b_i $(i = 1, 2, \ldots, m)$ follow normal distribution

Define a random variable h_i as

$$h_i = \sum_{j=1}^{n} a_{ij}x_j - b_i = \sum_{k=1}^{n+1} q_{ik}y_k,$$

where $q_{ik} = a_{ik}, \quad q_{i,n+1} = b_i.$
and $y_k = x_k, \quad y_{n+1} = -1; \quad k = 1, 2, \ldots, n + 1.$

The constraints of Eq. 1 can be rewritten as:

$$P_r[h_i \leq 0] \geq p_i \quad i = 1, 2, \ldots, m. \tag{6}$$

The mean $\bar{h}_i$ and variance of $Var(h_i)$ are given by

$$\bar{h}_i = \sum_{k=1}^{n+1} \bar{q}_{ik}y_k = \sum_{j=1}^{n} \bar{a}_{ij}x_j - \bar{b}_i,$$

and $Var(h_i) = X^T V_i X$ where V_i is given by

$$V_i = \begin{pmatrix} Var(a_{i1}) & cov(a_{i1}, a_{i2}) & \ldots & cov(a_{i1}, a_{in}) \\ cov(a_{i2}, a_{i1}) & Var(a_{i2}) & \ldots & cov(a_{i2}, a_{in}) \\ \ldots & \ldots & \ldots & \ldots \\ \ldots & \ldots & \ldots & \ldots \\ cov(a_{in}, a_{i1}) & cov(a_{in}, a_{i2}) & \ldots & Var(a_{in}) \end{pmatrix},$$

This can be also rewritten as:

$$Var(h_i) = \sum_{k=1}^{n+1}\left[y_k^2 Var(q_{ik}) + 2\sum_{l=k+1}^{n+1} y_k y_l cov(q_{ik}, q_{il})\right]$$

$$= \sum_{k=1}^{n}\left[y_k^2 Var(q_{ik}) + 2\sum_{l=k+1}^{n} y_k y_l cov(q_{ik}, q_{il})\right]$$

$$+ y_{n+1}^2 Var(q_{i,n+1}) + 2y_{n+1}^2 cov(q_{i,n+1}, q_{i,n+1})$$

$$+ \sum_{k=1}^{n}\left[2y_k y_{n+1} cov(q_{ik}, q_{i,n+1})\right]$$

$$= \sum_{k=1}^{n}\left[x_k^2 Var(a_{ik}) + 2\sum_{l=k+1}^{n} x_k x_l cov(a_{ik}, a_{il})\right]$$

$$+ Var(b_i) - 2\sum_{k=1}^{n} x_k cov(a_{ik}, b_i).$$

Thus, the constraints in Eq. 6 can be restated as follows:

$$P_r\left[\frac{h_i - \overline{h}_i}{\sqrt{Var(h_i)}} \leq \frac{-\overline{h}_i}{\sqrt{Var(h_i)}}\right] \geq p_i \quad i = 1, 2, \ldots, m.$$

$$\text{(7)}$$

Therefore, $P_r\left[h_i \leq 0\right] = \phi\left[\frac{-\overline{h}_i}{\sqrt{Var(h_i)}}\right].$

Defining e_i as $\phi(e_i) = p_i$, and then the constraints in equation (7) can be rewritten as follows:

$$\phi\left[\frac{-\overline{h}_i}{\sqrt{Var(h_i)}}\right] \geq \phi(e_i) \quad i = 1, 2, \ldots m.$$

This inequality will be satisfied only if

$$\frac{-\overline{h}_i}{\sqrt{Var(h_i)}} \geq e_i \quad i = 1, 2, \ldots m,$$

i.e., $\overline{h}_i + e_i\sqrt{Var(h_i)} \leq 0 \quad i = 1, 2, \ldots, m.$

Thus, finally, the probabilistic constraints (1) can be transformed into a deterministic constraints as $\sum_{j=1}^{n} \overline{a}_{ij}x_{ij} - \overline{b}_i + e_i\sqrt{X^T V_i X} \leq 0 \quad i = 1, 2, \ldots m.$ Thus, we obtain a multi-choice deterministic model (**Model 5**) as follows:

Model 5

$$\max_{\text{for DM}_{11}} Z_{11}(x) = \sum_{j=1}^{n}\left(c_{11j}^{(1)}, c_{11j}^{(2)}, \ldots, c_{11j}^{(k_j)}\right)x_j,$$

$$\max_{\text{for DM}_{2f}} Z_{2f}(x) = \sum_{j=1}^{n}\left(c_{2fj}^{(1)}, c_{2fj}^{(2)}, \ldots, c_{2fj}^{(k_j)}\right)x_j,$$

$$\text{subject to } \sum_{j=1}^{n} \overline{a}_{ij}x_j - \overline{b}_i + e_i\sqrt{X^T V_i X} \leq 0 \quad i = 1, 2, \ldots, m.$$

where $x_j \geq 0 \quad \forall j, f.$ $\quad\quad\quad$ (8)

Transformation of the objective functions involving multi-choice cost parameters

Now we present a transformation technique of MCSBLPP to formulate an equivalent mathematical model.

Step 1: Find the total number of choices from upper level and lower level decision maker's objective functions. Consider the total number of choices for upper level objective function is k_j. Suppose that $k_j \geq 2$.

Step 2: Find the number of binary variables, which is required to handle the multi-choice parameters in the following manner.

Find l_j, for which $2^{(l_j-1)} < k_j \leq 2^{l_j}$. Here l_j number of binary variables are needed. Let the binary variables be $z_j^{(1)}, z_j^{(2)}, \ldots, z_j^{(l_j)}$.

Step 3: Express $2^{l_j} = \begin{pmatrix} l_j \\ 0 \end{pmatrix} + \begin{pmatrix} l_j \\ 1 \end{pmatrix} + \cdots + \begin{pmatrix} l_j \\ r_{j_1} \end{pmatrix} + \cdots + \begin{pmatrix} l_j \\ r_{j_2} \end{pmatrix} + \cdots + \begin{pmatrix} l_j \\ l_j \end{pmatrix}$, and select the smallest number of consecutive terms whose sum is equal to or just greater than k_j from the expansion.

Let the terms be $\begin{pmatrix} l_j \\ r_{j_1} \end{pmatrix}, \begin{pmatrix} l_j \\ r_{j_1+1} \end{pmatrix}, \ldots, \begin{pmatrix} l_j \\ r_{j_2} \end{pmatrix}.$

Step 4: Set k_j binary codes to k_j number of choices as follows:

$$\max_{\text{for DM}_{11}} Z_{11}(x) = \sum_{j=1}^{n} \left[\sum_{t=1}^{\binom{l_j}{r_{j_1}}} P_t^{(r_{j_1})} Q_t^{(r_{j_1})} c_{11j}^{(t)} + \sum_{t=1}^{\binom{l_j}{r_{j_1+1}}} P_t^{(r_{j_1+1})} Q_t^{(r_{j_1+1})} c_{11j}^{(\binom{l_j}{r_{j_1}}+t)} \right.$$

$$\left. + \cdots + \sum_{t=1}^{\binom{l_j}{r_{j_2-1}}} P_t^{(r_{j_2-1})} Q_t^{(r_{j_2-1})} c_{11j}^{(\binom{l_j}{r_{j_1}}+\binom{l_j}{r_{j_1+1}}+\cdots+\binom{l_j}{r_{j_2-2}}+t)} + \sum_{t=1}^{(k_j-N_j^{(1)})} P_t^{(r_{j_2})} Q_t^{(r_{j_2})} c_{11j}^{(N_j^{(1)}+t)} \right] x_j$$

$$\max_{\text{for DM}_{2f}} Z_{2f}(x) = \sum_{j=1}^{n} \left[\sum_{t=1}^{\binom{l_j}{r_{j_1}}} P_t^{(r_{j_1})} Q_t^{(r_{j_1})} c_{2fj}^{(t)} + \sum_{t=1}^{\binom{l_j}{r_{j_1+1}}} P_t^{(r_{j_1+1})} Q_t^{(r_{j_1+1})} c_{2fj}^{(\binom{l_j}{r_{j_1}}+t)} \right.$$

$$\left. + \cdots + \sum_{t=1}^{\binom{l_j}{r_{j_2-1}}} P_t^{(r_{j_2-1})} Q_t^{(r_{j_2-1})} c_{2fj}^{(\binom{l_j}{r_{j_1}}+\binom{l_j}{r_{j_1+1}}+\cdots+\binom{l_j}{r_{j_2-2}}+t)} + \sum_{t=1}^{(k_j-N_j^{(1)})} P_t^{(r_{j_2})} Q_t^{(r_{j_2})} c_{2fj}^{(N_j^{(1)}+t)} \right] x_j$$

where $\quad N_j^{(1)} = \binom{l_j}{r_{j_1}} + \binom{l_j}{r_{j_1+1}} + \cdots + \binom{l_j}{r_{j_2-1}}$

$t_1 \in \{1, 2, 3, \ldots, (l_j-s)+1\}; \ t_2 \in \{2, 3, \ldots, (l_j-s)+2\}, \ldots, \ t_s \in \{s, s+1, \ldots, l_j\}$

$f = 1, 2, \ldots, n;$

$I_s(t) = \{\{t_1, t_2, \ldots, t_s\} | \ t_1 < t_2 < \cdots < t_s, \ s = r_{j_1}, r_{j_1}+1, \ldots, r_{j_2}\}$

$P_t^{s_j} = \{z_j^{(t_1)} z_j^{(t_2)} z_j^{(t_3)} \ldots z_j^{(t_s)} | \{t_1, t_2, \ldots, t_s\} \in I_s(t), \ s = r_{j_1}, r_{j_1}+1, \ldots, r_{j_2}\}$

$Q_t^{s_j} = \left\{\prod_{t=1}^{l_j}(1 - z_j(t)) | \ t \notin \{t_1, t_2, \ldots, t_s\}\right\}$

Step 5: Restrict $(2^{l_j} - k_j)$ number of binary codes to overcome repetitions as follows:

$z_j(1) + z_j(2) + \ldots + z_j(l_i) \geq r_{j_1}$

$z_j(1) + z_j(2) + \cdots + z_j(l_i) \leq r_{j_2}$

$z_j(t_1) + z_j(t_2) + \cdots + z_j(t_{r_{j_2}}) \leq r_{j_2-1},$

$t = (k_j - N_j(1)) + 1, (k_j - N_j(1)) + 2, \cdots, \binom{l_j}{r_{j_2}}$

Restrictions should be imposed on $z_j^{(t_1)} z_j^{(t_2)} z_j^{(t_3)} \ldots z_j^{(t_{r_{j_2}})} \in P_t^{(r_{j_2}j)}$

Step 6: Formulate the mathematical model and this model is denoted by **Model 6** as follows:

Model 6

$$
\begin{cases}
\max_{\text{for DM}_{11}} Z_{11}(x) = \sum_{j=1}^{n} \left[\sum_{t=1}^{\binom{l_j}{r_{j_1}}} P_t^{(r_{j_1})} Q_t^{(r_{j_1})} c_{11j}^{(t)} + \sum_{t=1}^{\binom{l_j}{r_{j_1+1}}} P_t^{(r_{j_1+1})} Q_t^{(r_{j_1+1})} c_{11j}^{(\binom{l_j}{r_{j_1}}+t)} + \cdots \right. \\[3ex]
\left. \sum_{t=1}^{\binom{l_j}{r_{j_2-1}}} P_t^{(r_{j_2-1})} Q_t^{(r_{j_2-1})} c_{11j}^{(\binom{l_j}{r_{j_1}}+\binom{l_j}{r_{j_1+1}}+\cdots+\binom{l_j}{r_{j_2-2}}+t)} + \sum_{t=1}^{(k_j-N_j^{(1)})} P_t^{(r_{j_2})} Q_t^{(r_{j_2})} c_{11j}^{(N_j^{(1)}+t)} \right] x_j
\end{cases}
$$

$$
\begin{cases}
\max_{\text{for DM}_{2f}} Z_{2f}(x) = \sum_{j=1}^{n} \left[\sum_{t=1}^{\binom{l_j}{r_{j_1}}} P_t^{(r_{j_1})} Q_t^{(r_{j_1})} c_{2fj}^{(t)} + \sum_{t=1}^{\binom{l_j}{r_{j_1+1}}} P_t^{(r_{j_1+1})} Q_t^{(r_{j_1+1})} c_{2fj}^{(\binom{l_j}{r_{j_1}}+t)} + \cdots \right. \\[3ex]
\left. \sum_{t=1}^{\binom{l_j}{r_{j2-1}}} P_t^{(r_{j2-1})} Q_t^{(r_{j2-1})} c_{2fj}^{(\binom{l_j}{r_{j1}}+\binom{l_j}{r_{j1+1}}+\cdots+\binom{l_j}{r_{j2-2}}+t)} + \sum_{t=1}^{(k_j-N_j^{(1)})} P_t^{(r_{j2})} Q_t^{(r_{j2})} c_{2fj}^{(N_j^{(1)}+t)} \right] x_j
\end{cases}
$$

$$
\begin{cases}
z_j(1) + z_j(2) + \cdots + z_j(l_j) \geq r_{j_1} \\[1ex]
z_j(1) + z_j(2) + \cdots + z_i(l_j) \leq r_{j_2} \\[1ex]
z_j(t_1) + z_j(t_2) + \cdots + z_j(t_{r_{j_2}}) \leq r_{j_2-1}\,, \\[1ex]
t = (k_j - N_j(1)) + 1, (k_j - N_j(1)) + 2, \ldots, \binom{l_j}{r_{j_2}} \\[2ex]
z_j^{(l_j)} = 0/1\,, l_j = 1, 2, \ldots, \lceil \frac{\ln k_j}{\ln 2} \rceil \quad j = 1, 2, \ldots, n \\[2ex]
\text{where } N_j^{(1)} = \binom{l_j}{r_{j1}} + \binom{l_j}{r_{j1+1}} + \cdots + \binom{l_j}{r_{j2-1}}
\end{cases}
\tag{9}
$$

subject to $\sum_{j=1}^{n} \overline{a}_{ij}x_j - \overline{b}_i + e_i\sqrt{X^T V_i X} \leq 0 \quad i = 1, 2, \ldots, m;$ $x_j \geq 0, \forall j$

or $\quad S = \{x_j, \forall j : \sum_{j=1}^{n} \overline{a}_{ij}x_j - \overline{b}_i + e_i\sqrt{X^T V_i X} \leq 0 \quad i = 1,$ $2, \ldots, m; \quad x_j \geq 0, \forall j\}$

Step 7: Mathematical **Model 6** is a mixed integer non-linear programming problem. Solve the model with the help of LINGO 13.0 packages.

Solution procedure

Basic concepts of fuzzy set and membership function

Fuzzy set was first introduced by Zadeh in 1965 on a mathematical way to represent impreciseness or vagueness in everyday life.

Fuzzy set: A fuzzy set A in a discourse X is defined as the following set of pairs $A = \{(x, \mu_A) : x \in A\}$, where $\mu_A : X \to [0, 1]$ is a mapping, called membership function of fuzzy set A and μ_A is called the membership value or degree of membership of $x \in X$ in the fuzzy set A. The larger μ_A is the stronger grade of membership form in A.

Normal fuzzy set: Let A be a fuzzy set in X. The height $h(A)$ of A is defined as

$$h(A) = Sup\{\mu_A(x)\}.$$

If $h(A) = 1$, then fuzzy set is called a normal fuzzy set, otherwise it is called subnormal.

$\alpha - $**cut:** Let A be a fuzzy set in X and $\alpha \in (0, 1]$. The $\alpha - cut$ of fuzzy set A in crisp set A_α given by

$$A_\alpha = \{x \in X : \mu_A(x) \geq \alpha\}.$$

Convex fuzzy set: A fuzzy set A in R^n is said to be a convex fuzzy set if its $\alpha - cut\, A_\alpha$ are (crisp) bounded sets, $\forall \alpha \in (0, 1]$.

Fuzzy number: Let A be a fuzzy set in **R** (set of real numbers). Then A is called a fuzzy number if

(i) A is normal,
(ii) A is convex,
(iii) μ_A is upper semicontinuous, and,
(iv) the support of A is bounded.

Triangular fuzzy number: A fuzzy number A is called a triangular fuzzy number(**TFN**) if its membership function μ_A is given by

$$\mu_A(x) = \begin{cases} 0, & \text{if } x < a_l, \quad x > a_u, \\ \dfrac{x - a_l}{a - a_l}, & \text{if } a_l \le x \le a, \\ \dfrac{a_u - x}{a_u - a}, & \text{if } a < x \le a_u. \end{cases} \quad (10)$$

The TFN A is denoted by the triplet $A = (a_l, a, a_u)$.

Fuzzy programming

In fuzzy programming, we construct the linear membership functions which are defined as:

$$\mu_{11}(Z_{11}(x)) = \begin{cases} 0, & \text{if } Z_{11}(x) > Z_{11}^0, \\ \dfrac{Z_{11}(x) - Z_{11}^1}{Z_{11}^0 - Z_{11}^1}, & \text{if } Z_{11}^1 < Z_{11}(x) \le Z_{11}^0, \\ 1, & \text{if } Z_{11}(x) \le Z_{11}^1, \end{cases}$$

$$(11)$$

$$\mu_{2f}(Z_{2f}(x)) = \begin{cases} 0, & \text{if } Z_{2f}(x) > Z_{2f}^0, \\ \dfrac{Z_{2f}(x) - Z_{2f}^1}{Z_{2f}^0 - Z_{2f}^1}, & \text{if } Z_{2f}^1 < Z_{2f}(x) \le Z_{2f}^0, \\ 1, & \text{if } Z_{2f}(x) \le Z_{2f}^1, \\ & \text{where } f = 1, 2, \ldots, n. \end{cases}$$

$$(12)$$

Zimmermann (1978) suggested a method for assessing the parameters of the membership function. In his method, the parameters Z_{11}^0, Z_{11}^1 and Z_{2f}^0, Z_{2f}^1 $\forall f$ are determined as:

$Z_{11}^0 = \max\limits_{x \in S} Z_{11}(x)$, $Z_{11}^1 = \min\limits_{x \in S} Z_{11}(x)$ and

$Z_{2f}^0 = \max\limits_{x \in S} Z_{2f}(x)$ $\forall f = 1, 2, \ldots, n$, $Z_{2f}^1 = \min\limits_{x \in S} Z_{2f}(x)$

$\forall f = 1, 2, \ldots, n$, where S is the feasible region of **Model 6**.

Now, to give an algorithm of fuzzy programming technique for deriving a compromise solution to **Model 2**, which is summarized in the following way:

Step 1: Solve the objective function of the upper level decision maker (i.e., leader) and the lower level decision makers (i.e., followers) with constraints (8) independently.

Step 2: Calculate the linear membership functions in equations (10) and (11) for DM_{11} and DM_{2f}, $\forall f$.

Step 3: Solve the problem which is defined as follows:

Model 7

max λ

subject to

$$\begin{cases} \mu_{11}(Z_{11}(x)) \ge \lambda \\ \mu_{2f}(Z_{2f}(x)) \ge \lambda \\ \text{and Eq. 8} \\ x \ge 0, f = 1, 2, \ldots, n \end{cases} \quad (13)$$

in which a smaller satisfactory degree between those of DM_{11} and DM_{2f} is maximized. If DM_{11} is satisfied with

the obtained optimal solution, the solution becomes a satisfactory solution. Otherwise, DM_{11} is to specify the minimal satisfactory level δ together with the lower and upper bounds $[\Delta_{min}, \Delta_{max}]$ of the ratio of satisfactory degree Δ, where $\Delta = \max\{\frac{\mu_{2f}(Z_{2f}(x))}{\mu_{11}(Z_{11}(x))}, \forall f\}$ with the satisfactory degree $\lambda^* (= \min\{\mu_{11}(Z_{11}(x^*)), \mu_{2f}(Z_{2f}(x^*))\}$, $\forall f$ and x^* is an optimal solution of **Model 6**) of DMs and the related information about the solution in mind.

Step 4: Solve the problem which is defined as follows:

Model 8

max $\mu_{2f}(Z_{2f}(x))$

subject to

$$\begin{cases} \mu_{11}(Z_{11}(x)) \ge \delta \\ \text{and Eq. 8} \\ x \ge 0, f = 1, 2, \ldots, n \end{cases} \quad (14)$$

in which the satisfactory degree of DM_{2f} is maximized under the condition that the satisfactory degree of DM_{11} is larger than or equal to the minimal satisfactory level δ, and then an optimal solution x in equation (12) is proposed to DM_{11} together with λ, $\mu_{11}(Z_{11})$, $\mu_{2f}(Z_{2f})$, $\forall f$ and Δ.

Step 5: If the solution x satisfies one of the following two conditions and DM_{11} accepts it, then goto **Step 7** and the solution x is determined to be the satisfactory solution.

5.1: DM_{11}'s satisfactory degree is larger than or equal to the minimal satisfactory level δ specified by DM_{11}'s self, i.e., $\mu_{11}(Z_{11}(x)) \ge \delta$.

5.2: The ratio Δ of satisfactory degrees lies in between the Δ_{min} and Δ_{max}, i.e., $\Delta \in [\Delta_{min}, \Delta_{max}]$.

Step 6: Ask DM_{11} to revise the minimal satisfactory level δ in accordance with the following procedure of updating the minimal satisfactory level.

6.1: If **Step 5.1** is not satisfied, then DM_{11} decreases the minimal satisfactory level δ.

6.2: If the ratio Δ exceeds its upper bound, then DM_{11} increases the minimal satisfactory level δ. Conversely, if Δ is below its lower bound, then DM_{11} decreases the minimal satisfactory level δ.

6.3: Although **Steps 5.1** and **5.2** are satisfied, if DM_{11} is not satisfied with the obtained solution and judges that it is desirable to increase the satisfactory degree of DM_{11} at the expense of the satisfactory degree of DM_{2f}, $\forall f$, then DM_{11} increases the minimal satisfactory level δ. Conversely, if DM_{11} judges that it is desirable to increase the satisfactory degree of DM_{2f}, $\forall f$ at the expense of the satisfactory degree of DM_{11}, then DM_{11} decreases the minimal satisfactory level δ.

Step 7: Stop.

Numerical example

A reputed industry farming organization operates six farms which are located at Bankura, Purulia, Bardwan, East Midnapur, West Midnapur and Nadia in West Bengal of India of comparable productivity. These farms planted two types of crops: rice and wheat, respectively. The output of each farm is limited both by the usable acreage and by the amount of water available for irrigation. Considering those parameters: usable acreage and water availability both follow a normal distribution. The data for the upcoming season is as shown below:

Farms	Usable Acreage		Minimum water available (in cubic feet)	
	Mean	Variance	Mean	Variance
Bankura	10	6	75	52
Purulia	13	7	78	55
Bardwan	15	8	95	62
Nadia	11	6.5	72	49
East Midnapur	12	6.6	80	59
West Midnapur	18	9	120	93

The organization is considering planting crops which differ primarily in their expected profits per acre and in their consumption of water and the profit to be considered in multi-choice type. Furthermore, the total acreage that can be devoted to each of the crops is limited by the amount of appropriate harvesting equipment available. The organization wishes to know how much of each crop should be planted at the respected farms to maximize the expected profit as well as the maximum revenue earned by the West Bengal Govt. Due to fluctuation of season in West Bengal, revenue is multi-choice type and also the expected revenue from these crops is (40, 42, 45) and (28, 30), respectively.

Crops	Maximum acreage		Water consumption (in cubic feet)		Expected profit (per acre)
	Mean	Variance	Mean	Variance	
Rice	70	52	65	48	(25000,26000,30000)
Wheat	50	41	50	41	(40000,45000)

Let $\mathbf{x_1}$, i.e., $x_{11}, x_{12}, x_{13}, x_{14}, x_{15}, x_{16}$ and $\mathbf{x_2}$, i.e., $x_{21}, x_{22}, x_{23}, x_{25}, x_{26}$ be the number of acres to be allocated for rice and wheat crops to the six firms located at Bankura, Purulia, Bardwan, East Midnapur, West

Midnapur and Nadia, respectively. There are two different level decision makers with respect to this problem, i.e., government (i.e., leader) and the manager of industry (i.e., follower) and each one controls only one decision variable. The government controls rice crop (i.e., $\mathbf{x_1}$) in the first level and the manager controls wheat crop (i.e., $\mathbf{x_2}$) in the second level. Two objectives are established, respectively: (i) revenue from the profits $Z_{11}(\mathbf{x_1}, \mathbf{x_2})$ and (ii) profit on the cultivation of crops $Z_{21}(\mathbf{x_1}, \mathbf{x_2})$.

This is clearly a multi-choice stochastic bi-level programming problem. The problem cannot be solved without using multi-choice programming and stochastic programming approaches.

Using the methodology presented in Sect. 3.2, first we convert the multi-choice objective functions into deterministic objective function and again using Sect. 3.1, we convert the probabilistic constraints into deterministic constraints and then the whole problem is transformed as follows:

$$\max_{\text{for DM}_{11}} : Z_{11}(\mathbf{x_1}, \mathbf{x_2}) = \left[40z_1z_2 + 42z_1 \ (1 - z_2) + 45 \right.$$
$$\left. (1 - z_1)z_2 \right] (x_{11} + x_{12} + x_{13} + x_{14} + x_{15} + x_{16}) + \left[28z_3 + \right.$$
$$\left. 30 \ (1 - z_3) \right] (x_{21} + x_{22} + x_{23} + x_{24} + x_{25} + x_{26}),$$

$$\max_{\text{for DM}_{21}} : Z_{21}(\mathbf{x_1}, \mathbf{x_2}) = \left[25000z_4z_5 + 26000z_4(1 - z_5) + \right.$$
$$\left. 30000(1 - z_4)z_5 \right] (x_{11} + x_{12} + x_{13} + x_{14} + x_{15} + x_{16}) +$$
$$\left[40000z_6 + 45000(1 - z_6) \right] (x_{21} + x_{22} + x_{23} + x_{24} + x_{25} + x_{26}),$$

subject to

$$x_{11} + x_{12} + x_{13} + x_{14} + x_{15} + x_{16} - 2.33(52)^{\frac{1}{2}} \leq 70,$$
$$x_{21} + x_{22} + x_{23} + x_{24} + x_{25} + x_{26} - 2.33(41)^{\frac{1}{2}} \leq 50,$$
$$x_{11} + x_{21} - 2.33(6)^{\frac{1}{2}} \leq 10,$$
$$x_{12} + x_{22} - 2.33(7)^{\frac{1}{2}} \leq 13,$$
$$x_{13} + x_{23} - 2.33(8)^{\frac{1}{2}} \leq 15,$$
$$x_{14} + x_{24} - 2.33(6.5)^{\frac{1}{2}} \leq 11,$$
$$x_{15} + x_{25} - 2.33(6.6)^{\frac{1}{2}} \leq 12,$$
$$x_{16} + x_{26} - 2.33(9)^{\frac{1}{2}} \leq 18,$$
$$65x_{11} + 50x_{21} + 2.33(48x_{11}^2 + 41x_{21}^2 + 52)^{\frac{1}{2}} \geq 75,$$
$$65x_{12} + 50x_{22} + 2.33(48x_{12}^2 + 41x_{22}^2 + 55)^{\frac{1}{2}} \geq 78,$$
$$65x_{13} + 50x_{23} + 2.33(48x_{13}^2 + 41x_{23}^2 + 62)^{\frac{1}{2}} \geq 95,$$
$$65x_{14} + 50x_{24} + 2.33(48x_{14}^2 + 41x_{24}^2 + 49)^{\frac{1}{2}} \geq 72,$$
$$65x_{15} + 50x_{25} + 2.33(48x_{15}^2 + 41x_{25}^2 + 59)^{\frac{1}{2}} \geq 80,$$
$$65x_{16} + 50x_{26} + 2.33(48x_{16}^2 + 41x_{26}^2 + 93)^{\frac{1}{2}} \geq 120,$$

$1 \leq z_1 + z_2 \leq 2,$

$1 \leq z_4 + z_5 \leq 2,$

where $x_{1j}, x_{2j} \geq 0 \quad j = 1, 2, \ldots, 6.$

Results and discussion

The above mathematical programming model is treated as non-linear programming problem and is solved by Lingo 13.0 packages. The results of the optimal solution to individual problems are obtained as:

$Z_{11}^1 = 204.82$ at $x_{1j} = 0 j = 1, 2, \ldots, 6, x_{21} = 1.04,$ $x_{22} = 1.09, x_{23} = 1.35, x_{24} = 1, x_{25} = 1.11, x_{26} = 1.72$ and the control variables are $z_1 = z_2 = z_3 = 1$; $Z_{11}^0 = 4532.97$ at $x_{11} = 13.14, x_{12} = 13.78, x_{13} = 7.94, x_{14} = 13.37, \quad x_{15} = 13.56, x_{16} = 24.99, x_{21} = 2.56, \quad x_{22} = 5.38, x_{23} = 3.57, x_{24} = 3.57, x_{25} = 4.42, x_{26} = 0$ and the control variables are $z_1 = 1, z_2 = z_3 = 0$;

$Z_{21}^1 = 15047.83$ at $x_{11} = 0.82, x_{12} = 0.86, x_{13} = 1.07,$ $x_{14} = 0.79, \ x_{15} = 0.88, x_{16} = 1.36, x_{2j} = 0 (j = 1, 2, \cdots, 6)$ and the control variables are $z_4 = 1, z_5 = z_6 = 0$; $Z_{21}^0 = 4465141$ at $x_{11} = 0, x_{12} = 19.16, x_{13} = 1.28, x_{14} = 9.23,$ $x_{15} = 9.75, x_{16} = 12.03, x_{21} = 15.71, x_{22} = 0, \ x_{23} = 20.31,$ $x_{24} = 7.71, x_{25} = 8.23, x_{26} = 12.96$ and the control variables are $z_5 = 1, z_4 = z_6 = 0.$

Next, we find the linear membership functions using the equations (10) and (11) by Zimmermann method and the maximin problem for this numerical example can be written as:

$\max \ \lambda,$

subject to $(Z_{11}(x) - 204.82)/(4532.97 - 204.82) \geq \lambda,$

$(Z_{21}(x) - 15047.83)/(4465141 - 15047.83) \geq \lambda,$

$x \in \mathbf{S},$

where $\mathbf{S}$ denotes the feasible region of **Model 6**.

Using the procedure described in Sect. 4, we derive the results after the first iteration and are shown in Table 1.

Suppose that DM_{11} is not satisfied with the solution obtained in Iteration 1 and then DM_{11} specifies the minimal satisfactory level $\delta = 0.9693$ and we see that the bounds of the ratio at the interval $[\Delta_{min}, \Delta_{max}] = [0.9693, 0.9852]$, taking into account of the result of the first iteration. Then, the problem with the minimal satisfactory level is rewritten as follows:

$\max \ \mu_{21}(Z_{21}(x)),$

subject to $(Z_{11}(x) - 204.82)/(4532.97 - 204.82) \geq 0.9693,$

where $x \in \mathbf{S}.$ \hfill (15)

The result of the second iteration including an optimal solution to problem (12) is shown in **Table 2** in a similar way as done in first iteration.

At the second iteration, the ratio $\Delta = 0.9999$ of satisfactory degree is not valid interval $[0.9693, 0.9852]$ of ratio. So, DM_{11} updates the minimal satisfactory level at $\delta = 0.9777$ Then, the problem with the revised minimum satisfactory level is solved, and the result of third iteration is shown in Table 3.

Table 1 Results from iteration 1

x_{1j}:	15.71	0	0	8.13	17.98	18.76
x_{2j}:	0	19.64	21.59	8.81	0	6.23
(Z_{11}, Z_{21}):	(4399.98, 4328401)					
$(z_1, z_2, z_3, z_4, z_5, z_6)$:	(0, 1, 0, 0, 1, 0)					
$\mu_{11}(Z_{11})$:	0.9692					
$\mu_{21}(Z_{21})$:	0.9692					
λ:	0.9692					
Δ:	0.9999					

Table 2 Results from iteration 2

x_{1j}:	15.71	19.16	21.59	1.16	1	1.5
x_{2j}:	0	0	0	15.32	16.98	23.49
(Z_{11}, Z_{21}):	(4400.1, 4328283)					
$(z_1, z_2, z_3, z_4, z_5, z_6)$:	(0, 1, 0, 0, 1, 0)					
$\mu_{11}(Z_{11})$:	0.9693					
$\mu_{21}(Z_{21})$:	0.9692					
λ:	0.9692					
Δ:	0.9999					

Table 3 Results from iteration
3

x_{1j}:	15.71	19.16	21.59	3.84	1	1.5
x_{2j}:	0	0	0	13.1	16.98	23.49
(Z_{11}, Z_{21}):	(4433.42, 4294956)					
$(z_1, z_2, z_3, z_4, z_5, z_6)$:	(0, 1, 0, 0, 1, 0)					
$\mu_{11}(Z_{11})$:	0.9769					
$\mu_{21}(Z_{21})$:	0.9617					
λ:	0.9618					
Δ:	0.9844					

At the third iteration, the satisfactory degree $\mu_{11}(Z_{11}) = 0.9777$ of DM_{11} which equals to the minimum satisfactory level $\delta = 0.9777$, and the ratio $\Delta = 0.9844$ of the satisfactory degree is in the valid interval [0.9693, 0.9852] of ratio. Therefore, this solution satisfies the termination condition of the interactive process, and it becomes a compromise solution for both DMs.

Sensitivity analysis

The main intention of this study is to formulate and solve the stochastic bi-level programming problem for cooperative game in multi-choice nature under fuzzy programming technique. Let us discuss why we have considered such a study and what is the contribution of this study compared to other research works carried out by many researchers in this direction. Bi-Level Programming Problem (BLPP) has been studied by several researchers, for example, (Anandalingam 1988; Sakawa et al. 2000; Roy 2006; Lachhwani and Poonia 2012; Dey et al. 2014) and many others. Most of them have not considered when the objective functions are in multi-choice nature. Due to globalization of the market or other real-life phenomena, we have assumed that the cost parameter of the objective functions is of multi-choice type and that non-linearity occurs in the BLPP. But here we have presented the parameters of constraints that follow a normal distribution. So, our proposed method treated non-linearity when the objective functions and the constraints are both non-linear.

By conventional method, we find that the objective functions of the upper level and lower level decision makers are 4393.38 and 4465141, respectively (using LINGO 13.0 pakages) but according to our findings the results are 4433.42 and 4294956. We see that only the upper level decision maker gives a better result with respect to our proposed method. Actually, in real-life situation we always give priority to the upper level decision maker while the lower level always remains secondary. In this situation too, our proposed model works well from this point of view. Hence, taking these observations into consideration, we feel that the proposed method is a better method for our study.

Conclusion

This paper has presented the solution procedure for solving the multi-choice stochastic bi-level programming problem with consideration of normal random variable. All the probabilistic constraints have been transferred into the equivalent deterministic constraints by stochastic programming approach and a general transformation technique has used for the multi-choice cost coefficients of the objective functions using fuzzy programming technique which provides a compromise solution. From our study, it has been concluded that in a cooperative environment there exists a compromise solution which governs by the upper level decision maker.

In the real-life decision-making problem, the cost coefficients of the objective functions and the constraints may not be known previously due to uncountable factors. For this reason, the cost coefficients of the objective functions are of multi-choice rather than by single choice and the constraints are followed random variables. In this paper, we have formulated the MCSBLPP model by considering both the factors. Finally, it is obvious that the formulated model is highly applicable for these types of bi-level programming problems such as supply chain planning problem, managerial decision-making problem, facility location, transportation problem, etc. and solving this model, the decision maker has provided the optimal planning for taking the right decision.

In future study, one can extend this work, i.e., to solve the multi-choice stochastic multi-level programming problem with interval programming using fuzzy goal programming technique.

Acknowledgments Authors are very much thankful to the anonymous reviewers for their comments to improve the quality of the paper.

References

Acharya S, Acharya MM (2013) Generalized transformation techniques for multi-choice linear programming problems. Int J Optimization Control Theor Appl 3:45–54

Anandalingam G (1988) A mathematical programming model of decentralized multi-level systems. J Operational Res Soc 39(11):1021–1033

Anandalingam G, Apprey V (1991) Multi-level programming and conflict resolution. Eur J Oper Res 51:233–247

Chang C-T (2007) Multi-choice goal programming. Omega Int J Manag Sci 35:389–396

Chang C-T (2008) Revised multi-choice goal programming. Appl Math Model 32:2587–2595

Charnes A, Cooper WW (1978) Chance constrained programming. Oper Res 16:576–586

Dempe S, Starostina T (2007) On the solution of fuzzy bi-level programming problems. http://www.optimization-online.org/DBFILE/2007/09/1778.pdf

Dey PP, Pramanik S, Giri BC (2014) TOPSIS approach to linear fractional bi-level MODM problem based on fuzzy goal programming. J Ind Eng Int 10:173–184

Lachhwani K, Poonia MP (2012) Mathematical solution of multi-level fractional programming problem with fuzzy goal programming approach. J Ind Eng Int 8(16):1–11

Lai YJ (1996) Hierarchical optimization: a satisfactory solution. Fuzzy Sets Syst 77:321–335

Leclercq JP (1982) Stochastic programming an interactive multi-criteria approach. Eur J Oper Res 10:33–41

Maity G, Roy SK (2014) Solving multi-choice multi-objective transportation problem: a utility function approach. J Uncertainty Anal Appl 2(11):1–20

Maity G, Roy SK (2015) Solving a Multi-Objective Transportation Problem with Nonlinear Cost and Multi-Choice Demand, International Journal Management Science and Engineering Management. Forthcoming. doi:10.1080/17509653.2014.988768

Mahapatra DR (2013) Roy, S.K and Biswal, M.P. Multi-choice stochastic transportation problem involving extreme value distribution. Appl Math Model 37:2230–2240

Roy SK, Mahapatra DR, Biswal MP (2012) Multi-choice stochastic transportation problem with exponential distribution. J Uncertain Syst 6(3):200–213

Roy SK (2006) Fuzzy programming techniques for stackelberg game. Tamsui Oxford J Manag Sci 22(3):43–56

Roy SK (2007) Fuzzy programming approach to two person multi-criteria bi-matrix games. J Fuzzy Math 15(1):141–153

Roy SK, Biswal MP, Tiwari RN (2001) An approach to multi-objective bi-matrix games for nash equilibrium solutions. Ricerca Operativa 30(95):56–63

Roy SK (2014) Multi-choice stochastic transportation problem involving weibull distribution. Int J Oper Res 21(1):38–58

Roy SK (2015) Langrange's interpolating polynomial approach to solve multi-choice transportation problem. Int J Appl Comput Math 1(4):639–649

Sakawa M, Nishizaki I, Oka Y (2000) Interactive fuzzy programming for multi-objective two-level linear programming problems with partial information of preference. Int J Fuzzy Syst 2:79–86

Shih HS, Lai YJ, Lee E (1996) Stanley fuzzy approach for multi-level programming problems, *Comput Oper Res* 23:73–91

Sengupta JK (1972) Stochastic programming methods and applications. North-Holland, Amsterdam

Sinha SB, Sinha SA (2004) Linear programming approach for linear multi-level programming problems. J Oper Res Soc 55:312–316

Zheng Y, Wana Z, Wang G (2011) A fuzzy interactive method for a class of bi-level multiobjective programming problem. Expert Syst Appl 38(8):10384–10388

Zimmermann HJ (1978) Fuzzy programming and linear programming with several objective functions. Fuzzy Sets Syst 1:45–55

16

Doctoral dissertations in logistics and supply chain management: a review of Nordic contributions from 2009 to 2014

Christopher Rajkumar[1] · Lone Kavin[1] · Xue Luo[1] · Jan Stentoft[1]

Abstract The purpose of this paper is to identify and analyze Nordic doctoral dissertations in logistics and supply chain management (SCM) published from the years 2009–2014. The paper is based on a detailed review of 150 doctoral dissertations. Compared with previous studies, this paper identifies a trend toward: more dissertations based on a collection of articles than monographs; more dissertations focusing on inter-organizational SCM issues; a shift from a focal company perspective to functional aspects and supply chain-related research; and finally, a continued decreased focus on the philosophy of science. A score for measuring the significance of article-based dissertations is also proposed.

Keywords Doctoral dissertations · Collection of articles · Monographs · Dissertation score · Logistics and supply chain management

1 Introduction

One way to keep track of the progress of logistics and supply chain management (SCM) discipline is to analyze the doctoral dissertations within the research area. By reviewing such dissertations, it will be possible to gain some interesting information regarding the development and direction of research within the discipline. Specifically, such a review will help us to understand the different approaches in relation to research framework, methodologies, theories applied and the empirical interpretations. Furthermore, the review could not only provide valuable insights into potential research gaps within the discipline, but also pave way for recognizing interesting topics for future research [13, 41]. Besides, given that PhD students are likely to form the next generation of established researchers, research conducted by them is important to the SCM discipline as it helps keeping the discipline on track with emerging topics as well as stimulate theory generation.

The number of PhD students in the Nordic countries has increased significantly during the last decades [24, 41]. Based on the rise in the number of dissertations as well as their varying content, it is interesting to investigate the requirements that are part of completing a PhD dissertation. There has been an escalation in dissertations that are based on collections of articles instead of a monograph [41]. A reason for this might be an increasing pressure to publish at the universities [21], which might be driving PhD students to learn and master the craft of publishing from the very start of their career. No matter what, the culture of many research departments is characterized by a high focus on performance in terms of publications in ranked journals [2, 20, 22, 27]. Therefore, by choosing an article-based dissertation, PhD students might have a better opportunity to work together with other PhD students and senior researchers; in doing so, they also learn how to "play the game."

The first two comprehensive digest of doctoral dissertations completed within the Nordic countries was

✉ Jan Stentoft
stentoft@sam.sdu.dk

Christopher Rajkumar
cra@sam.sdu.dk

Lone Kavin
loka@sam.sdu.dk

Xue Luo
xl@sam.sdu.dk

[1] Department of Entrepreneurship and Relationship Management, University of Southern Denmark, Kolding, Denmark

conducted by Gubi et al. [13] and Zachariassen and Arlbjørn [41]. These efforts provide PhD students, other academic staff as well as practitioners with an overview of what has been researched within the logistics and SCM area. These studies have also facilitated the comparison of Nordic dissertations themes to those in the USA (e.g., compiled by Stock [31] and Nakhata et al. [25]). Prior research has demonstrated that much confusion exists concerning SCM definitions and its overlap with logistics [32]. This paper extends the work of Zachariassen and Arlbjørn [41] that applies what Halldórsson et al. [14] call a relabeling approach between the terms of logistics and SCM. The purpose of this paper is to document the progress of doctoral work in logistics and SCM within the Nordic countries between 2009 and 2014. In all, 120 relevant dissertations were identified; however, we were unable to retrieve eight dissertations either in physical form or electronically; therefore, only 112 dissertations were reviewed. With a point of departure within the dimensions and classified categories of these two above-mentioned reviews, this paper provides two analyses:

1. An analysis of identified Nordic dissertations from the year 2009–2014; and
2. A longitudinal analysis that compares the above analysis with the result from Gubi et al. [13] and Zachariassen and Arlbjørn [41].

Accordingly, the analyses within this paper will not only reveal several important insights, but also identify new, potential research areas within the logistics and SCM discipline. Based on these insights, it will be possible to coordinate future research efforts and avoid any unnecessary replication or duplication of previous work.

The rest of the paper is organized as follows: Sect. 2 provides a brief literature review of earlier contributions dealing with doctoral dissertation reviews. Section 3 discusses the methodology used in this study; it also outlines the limitations concerning the chosen methodology. Subsequently, Sect. 4 discusses the results obtained from analyzing the Nordic dissertations. Finally, Sect. 5 concludes with the overall purpose of the paper and some directions for future research activities.

2 Extant literature on doctoral dissertations in logistics and supply chain management

Several authors have analyzed and classified doctoral dissertations in logistics and SCM, both within the Nordic countries as well as the USA. In this section, nine prior studies—seven American studies and two Nordic studies—are briefly mentioned so as to identify the trends in topical coverage through the years and to see whether there are any similarities between the topics chosen by PhD students across the Atlantic.

The first study of compendiums of PhD research in logistics conducted by Stock back in 1987 [28] examined 684 dissertations from the period 1970–1986 [29]. Subsequent reviews were conducted in (1) 1993 covering 422 American dissertations that were completed in the period of 1987–1991 [34] and (2) 2001 with an analysis of 317 PhD dissertations completed in the period of 1992–1998 [31]. In 2006, Stock completed his fourth review of PhD dissertations together with Broadus [33]. This study showed a distinct increasing trend in the number of dissertations within SCM- and/or logistic-related areas in the period from 1999 to 2004 when compared to the period covered by the 2001 study. But surprisingly, the count between 1999 and 2004 fell short of the overall levels of 1987–1991. A more interesting finding of the 2006 study was that the dissertations were more multifaceted due to the fact that they closely mirrored the cross-functional and boundary spanning nature of logistics; at the same time, their multifaceted nature made the classification task more difficult.

Another American study was conducted by Das and Handfield [8] wherein the authors investigated 117 PhD dissertations from the period 1987–1995 in order to evaluate the intellectual health of the purchasing discipline. The study was an extension of a previous research on purchasing dissertations conducted by Williams [39] to identify the key focus areas of research in the prior decade. In this study, Williams concluded that the key focus areas covered were supplier selection and development, information systems, organizational and measurement issues, negotiation and purchasing ethics.

The most recent review covering US dissertations was conducted by Nakhata et al. [25]. In this study, the authors reviewed 609 doctoral dissertations completed between 2005 and 2009. The number of identified dissertations in this study is significantly larger than the four reviews conducted by Stock and colleagues and clearly reflects a significant increase in colleges/universities graduating doctoral students within logistics- and supply chain-related areas. Nakhata et al. [25] also point out that a forthcoming retirement of academic "baby boomers" during the period 2005–2020 may explain the increase in the PhD production. The most prominent research methodologies employed by doctoral students in the study of Nakhata et al. [25] are modeling, simulation and empirical quantitative methods.

Two earlier studies of Nordic doctoral dissertations within logistics and SCM have been disseminated in academic journals. The first was developed by Gubi et al. [13], who reviewed 71 Nordic dissertations published between 1990 and 2001. Most of the dissertations were published as monographs with manufactures and carriers as the primary entity of analysis. The most recent Nordic contribution is by Zachariassen and Arlbjørn [41] wherein the authors analyze the development in Nordic doctoral research in logistics and SCM from the years 2002–2008. As opposed to Gubi et al. [13], Zachariassen and Arlbjørn [41] found more dissertations based on a collection of articles, which was reflective of a response to increase publication pressure. In contrast to the Gubi et al.'s [13] study, the primarily entity of analysis of most dissertations was the manufacturing companies. While prior reviews showed a focal company perspective, the review of Zachariassen and Arlbjørn [41] documented a shift toward an inter-organizational perspective covering dyadic and supply chain units of analysis. Additionally, according to Zachariassen and Arlbjørn [41], there has also been a decreased focus on the philosophy of science, since most dissertations were being published as collections of articles.

3 Method

The method applied for identifying, collecting and reviewing the doctoral dissertations in this paper follows a three-step process. These steps are explained in the following subsections.

3.1 Identifying and collecting Nordic doctoral dissertations

The first step consisted of sending e-mails to contact persons at different research institutions within the Nordic countries (i.e., Denmark, Finland, Iceland, Norway and Sweden) as outlined in [41]. Based on this e-mail contact, the list was further modified (e.g., adding University of Vaasa in Finland and Linnaeus University and Örebro University School of Business in Sweden). The final list included 39 research institutions which are presented in "Appendix 1." A contact person at each of the 39 research institution was identified and was contacted by e-mail. The e-mail provided a clear statement of the research project and requested the list of completed doctoral dissertations within logistics and SCM within the analysis period. This process provided an initial list of 120 dissertations. Majority of these dissertations were accessible in electronic form (we either received them by e-mail or downloaded them from the corresponding institution's Web site). The rest of the dissertations were available in hard copy form.

3.2 Validating the initial list of doctoral dissertations

The second step was concerned with the validation of the identified dissertations by senior researchers from each of the Nordic countries. This step was completed by e-mailing the initial list to these senior researchers and by attending the 27th annual NOFOMA conference in June 2015 at Molde University College. This process resulted in the inclusion of an additional 41 dissertations, thereby increasing the total count to 161.

3.3 Reviewing the received dissertations

In the third step, a detailed review of the 161 dissertations took place. During this process, three dissertations were excluded since they were judged as not being within the scope of the present analysis. Out of this net list of 158, it was possible to conduct reviews of 150 dissertations. As mentioned earlier, we were unable to retrieve eight dissertations as well as there was no response from the authors when we e-mailed them requesting for the copy of their dissertation. A complete list of the dissertations is included in "Appendix 2." The detailed review took place against a review framework as described in [13, 41]. Our analysis covered additional review elements as the dissertations were mostly based on a collection of articles. These elements were: (1) number of articles, (2) type of article (journal publication, book chapter, conference paper, working paper or unpublished paper), (3) year of publication, (4) ranking of the journal, (5) number of authors on each article and (6) the doctoral candidate's author number for the specific article.

3.4 Limitations

This dissertation review has some limitations that are worthwhile to mention. First, even though the gross list of dissertations was reviewed and validated by senior researchers within the Nordic countries, there is a possibility that some dissertations were not identified. Second, reviewer subjectivity could not be completely eliminated in the review of the 150 dissertations. However, in order to minimize subjectivity, an aligned interpretation of the review elements and their outcome was obtained by a common review of three different types of dissertations (one monograph and two article-based dissertations). Third, the list of dissertations that was composed and reviewed stems from the NOFOMA research community. Obviously, other Nordic researchers may produce doctoral dissertations that deal with topics under the scope of this review, but is outside the NOFOMA radar (e.g., researchers belonging to European Decision Sciences Institute (EDSI), European Logistics Association (ELA), European Operations Management (EurOMA), International Purchasing and Supply Education and Research Association (IPSERA) or Logistics Research Network (LRN) or Rencontres Internationales de Recherche en Logistique (RIRL)

(International Research Conference on Logistics and Supply Chain Management). This study can make observations only based on the dissertations reviewed under the NOFOMA umbrella. Fourth, since the contact persons were not provided with a definition of logistics and SCM, they might have excluded some dissertation that could have fallen within the scope of this analysis. Moreover, the senior researchers whom we e-mailed for dissertations might not be from the department of logistics and SCM. Therefore, including definitions for logistics and SCM will have no impact.

4 Analysis

This section is concerned with specific analyses of the 150 reviewed doctoral dissertations. For comparative purposes, the data from the present review are portrayed and analyzed with the categories that are similar to those used by Gubi et al. [13] and Zachariassen and Arlbjørn [41]. The results are displayed in tables and are followed with appropriate comments and interpretations.

4.1 Number and type of dissertations finalized in the period 2009–2014

Table 1 contains the PhD dissertations divided by country, year of publication and type of dissertation (monograph vs. a collection of articles). Compared with the earlier studies, these new figures show some interesting developments. First, the number of finalized dissertations in the period 2009–2014 is 158, which represents an average of 26 dissertations per year. Compared with averages numbers of 10 (from the period 2002–2008) and 6.25 (from the period 1990–2001), this shows that there has been a strong increase in PhD production in this research area. A similar pattern is identified by Nakhata et al. [25] in their study of doctoral dissertations published by Dissertation Abstracts International in the period of 2005 and 2009. They explain that one reason for such an increase might be an increased level of retiring academicians toward 2020 which creates a stronger market for Assistant Professors. Another explanation could be that there is a drive from the governments to boost the number of annual PhD production in order to support national social, economic and environmental well-being as well as to address major global challenges [12]. The majority of the Nordic logistics and SCM PhDs come from the Finnish, Norwegian and Swedish research environments. In Denmark, the production is stable with 17 dissertations in the period of 2009–2014. Iceland has reported their first PhD in this period of analysis. Twenty-five Nordic research institutions have produced within logistics and SCM in the period 2009–2014 (see "Appendix 2").

Another interesting finding in the current review is the increase in the share of dissertations that are based on a collection of articles. As given in Table 1, 92 out of the 150 dissertations (61 %) are based on a collection of articles. Thus, there is a much higher focus on the craft to write academic articles when compared to the share of article-based dissertations in previous periods (29 % in the period 2002–2008 and 21 % in the period 1990–2001). This trend supports the predictions previously made by Zachariassen and Arlbjørn [41]. One plausible explanation for this development might be an increased amount of public and private resource allocation to research environments based on publications in internationally recognized peer-reviewed journals as well as measures such as impact indicators and H-index (see, e.g., [27]).

4.2 Primary entity of analysis

In Table 2, all 150 dissertations are classified according to their entity of analysis; the classifications are also compared to previous results reported in [13, 14]. The study of the primary entity of analysis in the dissertations shows strong differences. First, the category others has increased to about one-third of the dissertations in the last reported period. This group consists of a variety of dissertations without a specific supply chain actors' perspective—such as fresh fish supply chains [26] and healthcare logistics [17]. Second, in absolute numbers, the manufacturer as the primary entity of analysis has increased when compared to the previous studies; but, if we measure the number as a percentage of the reviewed dissertations, there is a fall from 57 % in the dissertations from 2002 to 2008 to 33 % in the recent study.

Thus, it shows the more classical actors such as wholesalers, retailers and inventory hotels have obtained lesser research focus. Additionally, a remarkably low number of dissertations have focused on retail SCM within the Nordic countries. This is intriguing given the fact that the retail sector is well known for supply chain innovations such as quick response systems, efficient consumer response, distribution centers, reverse logistics, as well as collaborative planning, forecasting and replenishment [10].

4.3 Level of analysis arranged according to year of publication

Table 3 shows the analysis of the dissertations level of analysis arranged by the year of publication.

An interesting development evidenced in the above table is an increased focus on functional themes within dissertations (in the present analysis, this is about 26 % of the dissertations compared with 11 and 10 % in previous analyses). This development is primarily driven by Finnish

Table 1 Number and type of PhD dissertations finalized in the period 2009–2014

	2009	2010	2011	2012	2013	2014	Total (2009–2014)	Total (2002–2008)	Total (1990–2001)
All identified dissertations									
Danish	2	1	4	3	5	2	17	11	15
Finnish	10	9	7	8	6	8	48	22	20
Icelandic	–	1	–	–	–	–	1	–	–
Norwegian	6	5	–	5	6	5	27	17	12
Swedish	9	10	12	13	9	12	65	20	28
Total	27	26	23	29	26	27	158	70	75
Reviewed dissertations									
Danish reviewed									
Monograph	2	–	1	1	1	0	5	9	14
Collection of articles	–	1	3	2	4	2	12	2	1
Total Danish	2	1	4	3	5	2	17	11	15
Finnish reviewed									
Monograph	2	2	5	4	2	3	18	13	14
Collection of articles	8	6	2	3	4	4	27	9	3
Total Finnish	10	8	7	7	6	7	45	22	17
Icelandic reviewed									
Monograph	–	1	–	–	–	–	1	–	–
Collection of articles	–	–	–	–	–	–	0	–	–
Total Icelandic	0	1	0	0	0	0	1	0	0
Norwegian reviewed									
Monograph	5	1	–	1	2	2	11	9	9
Collection of articles	1	4	–	4	4	3	16	8	2
Total Norwegian	6	5	0	5	6	5	27	17	11
Swedish reviewed									
Monograph	2	6	4	5	1	5	23	13	19
Collection of articles	6	2	7	8	7	7	37	7	9
Total Swedish	8	8	11	13	8	12	60	20	28
Reviewed on total									
Monograph	11	10	10	11	6	10	58	44	56
Collection of articles	15	13	12	17	19	16	92	26	15
Total	26	23	22	28	25	26	150	70	71

Comparable data are included from the period 1990–2001 (Gubi et al. [13]) and 2002–2008 (Zachariassen and Arlbjørn [41])

dissertations and can indicate an emphasis on building stronger knowledge bases in certain sub-disciplines along the supply chain. Examples of such dissertations are [18, 19]. Another interesting development is the reduced focus on firm-level analysis and a subsequent increased focus on the supply chain or the network as the level of analysis, with the network level experiencing the highest increase. This increase is strongly evident within Swedish dissertations.

4.4 Research design, time frame and philosophy of science

Table 4 shows the classification of the dissertations according to research design, time frame and philosophy of science. The recent analysis reveals a decrease in share of dissertations that are purely theoretical in nature. Also, the share of dissertations founded on purely qualitative methods has also decreased compared with dissertations published in the period 2002–2008. The drop in share of these two categories has attributed to increases in shares of dissertations based on quantitative methods and on triangulations of qualitative and quantitative research methods. The increase in quantitative research methods can be explained by the increased pressure to publish, thereby favoring quantitative methods over time-consuming qualitative studies (see, e.g., [21]). While quantitative data collection can be automated, it is not possible for qualitative data collection. Qualitative data collection is, in general, more time-consuming and expensive when compared

Table 2 Primary entity of analysis

	Danish			Finnish			Icelandic			Norwegian			Swedish			Total		
	A	B	C	A	B	C	A	B	C	A	B	C	A	B	C	A	B	C
Manufacture	9	9	2	11	11	20	–	–	–	6	9	8	10	11	20	36	40	50
Carrier	5	1	2	3	4	5	–	–	–	3	1	14	10	3	15	21	9	36
Wholesaler	–	–	–	–	2	–	–	–	–	–	1	1	–	–	2	0	3	3
Retailer	–	–	–	–	2	3	–	–	–	–	1	1	–	1	4	0	4	8
Inventory hotel	–	–	–	–	–	1	–	–	–	–	–	1	1	–	3	1	0	5
Others	1	1	13	3	3	16	–	–	1	2	5	2	7	5	16	13	14	48
Total	15	11	17	17	22	45	0	0	1	11	17	27	28	20	60	71	70	150

A = time period 1999–2001 covered in Gubi et al. [13]

B = time period 2002–2008 covered in Zachariassen and Arlbjørn [41]

C = 2009–2014 covered in this paper

Table 3 Level of analysis arranged according to year of publication

	2009	2010	2011	2012	2013	2014	Total (2009–2014)	Total (2002–2008)	Total (1990–2001)
Total									
Function	6	4	4	7	11	7	39	8	7
Firm	1	4	1	0	4	5	15	19	33
Dyad	3	3	4	3	1	2	16	16	8
Supply chain	7	4	4	9	5	8	37	15	8
Network	5	7	7	8	5	4	36	8	5
Others	3	1	2	1	0	0	7	4	10
Total	25	23	22	28	26	25	150	70	71

Table 4 Research design, time frame and philosophy of science

	Danish	Finnish	Icelandic	Norwegian	Swedish	Total (2009–2014)	Total (2002–2008)	Total (1990–2001)
Research design								
Theoretical (desk research)	6	2	–	–	7	15	14	21
Empirical quantitative	4	9	–	11	1	25	9	6
Empirical qualitative	4	17	1	10	27	59	31	27
Empirical triangulation	3	17	–	6	25	51	16	17
Research design total	17	45	1	27	59	150	70	71
Time frame								
Snapshot	8	30	1	21	16	76	39	41
Longitudinal	1	8	–	2	22	33	11	6
Not specified	2	5	–	4	15	26	6	3
Time frame total	11	43	1	27	53	135	56	50
Containing philosophy								
Yes	9	13	–	7	11	40	20	39
No	8	32	1	21	47	110	50	32
Total	17	45	1	28	58	150	70	71

to quantitative research studies. Thus, it might be more cost-effective to slice one questionnaire survey into a number of articles than doing the same number of articles based on qualitative methodologies. This trend will

undoubtedly improve numerical performance metrics. At the same time, this trend need not necessarily deliver new knowledge that could move the discipline significantly forward. Particularly, quantitative surveys are exposed to the phenomenon of "salami-slicing" where the data from a particular project are disseminated in a number of articles that in fact are "sliced" so thinly that there might be overlap of the papers (e.g., text recycling in literature review sections, key findings and discussions ([9, 16]). We refer to the Committee on Publication Ethics (COPE) [7] for a discussion on different forms of text recycling.

Table 4 also shows an increased share of dissertations that apply a longitudinal perspective. Additionally, the share of snapshot time frames has decreased and the share of the category time frame not specified has increased. Examples of dissertations without any specific time frame are by (1) Mortensen [23], who investigate the concept of attraction and explain its role in initiation and development of buyer–seller relations, and (2) Tynjälä [35], who conceptually examines the methods and tools for supply chain decision making during new product development. A final remarkable finding from Table 4 is the continued decline of dissertations containing philosophy of science issues. The present analysis identifies 73 % of the dissertations not including such philosophical considerations in comparison with 71 and 45 % in the two previous periods of analysis. Twenty-nine percent of dissertations that are monographs do contain philosophy of science considerations, whereas 25 % have this content among the article-based dissertations. Zachariassen and Arlbjørn [41] provided four possible reasons for this decline: (1) There are no mandatory requirements for PhD students to attend philosophy of science courses, (2) article-based dissertations do not rely on philosophy of science argumentations to the same degree as the monographs that typically require more in-depth interaction of this issues and method considerations, (3) there is a lower prioritization of philosophy of science due to higher pressure from external funded projects that do not demand this theme, and (4) logistics and SCM are closely connected with industry that could lead to a perception that philosophy of science is less relevant. To this list, we add four more potential reasons. First, there are no requirements mentioned in Nordic countries' ministerial orders granting degrees of PhD about unfolding philosophical of science perspectives. Second, few, if any, journals within logistics and SCM demand such discussions. Third, philosophy of science has disappeared from PhD candidates' syllabi. Finally, only a few PhD advisors have the knowledge to join in such discussions with their PhD students; therefore, they do not send signals for offering such courses. Overall, this development is inexpedient if the discipline really has to move toward theory development [6, 30].

4.5 Dissertations distributed according to topic groups and country of origin

Table 5 displays the dissertations according to topic groups based on a title analysis of the 158 identified dissertations. For comparison reasons, the topic groups identified in previous studies were used [13, 41] and further supplemented with two new groups risk management and humanitarian logistics. The table shows a continued decline in the share of dissertations related to system design/structure/effectiveness, organizational development/competencies and material handling. In contrast, topics related to system integration/integration enablers and

Table 5 Division of dissertations according to topic groups and country of origin

	Danish			Finnish			Icelandic			Norwegian			Swedish			Total		
	A	B	C	A	B	C	A	B	C	A	B	C	A	B	C	A	B	C
System design/structure/effectiveness	5	4	1	10	6	7	–	–	1	5	–	6	5	2	7	25	12	22
Distribution/route planning	1	–	2	3	–	2	–	–	–	2	7	6	3	1	2	9	8	12
Organizational development/competencies	3	5	1	1	1	10	–	–	–	–	2	2	3	2	7	7	10	20
System integration/integration enablers	1	–	2	1	3	4	–	–	–	1	2	3	5	1	7	8	6	16
Environmental issues/CSR	–	1	2	–	1	1	–	–	–	–	–	1	2	1	3	2	3	7
Inter-organizational collaboration/third-party logistics	3	1	1	3	3	12	–	–	–	2	2	–	3	4	8	11	10	21
Material handling/material planning	–	–	5	–	3	–	–	–	–	–	3	3	3	8	6	3	14	14
Transport/transport systems	2	–	1	2	4	5	–	–	–	2	–	5	4	1	16	10	5	27
Risk management	–	–	–	–	–	2	–	–	–	–	–	1	–	–	6	0	–	9
Humanitarian logistics	–	–	–	–	–	2	–	–	–	–	–	–	–	–	1	0	–	3
Others	–	–	2	–	1	3	–	–	–	–	1	–	–	–	2	0	2	7
Total	15	11	17	20	22	48	–	–	1	12	17	27	28	20	65	75	70	158

A = Time period of 1999–2001 covered in Gubi et al. [13]

B = Time period of 2002–2008 covered in Zachariassen and Arlbjørn [41]

C = Time period of 2009–2014 covered in this paper

transport/transport systems have obtained increased awareness. The increase in these topics is primarily based on Swedish and Norwegian dissertations.

Additionally, while the study by [41] found that topics related to humanitarian logistics and risk management were absent, the present review of dissertations has remedied this with dissertations on this topic from Finland and Sweden.

4.6 Article-based dissertations

The study reported in this paper reveals an increased amount of article-based dissertations. This development confirms the expectations raised by Zachariassen and Arlbjørn [41]. However, an article-based dissertation can be composed in different ways—e.g., with respect to requirements regarding the type of articles, number of articles, co-author permission and the author order position among the co-authors in an article. The ministerial orders granting the degree of PhDs in the different countries do not provide any guidance and requirements concerning the format of a PhD. This opens up for various interpretations of the required workload to earn the PhD degree. Therefore, the increase in more article-based dissertations requires that we study how the practice of these types of dissertations is unfolded in the Nordic countries. This section takes a closer look at the 92 article-based dissertations and develops a measure to differentiate the various types of dissertations. The subsequent subsection proposes a measure for article-based dissertation; this measure is subsequently used in analyzing the identified Nordic article-based dissertations.

4.6.1 Measure for dissertation score

In order to develop a measure for an average article-based dissertation, we first need to recognize the fact that a specific article that is part of the dissertation can take different forms. Hence, the first element in this proposal for a dissertation measure is to differentiate between the different types of contributions that are part of the article-based dissertation and then to allocate different scores for the different types. The present analysis distinguishes between five different types of articles with specific scores as follows:

1. Peer-reviewed journal articles, score: 1
2. Peer-reviewed articles in form of book chapters (e.g., in an anthology), score: 0.8
3. Peer-reviewed conference articles, score: 0.8
4. Working papers, score: 0.5
5. Non-published papers, score: 0.5

The differences in scores are used to reflect different perceptions of workload as well as quality requirements. Thus, a peer-reviewed journal article obtains the highest score of 1 point followed by book chapters and conferences papers with scores of 0.8, and working papers and non-published manuscripts of 0.5 points.

The next step in developing a dissertation score is to propose a measure that takes into account the number of authors as well as the order of the authors. Thus, a sole authored paper by a PhD candidate counts more than a co-authored paper. And, in the present measure, a first-order author position counts more than a lower author position. Table 6 proposes scores for authorship indicators

Table 6 Authorship factor calculation	Number of authors (NoA)		Author order position (AOP)		Authorship factor
	NoA	Score	AOP	Score	NoA * AOP
	1	1	1	1	1
	2	0.9	1	1	0.9
	2	0.9	2	0.9	0.81
	3	0.8	1	1	0.8
	3	0.8	2	0.9	0.72
	3	0.8	3	0.8	0.64
	4	0.7	1	1	0.7
	4	0.7	2	0.9	0.63
	4	0.7	3	0.8	0.56
	4	0.7	4	0.7	0.49
	5	0.6	1	1	0.6
	5	0.6	2	0.9	0.54
	5	0.6	3	0.8	0.48
	5	0.6	4	0.7	0.42
	5	0.6	5	0.6	0.36

evaluating each of the articles that take into consideration various numbers of authors and author order positions.

Based on the above proposals of different types of articles and measures for different authorship indicators, we can now develop an overall dissertation score based on Formula 1:

Formula 1: overall dissertation score

$$S = \sum_{i=1}^{n} T_i * \text{Authorship factor}_i$$

where S = overall dissertation score, T = type of article and authorship factor = number of authors (NoA) * author order position (AOP).

The overall dissertation score is composed of multiplying the scores of the individual articles by the authorship factor. The basic idea is that the highest score per article of 1 point is reduced based on the lower the level of the perceived status of various channels (e.g., journal rankings; journal articles versus book chapters; conference articles versus journal articles/book chapters) and the number of co-authors as well as the author order the PhD student has for the given article. A conference paper that is included in an article-based dissertation, which have been through a double-blind review process (e.g., at a NOFOMA, LRN or a EurOMA conference), is valued 0.8 compared with an article that is published in a peer-reviewed journal. Thus, a conference article is considered as less mature when compared to an article that has been through perhaps several revisions in a journal before acceptance. However, we should avoid with the generalization of the different channels. The Danish Bibliometric Research Indicator, for example, value accepted conference articles as much as some journal papers, if they are accepted to be presented at some conferences (e.g., EURAM and AOM) [36]. Also, contributions to books are valued differently according to which publisher the work is published with [37].

The dissertation score is a measure for the PhD dissertation at hand—and thus, the status of it when it was judged and passed. Several article-based dissertations contain non-published papers, working papers and conference articles that find its way to peer-reviewed journals after the degree of PhD has been awarded and further workload is invested in those articles. Other articles of this nature never end in publications for various reasons. Thus, the status of the papers after the PhD evaluation is not included in the dissertation score presented in this article.

An article-based dissertation is evaluated not only on the enclosed articles, but also on the text (the frame) that bind the articles together. Also, herein there seems to be different practices—e.g., the scope and depths on positioning the thesis against extant research; the level and scope of methodological and philosophy of science discussions (see, e.g., [3]); independent literature reviews as well as discussions on the "red line" between the included articles.

The overall purpose with the dissertation score is to propose a measure to be used for discussing the scope and content-type of a PhD dissertation. What is enough? And how much of a dissertation can be co-authored with others? Is one dissertation better than another because it includes articles that are published? It is our experience that there are differences in what is needed in an article-based dissertation both within a department at a research institution and between research institutions. The present dissertation score can help in discussing what is needed and also in developing department guidelines.

The proposed overall dissertation score formula is not without limitations. First, the division of types of articles and their scores is subjective. The peer-reviewed journal category, for example, can be further divided into different scores by following specific journal ranking lists. Second, the scores for various numbers of authors and their author order position in the proposed authorship factor calculation are also subjective and can be altered. The order of authors of a paper does not necessarily display the true workload of different authors. The order of authors can be organized using different principles such as a simple alphabetical listing, organized after seniority (experience); listing the person first who got the idea to the article as the first author; or listing PhD students first because they need the credit more than their senior co-authors. Third, the dissertation score does not include a time perspective (i.e., the length of the PhD program)—a higher score can be obtained over a 5-year period when compared to a 3-year period. Fourth, the dissertation score does not take into consideration the extent of thesis frame (the text accompanying the articles such as scoping, positioning, methods, philosophy of science, contribution and implications). Fifth, the score of author order position decreases based on the position; this may not reflect the actual work load. In summary, as with any measures in general, this overall dissertation score also has its own shortcomings and these are important to be considered to ensure the practical use of the score. However, in spite of the sometime magical status of numbers—we should remember that "numbers are just number" and that they can be used to jump start discussions on how to compose an article-based dissertation.

4.6.2 Dissertation scores

This section provides an analysis of the 92 article-based dissertations using the dissertation score developed in the above subsection. This number is divided among 12

Table 7 Contents of article-based dissertations

		Art. 1	Art. 2	Art. 3	Art. 4	Art. 5	Art. 6	Art. 7	Art. 8	Total	Avg DS
3 Articles	JA	2	2	1	–	–	–	–	–	5	1.53
	BC	0	0	0	–	–	–	–	–	0	
	CP	0	0	0	–	–	–	–	–	0	
6	WP	0	0	0	–	–	–	–	–	0	
	NP	4	4	5	–	–	–	–	–	13	
4 Articles	JA	21	16	18	16	–	–	–	–	71	2.62
	BC	1	1	1	1	–	–	–	–	4	
	CP	6	7	6	5	–	–	–	–	24	
33	WP	1	3	2	3	–	–	–	–	9	
	NP	4	6	6	8	–	–	–	–	24	
5 Articles	JA	20	15	16	17	12	–	–	–	80	3.49
	BC	0	0	1	0	1	–	–	–	2	
	CP	2	6	3	4	8	–	–	–	23	
26	WP	1	1	1	2	1	–	–	–	6	
	NP	3	4	5	3	4	–	–	–	19	
6 Articles	JA	21	18	17	16	16	11	–	–	99	4.49
	BC	0	0	0	0	0	0	–	–	0	
	CP	3	5	5	5	4	5	–	–	27	
24	WP	0	1	1	2	2	6	–	–	12	
	NP	0	0	1	1	2	2	–	–	6	
7 Articles	JA	1	0	1	1	2	2	1	–	8	5.21
	BC	0	0	0	0	0	0	0	–	0	
	CP	1	2	1	1	0	0	1	–	6	
2	WP	0	0	0	0	0	0	0	–	0	
	NP	0	0	0	0	0	0	0	–	0	
8 Articles	JA	1	0	0	0	0	1	0	1	3	5.38
	BC	0	0	0	0	0	0	0	0	0	
	CP	0	1	1	0	1	0	0	0	3	
1	WP	0	0	0	0	0	0	0	0	0	
	NP	0	0	0	1	0	0	1	0	2	
Total		92	92	92	86	53	27	3	1	446	3.34

JA journal article, *BC* book chapter, *CP* conference paper, *WP* working papers, *NP* non-published paper

Danish, 27 Finnish, 37 Swedish and 16 Norwegian dissertations. As given in Table 7, the dissertations vary in the number of included articles spanning from three to eight articles. The majority of the dissertations are composed of four to six articles. The average number of articles counts to 4.84. Table 7 also shows that the average dissertation score increases from 1.53 with three articles to 5.38 with eight articles. This is not surprising given the design of the formula.

Furthermore, Table 7 shows that including journal articles is a well-established practice among the reviewed dissertations. The share of journal articles counts 54 % among dissertations with four articles (71/132); 62 % among dissertations with five articles (80/130); and 69 % among dissertations with six articles (99/144). After journal papers, conference papers and non-published papers are the second most typical types that are included in article-based dissertations. Finally, Table 7 shows that including reviewed books chapters is not that prevalent among the dissertations reviewed.

This review also shows that there are more PhD candidates at Chalmers University of Technology with highest number of contributions at 18 dissertations followed by Molde University College with 16 dissertations and Lund University with 11 dissertations (see Fig. 1). In all, the 92 article-based dissertations contain 446 articles divided into the five types discussed in Sect. 4.6.1. Out of this, 266 are peer-reviewed journal articles of which 68 are from Chalmers University of Technology; 39 are from Molde University College; and 39 are from Lappeenranta University of Technology. In total, 83 articles are included as conference papers; 64 are included as non-published papers; 27 are included as working papers; and only six articles take the form of book chapters.

Fig. 1 Average dissertation scores divided by research institution. *AU* Aalto University, *ASB/AU* Aarhus School of Business/Aarhus University, *CUT* Chalmers University of Technology, *HSE* Hanken School of Economics, *JIBS* Jönköping International Business School, *KI* Karolinska Institutet, *LiU* Linköping University, *LU* Lund University, *LUT* Lappeenranta University of Technology, *MUC* Molde University College, *SU* Stockholm University, *TSE* Turku School of Economics, *TUD* Technical University Denmark, *UO* University of Oulu, *UV* University of Vaasa, *USD* University of Southern Denmark

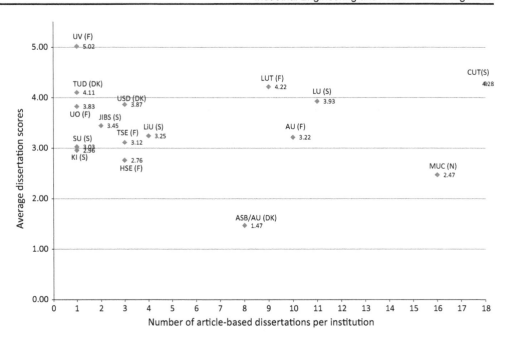

Figure 1 demonstrates a relatively high variety among the dissertations across the different research institutions with respect to the average institutional dissertations (ranging from 1.47, 2.47, 2.76, 2.96, 3.03, 3.12, 3.22, 3.25, 3.45, 3.83, 3.87, 3.93, 4.11, 4.22, 4.28, 5.02). Four of the six dissertations composed of three articles (see Table 7) are Danish dissertations from Aarhus School of Business/Aarhus University. In contrast, eight out of the 24 dissertations holding six articles are composed of 4–6 peer-reviewed journal articles. The dissertation score is here suggested as an instrument that can be used in discussions on what is required to earn the degree of PhD at the specific research institutions and, through benchmarks, facilitate alignment processes if needed. With the limitations in mind, a PhD student can discuss with her or his advisor(s) as of what type of articles are required; whether co-authored articles with advisors or other senior researchers or PhD students can be included; and how a PhD student documents her or his contribution in each article if they are co-authored. The dissertation score can also stimulate further discussions on what should be included in the binder of the dissertation—e.g., prioritizes theme to be discussed in the binder instead of adding another article.

4.7 Potential topics for inclusion in the dissertations

The review process of the 150 dissertations reveals that five research areas seem to be under-prioritized among Nordic scholars within logistics and SCM. The topics are as follows (not prioritized):

(1) *Cloud technology* Although there is a focus on maximizing the effectiveness of shared resources [5],

the increased availability of high-capacity network, low-cost computers and storage devices as well as the widespread adoption of hardware virtualization, service-oriented architecture, and autonomic and utility computing have led to a growth in the use of cloud technology. In spite of its increased importance, it is distinct that this subject has not received more attention in the dissertations reviewed.

(2) *Globalization of SC* As mentioned previously, this topic was still largely ignored in the dissertation reviewed. With the natural outcome of expanding growing market and sustaining competitive advantage, companies have to make key decisions about managing costs and complexity through globalized supply chain (see, e.g., [11]).

(3) *Supply chain innovation (SCI)* The discipline of innovations has branched out into the supply chain context and should be regarded as an important topic. While firms are applying their assets, operating resources and capabilities to develop new ways of improve performance, they cannot ignore the importance of their supply chain partners in enabling product and process innovations (see, e.g., [1]). Accordingly, more attention needs to be paid to this topic.

(4) *Security* This topic was also mentioned in the previous study by Zachariassen and Arlbjørn [41]. Yet, this topic has only received limited attention among the 112 dissertations published between 2009 and 2014. Security is increasingly becoming a major concern to both private and public sector organizations. Security searches for cooperative arrangements between businesses as well as identifies risks

before the goods move. It is also concerned with controlling theft and reducing illegal import and export of stolen goods. There are many areas of research within the topic of security from formal aspects to empirical research (see, e.g., [4, 40]). Hence, it is essential that researchers start focusing on the different aspects of this topic in the future. This is one of the key topics of research within supply chain now as well as in the distant future.

(5) *Big data* This is currently drowning the world. The huge amount of data is an invaluable asset in the context of supply chains. The quality of the evidence extracted significantly benefits from the availability of broad datasets. On the one hand, extensive vision is more promising when extra data are available. And, on the other hand, it is a big challenge. The current approach is not appropriate to handle large data, and therefore, there is a need for new solutions to handle large datasets. This research field is new and rapidly evolving [38], and there is also a lead-time issue before we see the first dissertations within this area. As a result, any initial effort taken will be a strong contribution from both experimental and theoretical perspective (see, e.g., [15]).

result, there is an extreme decrease (110/150). Finally, there is clear shift toward disseminating doctoral research as an article-based dissertation (2009–2014: 92/150; 2002–2008: 26/70; 1990–2001: 15/71, respectively). However, this piece of research has demonstrated that the content of the Nordic article-based dissertations varies along dimensions such as types of articles included, the number of articles included and the number of co-authors at the papers. A dissertation score has been suggested as a measure for initiating discussions about such dissertations at a single research institution and as a benchmark between different institutions. As discussed earlier, such a score is not without limitations and must not stand alone. Therefore, the next logical step might be to discuss how far should we go with the number and the type of contributions. What is enough in order to earn a PhD degree? The content of this paper can be used to match the expectations of PhD students and their supervisors so as to find the right path to learn the craft of conducting and disseminating logistics and SCM research.

5 Conclusion

This paper has set out to analyze the development of Nordic doctoral dissertations in logistics and SCM from the years 2009–2014. The paper identified 158 dissertations relevant for review; out of this, 150 dissertations were reviewed in depth based on different criteria. Compared with previous studies, this research found several important developments in Nordic doctoral research. First, there has been an increase in the average annual number of finalized dissertations when compared to previous Nordic studies. The results predominantly show the remarkable increase in dissertations between 2009 and 2014 (158 dissertations) wherein Sweden ranks highest followed by Finland, Norway, Denmark and Iceland, respectively. Second, there has been a decrease in dissertations that focus on classical entities of analysis such as carriers, wholesalers, retailers and inventories. Therefore, it is clear that the PhD dissertation's focal point is still more on manufacturing firms (50/150) rather than on other entities. Third, there has been an increase in functional subject areas of logistics and SCM and the supply chain/network level. Fourth, the number of dissertations containing philosophy of science discussions is continuing to decline. As mentioned earlier, there is a significant increase in article-based dissertations and these dissertations do not adopt philosophy of science; as a

Appendix 1: Research institutions contacted

Denmark

- Aalborg University
- Aarhus School of Business/Aarhus University
- Copenhagen Business School
- Danish Technical University/Technical University of Denmark
- Roskilde University
- University of Southern Denmark

Finland

- Aalto University
- Åbo Akademi University
- Hanken School of Economic
- Helsinki University
- Lappeenranta University of Technology
- National Defence University
- Swedish School of Economics and Business Administration
- Tampere University of Technology
- Technical Research Center of Finland
- Turku School of Economics

- University of Oulu Business School
- University of Vaasa

Iceland

- University of Iceland

Norway

- BI Norwegian School of Management
- Institute of Transport Economics
- Molde University College
- Norwegian School of Economics and Business Administration
- Norwegian University of Science and Technology
- SINTEF Industrial Management
- University of Oslo Business School
- University of Nordland

Sweden

- Chalmers University of Technology
- Gothenburg University
- Jönköping International Business School
- Karolinska Institute Department of Public Health Sciences
- Linköping University
- Linnæus University
- Lund University
- Örebro University
- Stockholm School of Economics
- Stockholm University
- Swedish National Road and Transport Research Institute
- University College of Borås

Appendix 2: Doctoral dissertations identified

Danish dissertations

Aarhus School of Business/Aarhus University:

- Abginehchi, S. (2012), *Essays on Inventory Control in Presence of Multiple Sourcing*, Aarhus.
- Bach, L. (2014), *Routing and Scheduling Problems—Optimization using Exact and Heuristic Methods*, Aarhus.
- Bendre, A.B. (2010), *Numerical Studies of Single-stage, Single-item Inventory Systems with Lost Sales*, Aarhus.
- Bodnar, P. (2013), *Essays on Warehouse Operations*, Aarhus.
- Christensen, T.R.L. (2013), *Network Design Problems with Piecewise Linear Cost Functions*, Aarhus.

- Du, B. (2011), *Essays on Advance Demand Information, Prioritization and Real Options in Inventory Management*, Aarhus.
- Hanghøj, A. (2014), *Papers in Purchasing and Supply Management: A Capability-Based Perspective*, Aarhus.
- Kjeldsen, K.H. (2012), *Routing and Scheduling in Liner Shipping*, Aarhus.

Copenhagen Business School:

- Andreasen, P.H. (2012), *The Dynamics of Procurement Management—A Complexity Approach*, Frederiksberg.
- Kinra, A. (2009), *Supply Chain (Logistics) Environmental Complexity*, Frederiksberg.
- Nøkkentved, C. (2009), *Enabling Supply Networks with Collaborative Information Infrastructures: An Empirical Investigation of Business Model Innovation in Supplier Relationship Management*, Frederiksberg.
- Yu, L.A. (2012), *Fabricating an S&OP Process: Circulating References and Matters of Concern*, Frederiksberg.

University of Southern Denmark:

- Jensen, J.K. (2013), *Development of Environmentally Sustainable Food Supply Chains*, Kolding.
- Mikkelsen, O.S. (2011), *Strategic Sourcing in a Global Organizational Context*, Kolding.
- Mortensen, M.H. (2011), *Towards Understanding Attractiveness in Industrial Relationships*, Kolding.
- Zachariassen, F. (2011), *Supply Chain Management and Critical Theory: Meta-Theoretical, Disciplinary and Practical Contributions to the Supply Chain Management Discipline Based on Insights from the Management Accounting Discipline*, Kolding.

Technical University of Denmark:

- Jørgensen, P. (2013), *Technology in Health Care Logistics*, Lyngby.

Finnish dissertations

Aalto University School of Business:

- Aaltonen, K. (2010), *Stakeholder Management in International Projects*, Finland.
- Ahola, T. (2009), *Efficiency in Project Networks: The Role of Inter-Organizational Relationships in Project Implementation*, Finland.
- Ala-Risku, T. (2009), *Installed Base Information: Ensuring Customer Value and Profitability after the Sale*, Finland.

- Groop, J. (2012), *Theory of Constraints in Field Service: Factors Limiting Productivity in Home Care Operations*, Finland.
- Helkiö, P. (2013), *Developing Explorative and Exploitative Strategic Intentions—Towards a Practice Theory of Operations Strategy*, Finland.
- Hinkka, V. (2013), *Implementation of RFID Tracking across the Entire Supply Chain*, Finland.
- Karjalainen, K. (2009), *Challenges of Purchasing Centralization—Empirical Evidence from Public Procurement*, Finland.
- Karrus, K. (2011), *Policy Variants for Coordinating Supply Chain Inventory Replenishments*, Finland.
- Kauremaa, J. (2010), *Studies on the Utilization of Electronic Trading Systems in Supply Chain Management*, Finland.
- Nieminen, S. (2011), *Supplier Relational Effort in the Buyer–Supplier Relationship*, Finland.
- Peltokorpi, A. (2010), *Improving Efficiency in Surgical Services: A Production Planning and Control Approach*, Finland.
- Porkka, P. (2010), *Capacitated Timing of Mobile and Flexible Service Resources*, Finland.
- Rajahonka, M. (2013), *Towards Service Modularity—Service and Business Model Development*, Finland.
- Ristola, P. (2012), *Impact of Waste-to-Energy on the Demand and Supply Relationships of Recycled Fibre*, Finland.
- Seppälä, T. (2014), *Contemporary Determinants and Geographical Economy of Added Value, Cost of Inputs, and Profits in Global Supply Chains: An Empirical Analysis*, Finland.
- Tenhiälä, A. (2009), *Contingency Theories of Order Management, Capacity Planning, and Exception Processing in Complex Manufacturing Environments*, Finland.
- Torkki, P. (2012), *Best Practice Processes—What are the Reasons for Differences in Productivity between Surgery Units*, Finland. **NOT reviewed**
- Turunen, T. (2013), *Organizing Service Operations in Manufacturing*, Finland.
- Tynjälä, T. (2011), *An Effective Tool for Supply Chain Decision Support During New Product Development Process*, Finland.
- Viitamo, E. (2012), *Productivity as a Competitive Edge of a Service Firm: Theoretical Analysis and a Case Study of the Finnish Banking Industry*, Finland.
- Voutilainen, J. (2014), *Factory Positioning in an Unpredictable Environment: A Managerial View of Manufacturing Strategy Formation*, Finland.

Åbo Akademi University:

- Nyholm, M. (2011), *Activation of Supply Relationships*, Turku.

Hanken School of Economic:

- Antai, I. (2011), *Operationalizing Supply Chain vs. Supply Chain Completion*, Finland.
- Haavisto, I. (2014), *Performance in Humanitarian Supply Chains*, Finland.
- Harilainen, H. (2014), *Managing Supplier Sustainability Risk*, Finland.
- Tomasini Ponce, R. (2012), *Informal Learning Framework for Secondment: Logistics Lessons from Disaster Relief Operations*, Finland.
- Vainionpää, M. (2010), *Tiering Effects in Third-party Logistics: A First-Tier Buyer Perspective*, Finland.

University of Oulu Business School:

- Juntunen, J. (2010): *Logistics Outsourcing for Economies in Business Network*, Finland

Lappeenranta University of Technology—Industrial Engineering and Management:

- Karppinen, H. (2014), *Reframing the Relationship between Service Design and Operations: A Service Engineering Approach*, Lappeenranta.
- Kerkkänen, A. (2010), *Improving Demand Forecasting Practices in the Industrial Context*, Lappeenranta.
- Laisi, M. (2013), *Deregulation's Impact on the Railway Freight Transport Sector's Future in the Baltic Sea Region*, Lappeenranta.
- Lättilä, L. (2012), *Improving Transportation and Warehousing Efficiency with Simulation-Based Decision Support Systems*, Lappeenranta.
- Niemi, P. (2009), *Improving the Effectiveness of Supply Chain Development Work—An Expert Role Perspective*, Lappeenranta.
- Pekkanen, P. (2011), *Delay Reduction in Courts of Justice—Possibilities and Challenges of Process Improvement in Professional Public Organizations*, Lappeenranta.
- Salmela, E. (2014), *Kysyntä-Toimitusketjun Synkronointi Epävarman Kysynnän ja Tarjonnan Toimintaympäristössä*, Lappeenranta. **NOT reviewed.**
- Saranen, J. (2009), *Enhancing the Efficiency of Freight Transport by Using Simulation*, Lappeenranta.

Lappeenranta University of Technology—School of Business:

- Kähkönen, A. (2010), *The Role of Power Relations in Strategic Supply Management—A Value Net Approach*, Lappeenranta.

- Lintukangas Annaliisa, K. (2009), *Supplier Relationship Management Capability in the Firm's Global Integration*, Lappeenranta.
- Vilko, J. (2012), *Approaches to Supply Chain Risk Management: Identification, Analysis and Control*, Lappeenranta.

Turku School of Economic:

- Koskinen, P. (2009), *Supply Chain Challenges and Strategies of a Global Paper Manufacturing Company*, Turku.
- Lorentz, H. (2009), *Contextual Supply Chain Constraints in Emerging Markets—Exploring the Implications for Foreign Firms*, Turku.
- Rantasila, K. (2013), *Measuring Logistics Costs. Designing a Generic Model for Assessing Macro Logistics Costs in a Global Context with Empirical Evidence from the Manufacturing and Trading Industries*, Turku.
- Solakivi, T. (2014), *The Connection between Supply Chain Practices and Firm Performance—Evidence from Multiple Surveys and Financial Reporting Data*, Turku.

Tampere University of Technology:

- Jokinen, J. (2010), *Multi-Agent Control of Reconfigurable Pallet Transport Systems*, Tampere. **NOT reviewed**

University of Vaasa:

- Addo-Tenkorang, R. (2014), *Conceptual Framework for Large-Scale Complex Engineering- Design & Delivery Processes. A Case of Enterprise SCM Network Activities and Analysis*, Finland.
- Kärki, P. (2012), *The Impact of Customer Order Lead Time-Based Decisions on the Firm's Ability to Make Money—Case Study: Build to Order Manufacturing of Electrical Equipment and Appliances*, Finland.
- Moilanen, V. (2011), *Case study: Developing a Framework for Supply Network Management*, Finland.
- Nugroho Widhi, Y K. (2009), *Structuring Postponement Strategies in the Supply Chain by Analytical Modeling*, Finland.

Icelandic dissertations

University of Iceland—School of Health Sciences:

- Nga, M. (2010), *Enhancing Quality Management of Fresh Fish Supply Chains through Improved Logistics and Ensured Traceability*, Reykjavik.

Norwegian dissertations

Norwegian University of Science and Technology:

- Bai, Y. (2013), *Reliability of International Freight Trains, An Exploratory Study Drawing on Three Mainstream Theories*, Trondheim.

BI Norwegian School of Management:

- Hatteland, C.J. (2010), *Ports as Actors in Industrial Networks*, Norway.
- Bjørnstad, S. (2009), *Shipshaped Kongsberg Industry and Innovations in Deepwater Technology, 1975–2000*, Norway.
- Hoholm, T. (2009), *The Contrary Forces of Innovation-An Ethnography of Innovation Processes in the Food Industr*, Norway.
- Zhovtobryukh, Y (2014), *The Role of Technology, Ownership and Origin in M&A Performance*, Norway.

Molde University College:

- Halse, L.L. (2014), *Walking the Path of Change. Globalization of the Maritime Cluster in North West Norway*, Norway.
- Salema, G.L. (2014), *The Antecedents of Supplier Logistics Performance: an Empirical Study of the Essential Medicines Supply in Tanzania*, Norway.
- Chaudhry, M.O. (2014), *An Assessment of Linkages between Investment in Transport Infrastructure and Economic Development*, Norway.
- Søvde, N.E. (2014) *Optimization of Terrain Transportation Problems in Forestry*, Norway.
- Iversen, H.P. (2013) *Logististikkerfaringer i Psykiatri og Psykisk Helsearbeid: Om Forståelse, Organisering og Ledelsed av Relasjoner i en Profesjonell Organisasjon i Omstilling*, Norway.
- Bottolfsen, T. (2013) *The Impact of Internal, Customer and Supplier Integration on Store Performance*, Norway.
- Schøyen, H. (2013) *Identifying Efficiency Potentials in Maritime Logistics: Investigations from Container and Bulk Trades*, Nowary.
- Jin, J.Y. (2013) *Cooperative Parallel Metaheuristics for Large Scale Vehicle Routing Problems*, Norway.
- Regmi, U.K. (2013) *Essays on Air Transport Marketing and Economics*, Norway.
- Glavee-Geo, R. (2012) *The Antecedents and Consequences of Supplier Satisfaction in Agro Commodity Value Chain: an Empirical Study of Smallholder Cocoa Growers of Ghana*, Norway.
- Lanquepin, G. (2012) *Algorithms for Dynamic Pricing and Lot Sizing*, Norway.

- Qin, F.F. (2012) *Essays on Efficient Operational Strategy of Urban Rail Transit,* Norway.
- Qian, F.B. (2012) *Passenger Risk Minimization in Helicopter Transportation for the Offshore Petroleum Industry,* Norwary.
- Bø, O. (2012) *Aspects of Production Tracking Systems in the Supply Network for Caught Seafood,* Norway.
- Yue, X. (2010) *Competition and Cooperation: a Game theoretic Analysis on the Development of Norwegian Continental shelf,* Norway.
- Shyshou, A. (2010) *Vessel Planning in Offshore Oil and Gas Operations,* Norway.
- Thapalia, B.K. (2010) *Stochastic Single-commodity Network Design,* Norway.
- Bakhrankova, K. (2010) *Production Planning in Continuous Process Industries: Theoretical and Optimization Issues,* Norway.
- Bhatta, B.P. (2009) *Discrete Choice Analysis with Emphasis on Problems of Network-based Level of Service Attributes in Travel Demand Modeling,* Norway.
- Burki, U. (2009) *Cross Cultural Effects on the Relational Governance of Buyer–Supplier Relationships: an Empirical Study of the Textile Exporting Firms of Pakistan,* Norway.
- Vaagen, H. (2009) *Assortment Planning under Uncertainty,* Norway.
- Saeed, N. (2009) *Competition and Cooperation among Container Terminals in Pakistan: with Emphasis on Game Theoretical Analysis,* Norway.

Swedish dissertations

Chalmers University of Technology:

- Almotairi, B. (2012), *Integrated Logistics Platform the Context of the Port Relational Exchanges and Systematic Integration,* Gothenburg.
- Andersson, R. (2009), *Supply Chain Resilience through Quality Management,* Gothenburg.
- Bankvall, L. (2011), *Activity Linking in Industrial Networks,* Gothenburg.
- Behrends, S. (2011), *Urban Freight Transport Sustainability—The Interaction of Urban Freight and Intermodal Transport,* Gothenburg.
- Ekwall, D. (2009), *Managing Risk for Antagonistic Threats against Transport Network,* Gothenburg.
- Ellis, J. (2011), *Assessing Safety Risks for the Sea Transport Link of a Multimodal Dangerous Goods Transport Chain,* Gothenburg.
- Finnsgård, C. (2013), *Materials Exposure: The Interface between Materials Supply and Assembly,* Gothenburg.

- Fredriksson, A. (2011), *Materials Supply and Production Outsourcing,* Gothenburg.
- Hanson, R. (2012), *In-Plant Materials Supply: Supporting the Choice between Kitting and Continuous Supply,* Gothenburg.
- Hilletofth, P. (2010), *Demand–Supply Chain Management,* Gothenburg.
- Hjort, K. (2013), *On Aligning Returns Management with the Ecommerce Strategy to Increase Effectiveness,* Gothenburg.
- Ingrid, H. (2014), *Organizing Purchasing and Supply Management across Company Boundaries,* Gothenburg.
- Ivert, L. (2012), *Use of Advanced Planning and Scheduling (APS) Systems to Support Manufacturing Planning and Control Processes,* Gothenburg.
- Kalantari, J. (2012) *Foliated Transportation Networks-Evaluating Feasibility and Potential,* Gothenburg.
- Kharrazi, S. (2012), *Steering Based Lateral Performance Control of Long Heavy Vehicle Combinations,* Gothenburg.
- Lindholm, M.E., (2012) *Enabling Sustainable Development of Urban Freight from a Local Authority Perspective,* Gothenburg.
- Mirzabeiki, V. (2013), *Collaborative Tracking and Tracing—A Supply Chain Perspective,* Gothenburg.
- Roso, V. (2009), *The Dry Port Concept,* Gothenburg.
- Sternberg, H. (2011), *Waste in Road Transport Operations—Using Information Sharing to Increase Efficiency,* Gothenburg.
- Styhre, L. (2010), *Capacity Utilization in Short Sea Shipping,* Gothenburg.
- Sundquist, V. (2014), *The Role of Intermediation in Business Networks,* Gothenburg.
- Thörnblad, K. (2013), *Mathematical Optimization in Flexible Job Shop Scheduling: Modelling, Analysis, and Case Studies,* Gothenburg.

Jönköping International Business School:

- Borgström, B. (2010), *Supply chain strategising: Integration in practice,* Jönköping.
- Cui, L.G., (2012), *Innovation and network development of logistics firms,* Jönköping.
- Skoglund, P. (2012), *Sourcing decisions for military logistics in Peace Support Operations: A case study of the Swedish armed forces,* Jönköping.
- Jafari, H. (2014), *Postponement and Logistics Flexibility in Retailing,* Jönköping.
- Wikner, S. (2011), *Value Co-creation as Practice—On a supplier's capabilities in the value generation process,* Jönköping.

Karolinska Institute Department of Public Health Sciences:

- Anund, A. (2009), *Sleepiness at the Wheel,* Sweden.

Linköping University:

- Ekström, J. (2012), *Optimization Approaches for Design of Congestion Pricing Schemes*, Linköping.
- Feldmann, A. (2011), *A Strategic Perspective on Plants in Manufacturing Networks*, Linköping.
- Hansson, L. (2010), *Public Procurement at the Local Government Level: Actor Roles, Discretion and Constraints in the Implementation of Public Transport Goals*, Linköping.
- Isaksson, K. (2014), *Logistics Service Providers Going Green—A Framework for Developing Green Service Offerings*, Linköping.
- Lindskog, M. (2012), *On Systems Thinking in Logistics Management—A Critical Perspective*, Linköping.
- Malmgren, M. (2010), *Managing Risks in Business Critical Outsourcing: A Perspective from the Outsourcer and the Supplier*, Linköping.
- Martinsen, U. (2014), *Towards Greener Supply Chains: Inclusion of Environmental Activities in Relationships between Logistics Service Providers and Shippers*, Linköping.
- Mårdh, S. (2013), *Cognitive Erosion and its Implications in Alzheimer's Disease*, Linköping.
- Musa, S.N. (2012), *Supply Chain Risk Management: Identification, Evaluation and Mitigation Techniques*, Linköping.
- Olstam, J. (2009), *Simulation of Surrounding Vehicles in Driving Simulators*, Linköping.
- Thoresson, K. (2011), *To Calculate the Good Society: Cost-Benefit Analysis and the Border between Expertise and Policy in the Transport Sector*, Linköping. **NOT reviewed.**

Linnaeus University:

- Farvid, S.M. (2014), *Essays on Inventory Theory*, Sweden.
- Samadi, R. (2010), *Supply Chain Optimization and Market Coordinated Inventory*, Sweden. **NOT Reviewed.**

Lund University:

- Bagdadi, O. (2012), *The Development of Methods for Detection and Assessment of Safety Critical Events in Car Driving*, Lund.
- Eng Larsson, F. (2014), *On the Incentives to Shift to Low-Carbon Freight Transport*, Lund.
- Howard, C. (2013), *Real-Time Allocation Decisions in Multi-Echelon Inventory Control*, Lund.
- Lundin, J. (2011), *On Supply Chain Incentive Alignment: Insight from a Cash Supply Chain and a Trucking Service Supply Chain*, Lund.

- Olander, M. (2010), *Logistik och Juridik. Moderna Affärsförbindelser och Kontrakt som Utmaningar för Förmögenhetsrätten*, Lund. **NOT Reviewed**
- Pazirandeh, A. (2014), *Purchasing Power and Purchasing Strategies—Insights from the Humanitarian Sector*, Lund.
- Urciuoli, L. (2011), *Security in Physical Distribution Networks: A Survey Study of Swedish Transport Operators*, Lund.
- Olander Roses, K. (2014), *From PowerPoints to Reality-managing Strategic Change in the Paper Packaging Industry*, Lund.
- Sohrabpour, V. (2014), *Packaging Design and Development for Supply Chain Efficiency and Effectiveness,* Lund.
- Abbasi, M. (2014), *Exploring Themes and Challenges in Developing Sustainable Supply Chains-A Complexity Theory Perspective*, Lund.
- Ringsberg, H. (2013), *Food Traceability in Regulated Fresh Food Supply Chains with an Emphasis on the Swedish Fishing Industry*, Lund.
- Beckeman, M. (2011), *The Potential for Innovation in the Swedish Food Sector*, Lund.
- Dominic, C. (2011), *Packaging Logistics Performance*, Lund.
- Johansson, O. (2009), *On the Value of Intelligent Packaging-A Packaging Logistics Perspective*, Lund.
- Pålsson, H. (2009), *Logistics value of using tracking data from uniquely labelled goods*, Lund.

Stockholm University, Department of Psychology:

- Eriksson, G. (2014), *On Physical Relations in Driving: Judgments, Cognition and Perception*, Stockholm.

Örebro University School of Business:

- Ahlberg, J. (2012), *Multi-Unit Common Value Auctions: Theory and Experiments*, Örebro.
- Arvidsson, S. (2010), *Essays on Asymmetric Information in the Automobile Insurance Market*, Örebro.
- Bohlin, L. (2010), *Taxation of Intermediate Goods—A CGE Analysis*, Örebro.
- Krüger, N. (2009), *Infrastructure Investment Planning under Uncertainty*, Örebro. **NOT reviewed**
- Liu, X. (2013), *Transport and Environmental Incentive Policy Instruments—Effects and Interactions*, Örebro. **NOT reviewed**
- Lodefalk, M. (2013), *Tackling Barriers to Firm Trade. Liberalisation, Migration, and Servicification*, Örebro.
- Sund, B. (2010), *Economic Evaluation, Value of Life, Stated Preference Methodology and Determinants of Risk*, Örebro.
- Swärdh, J. (2009), *Commuting Time Choice and the Value of Travel Time*, Örebro.

References

1. Arlbjørn JS, de Haas H, Munksgaard KB (2011) Exploring supply chain innovation. Logist Res 3(1):3–18
2. Arlbjørn JS, Freytag PV, Damgaard T (2008) The beauty of measurement. Eur Bus Rev 20(2):112–127
3. Arlbjørn JS, Halldórsson Á (2002) Logistics knowledge creation: reflections on content, context and processes. Int J Phys Distr Log 32(1):22–40
4. Blackhurst J, Ekwall D, Martens BJ (2015) Special issue on supply chain security. Int J Phys Distr Log 45(7). doi:10.1108/IJPDLM-04-2015-0104
5. Cegielski CG, Allison Jones-Farmer L, Wu Y, Hazen BT (2012) Adoption of cloud computing technologies in supply chains: an organizational information processing theory approach. Int J Logist Manag 23(2):184–211
6. Choi TY, Wacker JG (2011) Theory building in the OM/SCM field: pointing to the future by looking at the past. Supply Chain Manag Int J 47(2):8–11
7. Committee on Publication Ethics (COPE) (2014) How to deal with text recycling. http://publicationethics.org/files/BioMed%20Central_text_recycling_editorial_guidelines_1.pdf. Accessed 09 Jan 2016
8. Das A, Handfield RB (1997) A meta-analysis of doctoral dissertations in purchasing. J Oper Manag 15(2):101–121
9. Dyrud MA (2015) Ethics and text recycling. In: 122nd ASEE annual conference & exposition, 14–17 June 2015, Seattle, paper ID#11150
10. Fernie J, Sparks L, McKinnon AC (2010) Retail logistics in the UK: past, present and future. Int J Retail Distr Manag 38(11/12):894–914
11. Gereffi G, Lee J (2012) Why the world suddenly cares about global supply chains. Supply Chain Manag Int J 48(3):24–32
12. Group of Eight (2013) The changing. In: Ph.D.: discussion paper. O'Conner ACT Australia. https://go8.edu.au/sites/default/files/docs/the-changing-phd_final.pdf. Accessed 01 March 2015
13. Gubi E, Stentoft Arlbjørn J, Johansen J (2003) Doctoral dissertations in logistics and supply chain management: a review of Scandinavian contributions from 1990 to 2001. Int J Phys Distr Log 33(10):854–885
14. Halldórsson Á, Larson PD, Poist RF (2008) Supply chain management: a comparison of Scandinavian and American perspectives. Int J Phys Distr Log 38(2):126–142
15. Huang YY, Handfield RB (2015) Measuring the benefits of ERP on supply management maturity model: a "big data" method. Int J Oper Prod Man 35(1):2–25
16. Hughes M (2014) Editorial: reflecting on ethical questions and peer reviewing. Aust Soc Work 67(4):463–466
17. Jørgensen P (2013) Technology in health care logistics. Dissertation, Technical University of Denmark
18. Karjalainen K (2009) Challenges of purchasing centralization—empirical evidence from public procurement. Dissertation, Aalto University School of Business
19. Karrus KE (2011) Policy variants for coordinating supply chain inventory replenishments. Dissertation, Aalto University School of Business
20. Macdonald S, Kam J (2007) Ring a Ring o'Roses: quality journals and gamesmanship in management studies. J Manag Stud 44(4):640–655
21. McKinnon AC (2013) Starry-eyed: journal rankings and the future of logistics research. Int J Phys Distr Log 43(1):6–17
22. Menachof DA, Gibson BJ, Hanna JB, Whiteing AE (2009) An analysis of the value of supply chain management periodicals. Int J Phys Distr Log 39(2):145–165
23. Mortensen, MH (2011) Towards understanding attractiveness in industrial relationships. Dissertation, University of Southern Denmark
24. Myklebust, JP (2013). Sharp rise in foreign. In: Ph.D. enrolments in Scandinavia. University World News 18
25. Nakhata C, Stock JR, Texiera TB (2013) Doctoral dissertations in logistics and supply chain-related areas: 2005–2009. Logist Res 6(4):119–131
26. Nga, MTT (2010) Enhancing quality management of fresh fish supply chains through improved logistics and ensured traceability. Dissertation, University of Iceland
27. Rao S, Iyengar D, Goldsby JT (2013) On the measurement and benchmarking of research impact among active logistics scholars. Int J Phys Distr Log 43(10):814–832
28. Stock JR (1987) A compendium of doctoral research in logistics: 1970–1986. J Bus Logist 8(2):123–136
29. Stock JR (1988) A compendium of doctoral research in logistics: 1970–1986. J Bus Logist 9(1):125–233
30. Stock JR (1997) Applying theories from other disciplines to logistics. Int J Phys Distr Log 27(9/10):515–539
31. Stock JR (2001) Doctoral research in logistics and logistics-related areas 1992–1998. J Bus Logist 22(1):125–256
32. Stock JR, Boyer SL (2009) Developing a consensus definition of supply chain management: a qualitative study. Int J Phys Distr Log 39(8):690–711
33. Stock JR, Broadus CJ (2006) Doctoral research in supply chain management and/or logistics-related areas: 1999–2004. J Bus Logist 27(1):139–151
34. Stock JR, Luhrsen DA (1993) Doctoral research in logistics-related areas 1987–1991. J Bus Logist 14(2):197–210
35. Tynjälä T (2011) An effective tool for supply chain decision support during new product development process. Dissertation, Aalto University School of Business
36. Undervisnings & Forsknings Ministeriet (Ministry of Teaching and Research) (UFM) (2015a) http://ufm.dk/forskning-og-innovation/statistik-og-analyser/den-bibliometriske-forskningsindikator/autoritetslister/list-of-series-18112015-xlsx.pdf. Accessed 31 Dec 2015
37. Undervisnings & Forsknings Ministeriet (Ministry of Teaching and Research) (UFM) (2015b) http://ufm.dk/forskning-og-innovation/statistik-og-analyser/den-bibliometriske-forskningsindikator/autoritetslister/autoritetslisten-for-forlag-2015-november-xlsx.pdf. Accessed 31 Dec 2015
38. Waller M, Fawcett S (2013) Data science, predictive analytics, and big data: a revolution that will transform supply chain design and management. J Bus Logist 34(2):77–84
39. Williams AJ (1986) Doctoral research in purchasing and materials management: an assessment. J Purch Mater Manage 22(1):13–16
40. Williams Z, Lueg JE, Taylor RD, Cook RL (2009) Why all the changes? An institutional theory approach to exploring the drivers of supply chain security (SCS). Int J Phys Distr Log 39(7):595–618
41. Zachariassen F, Arlbjørn JS (2010) Doctoral dissertations in logistics and supply chain management: a review of Nordic contributions from 2002 to 2008. Int J Phys Distr Log 40(4):332–352

Permissions

The contributors of this book come from diverse backgrounds, making this book a truly international effort. This book will bring forth new frontiers with its revolutionizing research information and detailed analysis of the nascent developments around the world.

We would like to thank all the contributing authors for lending their expertise to make the book truly unique. They have played a crucial role in the development of this book. Without their invaluable contributions this book wouldn't have been possible. They have made vital efforts to compile up to date information on the varied aspects of this subject to make this book a valuable addition to the collection of many professionals and students.

This book was conceptualized with the vision of imparting up-to-date information and advanced data in this field. To ensure the same, a matchless editorial board was set up. Every individual on the board went through rigorous rounds of assessment to prove their worth. After which they invested a large part of their time researching and compiling the most relevant data for our readers.

The editorial board has been involved in producing this book since its inception. They have spent rigorous hours researching and exploring the diverse topics which have resulted in the successful publishing of this book. They have passed on their knowledge of decades through this book. To expedite this challenging task, the publisher supported the team at every step. A small team of assistant editors was also appointed to further simplify the editing procedure and attain best results for the readers.

Apart from the editorial board, the designing team has also invested a significant amount of their time in understanding the subject and creating the most relevant covers. They scrutinized every image to scout for the most suitable representation of the subject and create an appropriate cover for the book.

The publishing team has been an ardent support to the editorial, designing and production team. Their endless efforts to recruit the best for this project, has resulted in the accomplishment of this book. They are a veteran in the field of academics and their pool of knowledge is as vast as their experience in printing. Their expertise and guidance has proved useful at every step. Their uncompromising quality standards have made this book an exceptional effort. Their encouragement from time to time has been an inspiration for everyone.

The publisher and the editorial board hope that this book will prove to be a valuable piece of knowledge for researchers, students, practitioners and scholars across the globe.

List of Contributors

D. Mourtzis
Laboratory for Manufacturing Systems and Automation, University of Patras, 26500 Patras, Greece

Nafissa Rezki, Leila Hayet Mouss and Djamil Rezki
LAP: Laboratory, Industrial Engineering Department, University of Batna, 05000 Batna, Algeria

Okba Kazar and Laid Kahloul
LINFI Laboratory, Computer Science Department, University of Biskra, 07000 Biskra, Algeria

G. V. S. S. Sharma
Department of Mechanical Engineering, GMR Institute of Technology, GMR Nagar, Rajam 532127, Andhra Pradesh, India

P. Srinivasa Rao
Department of Mechanical Engineering, Centurion University, Parlakhemundi 761211, Odisha, India

B. Surendra Babu
Department of Industrial Engineering, GITAM Institute of Technology, GITAM University, Visakhapatnam 530045, Andhra Pradesh, India

Maghsoud Amiri
Department of Industrial Management, Allameh Tabataba'I University, Tehran, Iran

Mostafa Khajeh
Department of Industrial Management, Qom Branch, Islamic Azad University, P.O. Box 3749113191, Qom, Iran

Masood Rabieh
Department of Industrial Management, Shahid Beheshti University, Tehran, Iran

Mohammad Ali Soukhakian and Ali Naghi Mosleh Shirazi
Department of Management, Shiraz University, Shiraz, Iran

Rajeev Rathi, Dinesh Khanduja and S. K. Sharma
Department of Mechanical Engineering, National Institute of Technology, Kurukshetra, Haryana 136119, India

Shokrollah Ziari
Department of Mathematics, Firoozkooh Branch, Islamic Azad University, Firoozkooh, Iran

Sadigh Raissi
School of Industrial Engineering, Islamic Azad University, South Tehran Branch, Tehran, Iran

Mahtab Sherafati
Department of Industrial Engineering, South Tehran Branch, Islamic Azad University, Tehran, Iran

Mahdi Bashiri
Department of Industrial Engineering, Shahed University, Tehran, Iran

Angelos P. Markopoulos, Sotirios Georgiopoulos and Dimitrios E. Manolakos
Section of Manufacturing Technology, School of Mechanical Engineering, National Technical University of Athens, Heroon Polytechniou 9, 15780 Athens, Greece

Abdollah Arasteh
Industrial Engineering Department, Babol Noshirvani University of Technology, Shariati Av., P.O. Box: 484, Babol, Mazandaran, Iran

Masoud Rabbani, Farshad Ramezankhani, Ramin Giahi and Amir Farshbaf-Geranmayeh
School of Industrial and Systems Engineering, College of Engineering, University of Tehran, P.O. Box 11155-45632, Tehran, Iran

Mostafa Moradgholi and Iraj Mahdavi
Department of Industrial Engineering, Mazandaran University of Science and Technology, Babol, Iran

Mohammad Mahdi Paydar
Department of Industrial Engineering, Babol University of Technology, Babol, Iran

Javid Jouzdani
Department of Industrial Engineering, Najafabad Branch, Islamic Azad University, Najafabad, Iran

Mohammadreza Shahriari
Faculty of Management, South Tehran Branch, Islamic Azad University, Tehran, Iran

Mohammadreza Shahriari
Faculty of Management, South Tehran Branch, Islamic Azad University, Tehran, Iran

Naghi Shoja
Department of Industrial Engineering, Firoozkooh Branch, Islamic Azad University, Firoozkooh, Iran

Amir Ebrahimi Zade
Department of Industrial Engineering, Amirkabir University of Technology, Tehran, Iran

Sasan Barak
Faculty of Economics, Technical University of Ostrava, Ostrava, Czech Republic

Mani Sharifi
Faculty of Industrial and Mechanical Engineering, Qazvin Branch, Islamic Azad University, Qazvin, Iran

Sumit Kumar Maiti
School of Applied Sciences and Humanities, Haldia Institute of Technology, Purba Midnapore, 721 157 Haldia, India

Sankar Kumar Roy
Department of Applied Mathematics with Oceanology and Computer Programming, Vidyasagar University, 721 102 Midnapore, India

Christopher Rajkumar, Lone Kavin, Xue Luo and Jan Stentoft
Department of Entrepreneurship and Relationship Management, University of Southern Denmark, Kolding, Denmark

Index

Printed in the USA
CPSIA information can be obtained
at www.ICGtesting.com
JSHW051438221024
72173JS00006B/1514

9 781632 405845